Antioxidant Activity of Foods and Natural Products

Antioxidant Activity of Foods and Natural Products

Antioxidant Activity of Foods and Natural Products

Editors

José Virgílio Santulhão Pinela
Maria Inês Moreira Figueiredo Dias
Carla Susana Correia Pereira
José Ignacio Alonso-Esteban

MDPI

Basel • Beijing • Wuhan • Barcelona • Belgrade • Novi Sad • Cluj • Manchester

Editors
José Virgílio Santulhão Pinela
Mountain Research Centre (CIMO)
Polytechnic Institute of Bragança
Bragança
Portugal

Maria Inês Moreira Figueiredo Dias
Mountain Research Centre (CIMO)
Polytechnic Institute of Bragança
Bragança
Portugal

Carla Susana Correia Pereira
Mountain Research Centre (CIMO)
Polytechnic Institute of Bragança
Bragança
Portugal

José Ignacio Alonso-Esteban
Department of Biomedical Sciences
Faculty of Pharmacy, University of Alcalá
Alcalá de Henares (Madrid)
Spain

Editorial Office
MDPI
St. Alban-Anlage 66
4052 Basel, Switzerland

This is a reprint of articles from the Special Issue published online in the open access journal *Molecules* (ISSN 1420-3049) (available at: www.mdpi.com/journal/molecules/special_issues/Foods_Antioxidant).

For citation purposes, cite each article independently as indicated on the article page online and as indicated below:

Lastname, A.A.; Lastname, B.B. Article Title. *Journal Name* **Year**, *Volume Number*, Page Range.

ISBN 978-3-7258-1116-8 (Hbk)
ISBN 978-3-7258-1115-1 (PDF)
doi.org/10.3390/books978-3-7258-1115-1

Contents

Contents

About the Editors

José Virgílio Santulhão Pinela

José Pinela is a junior researcher at the Mountain Research Centre (CIMO) of the Polytechnic Institute of Bragança (IPB), Portugal, where he teaches subjects related to chemistry and food science. He completed his degree in Phytochemistry and Phytopharmacology at the School of Agriculture of IPB in 2010, followed by his M.Sc. in Biotechnology at the same institution in 2012. In 2017, he received his Ph.D. in Sustainable Chemistry from the Faculty of Sciences of the University of Porto, Portugal, during which he undertook a three-month research visit to the Faculty of Pharmacy of the Complutense University of Madrid, Spain, to obtain a European Doctorate. He also studied at the University of Ljubljana, Slovenia, under the Erasmus program. After receiving his Ph.D., he worked on the development of bio-based ingredients with bioactive and technological properties from agri-food by-products and on the chemical characterization of tomato germplasm and other plant-based foods. Currently, his attention is focused on sustainable methodologies for the production and biofortification of food crops.

Maria Inês Moreira Figueiredo Dias

Maria Inês Dias is a junior researcher at the Mountain Research Centre (CIMO) of the Polytechnic Institute of Bragança (IPB), Portugal, assuming pedagogical functions related to biology, biochemistry, and analytical chemistry. She holds a Ph.D. in Sustainable Chemistry from the Faculty of Sciences of the University of Porto (2017), with a European Doctorate mention after a three-month research visit to the Faculty of Pharmacy of the Complutense University of Madrid, Spain. She also holds a master's degree in Biotechnology (2011), and a Bachelor's degree in Biotechnological Engineering (2009), both from the School of Agriculture of IPB. The focus of her research has been the nutritional, chemical, and bioactive characterization of edible medicinal and aromatic plants, with a particular emphasis on utilizing/developing chromatographic methodologies. Her scientific endeavors have primarily revolved around investigating the increased production of plant secondary metabolites through in vitro cell and tissue culture techniques. Recently, she has shifted her focus to crop production via hydroponic methods, utilizing fish production wastewater (decoupled aquaponics) for irrigation.

Carla Susana Correia Pereira

Carla Pereira is a junior researcher at the Mountain Research Centre (CIMO) of the Polytechnic Institute of Bragança (IPB), Portugal. She holds a Ph.D. in Pharmacy and Health from the Faculty of Pharmacy of the University of Salamanca, Spain (2016), an M.Sc. in Food Quality and Safety from the School of Agriculture of IPB (2011), and a degree in Chemical Engineering from the School of Technology and Management of IPB (2007). She teaches chemistry, biochemistry, and related subjects and develops research activities in natural product chemistry and bioactive properties. The main focus of her research projects is related to plants and mushrooms production, characterization, and application in the food and pharmaceutical industries.

José Ignacio Alonso-Esteban

José Ignacio Alonso-Esteban is an Assistant Professor of Nutrition and Food Science at the Department of Biomedical Sciences of the Faculty of Pharmacy of the University of Alcalá (UAH), Spain. He holds a Bachelor's Degree at Master's level in Pharmacy (2015) and a PhD in Pharmacy (2021) from the Complutense University of Madrid (UCM), Spain. After that, he carried out his postdoctoral training (2021-2023) at the Mountain Research Centre (CIMO) of the Polytechnic

Institute of Bragança (IPB), Portugal. His main lines of research are the study of nutrients, anti-nutrients, and bioactive compounds and properties in plant foods, especially of the Cannabaceae family, such as hemp and hop, as well as in by-products from agri-food and forestry industries.

Preface

Antioxidants are a point of focus for different research fields due to their potential oxidative stress prevention and multiple biological properties, potentiating several industrial applications. Oxidative stress plays a major role in the pathogenesis of chronic diseases such as cardiovascular and neurodegenerative disorders, diabetes, and cancer. Antioxidants are capable of inhibiting oxidation and protecting the human body from free radicals and reactive species, reducing oxidative stress and, consequently, preventing or delaying many diseases. In foods, antioxidants naturally present or intentionally added as additives or bioactive compounds can delay or prevent degradation and rancidity, preserving food quality and extending shelf-life. Despite the growing demand for food supplements and nutraceuticals, today's consumers are increasingly aware that diets rich in fruits and vegetables, among other foods, are directly related to health and well-being due to their high content of antioxidant compounds. Therefore, the role of antioxidants from foods and natural products in human health is being extensively investigated, as is the development of new functional ingredients and food products containing these bioactive molecules. This reprint provides new insights into the antioxidant properties of foods and natural products, important structural features and the mechanisms of action of antioxidants and pro-oxidants, as well as antioxidants and pro-oxidants' potential health-promoting effects.

José Virgílio Santulhão Pinela, Maria Inês Moreira Figueiredo Dias, Carla Susana Correia Pereira, and José Ignacio Alonso-Esteban

Editors

molecules

MDPI

Editorial

Antioxidant Activity of Foods and Natural Products

José Pinela [1,2,*], Maria Inês Dias [1,2], Carla Pereira [1,2] and José Ignacio Alonso-Esteban [3]

1 Centro de Investigação de Montanha (CIMO), Instituto Politécnico de Bragança, Campus de Santa Apolónia, 5300-253 Bragança, Portugal; maria.ines@ipb.pt (M.I.D.); carlap@ipb.pt (C.P.)
2 Laboratório Associado para a Sustentabilidade e Tecnologia em Regiões de Montanha (SusTEC), Instituto Politécnico de Bragança, Campus de Santa Apolónia, 5300-253 Bragança, Portugal
3 Departamento de Ciencias Biomédicas, Facultad de Farmacia, Universidad de Alcalá, Carretera Madrid-Barcelona, Km 33.600, 28871 Madrid, Spain; jignacio.alonso@uah.es
* Correspondence: jpinela@ipb.pt

1. Introduction

Today, there is growing recognition of the importance of antioxidants in promoting human health and well-being. These compounds play a vital role in protecting the organism from oxidative stress, which is implicated in the pathogenesis of various chronic degenerative diseases, including cardiovascular and neurodegenerative disorders, diabetes, and different types of cancer [1]. Furthermore, antioxidants serve as indispensable agents in food preservation by preventing or slowing down oxidation processes, which could otherwise result in the deterioration of food quality, flavor, color, texture, and nutritional value [2]. This underscores ongoing research endeavors and the emergence of novel applications for antioxidant compounds across different disciplines, including chemistry, food science, nutrition, pharmacology, and medicine.

This Special Issue, titled "Antioxidant Activity of Foods and Natural Products", has brought together an interesting collection of cutting-edge research and development contributions. A total of 30 manuscripts were submitted for consideration, each undergoing the *Molecules* journal's rigorous pre-check and peer review process. Ultimately, 12 articles were accepted for publication in this Special Issue, comprising 1 review article and 11 original research articles. These studies address current challenges and topics related to the antioxidant properties of foods and natural products, including the antioxidant activity of plant extracts, botanical preparations, and isolated compounds, the mechanisms of action of antioxidants and pro-oxidants, and their potential therapeutic effects in vitro and in vivo.

Citation: Pinela, J.; Dias, M.I.; Pereira, C.; Alonso-Esteban, J.I. Antioxidant Activity of Foods and Natural Products. *Molecules* **2024**, *29*, 1814. https://doi.org/10.3390/molecules29081814

Received: 4 April 2024
Revised: 12 April 2024
Accepted: 15 April 2024
Published: 17 April 2024

2. An Overview of Published Articles

Bērziņa and Mieriņa (contribution 1) discussed the role of compounds containing the 1,3-dicarbonyl moiety in preventing oxidative damage and their potential as antioxidants. The study suggested that while some research points to the importance of substituents in the benzene ring, others indicate the crucial role of the 1,3-dicarbonyl moiety itself. Additionally, structural elements such as α-monosubstituted compounds and cyclic structures were identified as important factors influencing antiradical and antioxidant activity. Overall, the findings suggest that 1,3-dicarbonyl compounds hold potential for the development of effective antioxidants.

The study by Altuntaş et al. (contribution 2) explored the phenolic composition and bioactivity of 24 Anatolian propolis samples from Türkiye, aiming to classify them by origin. Phenolic and aromatic acids were identified, and propolis's antioxidant and antimicrobial properties were highlighted. While no single compound was solely responsible for these effects, the natural mixture as a whole demonstrated therapeutic significance. The study's use of principal component analysis effectively clustered samples based on their biochemical properties, providing insights into propolis's geographical variations.

Olędzki and Harasym (contribution 3) examined the impact of heat treatments on the bioactive properties and color–structural characteristics of bell peppers at different maturity stages (green, yellow, and red). Heat treatments such as contact grilling and roasting combined with microwaving increased the total phenolic content (TPC) of green peppers, while roasting and steaming decreased the antioxidant activity of yellow bell peppers. Moreover, certain methods significantly reduced the content of reducing sugars in red bell peppers. The study underscored the importance of selecting appropriate heat treatments to preserve the antioxidant and bioactive properties of bell peppers, thus enhancing the digestibility and bioavailability of these bioactive constituents.

Sousa et al. (contribution 4) investigated how different time/temperature combinations (65 °C/30 min; 77 °C/1 min; 88 °C/15 s; and 120 °C/20 min) affect the chemical composition and microbial load of olive pomace paste, a by-product of olive oil production. They found that while there were significant changes in components like ash, fat, vitamin E, TPC, total flavonoid content (TFC), and hydroxytyrosol and in antioxidant activity, the fatty acid profile remained constant. The 88 °C/15 s combination was the most effective in preserving the beneficial properties of the paste, making it a feasible industrial-scale solution to provide a safe and sustainable functional ingredient for novel food products.

In a different study, Ramarao et al. (contribution 5) explored the effects of convective air-drying, oven-drying, and sun-drying methods on the antioxidant properties of betel (*Piper betle* L.) leaves, with fresh leaves serving as a control. The findings revealed that sun-dried leaves exhibited superior antioxidant properties, including better total antioxidant and DPPH radical scavenging activities and higher TPC and TFC, along with a lower alkaloid concentration. Additionally, aqueous extracts from fresh and sun-dried leaves contained constituents with antioxidant and anti-inflammatory effects. These findings suggest that sun-drying maintains the antioxidant potential of betel leaves while augmenting the presence of biologically active phytoconstituents, suggesting potential applications in the food and pharmaceutical industries. Further research was recommended to assess safety aspects and individual composition variations.

Núñez-Gómez et al. (contribution 6) explored the chemical composition, functional properties, and antioxidant capacity of dietary fiber extracted from lemon peel by drying with warm air and enzymatic hydrolysis with pectinesterase. The enzymatic treatment caused a reduction in soluble fiber and an increase in insoluble fiber, along with changes in the pectin structure, resulting in diminished water and fat absorption capacities; it also decreased the TPC and antioxidant activity. Conversely, warm air drying had higher potential for producing high-quality fiber with antioxidant properties from lemon peel. These findings hold significant implications for the development of novel ingredients rich in dietary fiber and (poly)phenols with antioxidant capacity, as well as for the valorization of lemon peels and mitigation of the environmental impact associated with their disposal.

The study by dos Santos et al. (contribution 7) explored the in vitro antidiabetic, antiglycation, and antioxidant properties of an ethanolic seed extract of *Passiflora edulis* and piceatannol. Both samples significantly inhibited the α-amylase, α-glucosidase, and dipeptidyl-peptidase-4 enzymes, with IC_{50} values indicating their effectiveness compared to standard drugs. The formation of advanced glycation end-products (AGEs) and β-amyloid fibrils was also inhibited, demonstrating antiglycation properties. The samples also showed strong antioxidant activity. These findings suggested the potential of both *P. edulis* seed extract and piceatannol as antidiabetic agents, warranting further investigation.

The study conducted by Maliar et al. (contribution 8) addressed the need for a better understanding of both antioxidant and pro-oxidant effects. A methodology for the simultaneous determination of antioxidant and pro-oxidant activity on a single microplate was developed, assuming that the FRAP method could measure both effects due to the generation of pro-oxidant Fe^{2+} ions in the Fenton reaction. The study suggested that compounds with higher numbers of oxygen heteroatoms, particularly sp2-hybridized compounds like flavonoids, exhibit dominant pro-oxidant effects. Conversely, catechins, carotenoids, and certain plant extracts, such as those from birch and chestnut leaves, demonstrate dominant antioxidant activity over pro-oxidant. These initial findings prompt further systematic evaluation of a broader range of compounds and plant extracts using this method.

The potential of a standardized botanical composition from *Scutellaria baicalensis* and *Acacia catechu* in mitigating acute inflammatory lung injury and reducing mortality by reducing extracellular HMGB1 levels was investigated by Yimam et al. (contribution 9). HMGB1 is a crucial late inflammatory mediator associated with air pollution-induced oxidative stress and lung injury. The botanical composition was tested using murine models of acute lung injury and sepsis. Significant reductions in mortality, pro-inflammatory cytokines and chemokines, bacterial counts in the lungs and airways, and extracellular HMGB1 levels were observed. Moreover, cultured macrophages treated with the botanical product exhibited increased phagocytic activity and decreased extracellular HMGB1 levels. These findings suggested that the botanical could potentially protect against oxidative stress-induced lung damage by reducing extracellular HMGB1 accumulation.

Regarding in vivo studies, Truong et al. (contribution 10) investigated the protective effects of orange sweet pepper (*Capsicum annuum* L.) juices, prepared by both high-speed blender and low-speed masticating juicer, against UVB-induced skin damage in SKH-1 hairless mice. Oral administration of these juices reduced UVB-induced skin photoaging by regulating genes involved in dermal matrix production and maintenance, such as collagen type I α 1 and matrix metalloproteinases-2, 3, and 9. The juices also restored total collagen levels in UVB-exposed mice and downregulated the expression of pro-inflammatory proteins, including cyclooxygenase-2, interleukin (IL)-1β, IL-17, and IL-23, likely via inhibiting the NF-κB pathway. Additionally, the juices enhanced primary antioxidant enzymes in the skin, such as catalase, glutathione peroxidase, and superoxide dismutase-2, and reduced UVB-induced DNA damage by preventing 8-OHdG formation. These findings suggest that sweet pepper juices offer a protective effect against photoaging by inhibiting dermal matrix breakdown, inflammatory response, and DNA damage, while enhancing antioxidant defense, ultimately leading to a reduction in skin damage.

In another work, Yang et al. (contribution 11) investigated the protective effects and mechanisms of salidroside in age-related renal fibrosis using SAMP8 mice. The administration of salidroside for 12 weeks led to improvements in renal function, with reduced levels of blood urea nitrogen and serum creatinine, and increased serum albumin levels. Additionally, this phenylpropanoid glycoside reduced mesangial hyperplasia and levels of transforming growth factor-β and α-smooth muscle actin, indicating the mitigation of renal fibrosis. The treatment also decreased lipid peroxidation in the kidneys and regulated iron transport-related proteins and ferroptosis-related proteins. These findings suggested that salidroside delays renal aging and inhibits age-related glomerular fibrosis by suppressing ferroptosis in SAMP8 mice.

Lastly, the study by Mattioli et al. (contribution 12) investigated the impact of dietary *n*-3 polyunsaturated fatty acids (PUFA) on apelin and resolvin D1 (RvD1) levels in rabbit testis and sperm. Apelin is an endogenous peptide known for its involvement in both the release of inflammatory mediators and the expression of antioxidant enzymes, whereas RvD1 is a specialized pro-resolving mediator derived from *n*-3 PUFA [3]. The authors fed rabbits diets enriched with either flaxseed (rich in α-linolenic acid) or fish oil (containing eicosapentaenoic acid, docosapentaenoic acid, and docosahexaenoic acid) and observed increased apelin levels in testes of both groups, particularly in the interstitial tissue of the flaxseed-fed rabbits. The flaxseed-enriched diet also enhanced serum testosterone levels, while both diets led to higher malondialdehyde and RvD1 levels in the testis. In ejaculated sperm, apelin and RvD1 were mainly located in the tail, with positive correlations observed between apelin, sperm motility, and RvD1 levels, suggesting their potential involvement in male reproduction and inflammation resolution. These findings underscore the potential benefits of a flaxseed-enriched diet in increasing testicular apelin levels, thereby potentially ameliorating male reproductive health and inflammatory status.

3. Conclusions

The articles comprising this Special Issue highlight the pivotal role of antioxidants in promoting human health and well-being, alongside their versatile applications across diverse industrial sectors such as food, pharmaceuticals, and medicine. Through deeper exploration of antioxidant compounds from different foods and natural products and their underlying mechanisms of action, these studies pave the way for leveraging their therapeutic and technological potential in combating oxidative stress and its related diseases. Furthermore, this advancement will enable the development of innovative strategies for the preservation and functionalization of foods and dietary supplements, with the ultimate goal of improving the overall quality of life for individuals worldwide.

Author Contributions: Conceptualization and writing—original draft preparation and review and editing, J.P., M.I.D., C.P. and J.I.A.-E. All authors have read and agreed to the published version of the manuscript.

Funding: CIMO (UIDB/00690/2020 and UIDP/00690/2020) and SusTEC (LA/P/0007/2020) received financial support from Foundation for Science and Technology (FCT, Portugal) through national funds FCT/MCTES (PIDDAC). National funding was provided by FCT, through the scientific employment program contract with J. Pinela (DOI: 10.54499/CEECIND/01011/2018/CP1578/CT0002), M.I. Dias (DOI: 10.54499/CEECINST/00016/2018/CP1505/CT0004), and C. Pereira (DOI: 10.54499/CEECINST/00016/2018/CP1505/CT0010).

Acknowledgments: The Guest Editors would like to thank all the authors whose valuable work was published under this Special Issue and contributed to its success.

Conflicts of Interest: The authors declare no conflicts of interest.

List of Contributions

1. Bērziņa, L.; Mieriņa, I. Antiradical and Antioxidant Activity of Compounds Containing 1,3-Dicarbonyl Moiety: An Overview. *Molecules* **2023**, *28*, 6203. https://doi.org/10.3390/molecules28176203.
2. Altuntaş, Ü.; Güzel, İ.; Özçelik, B. Phenolic Constituents, Antioxidant and Antimicrobial Activity and Clustering Analysis of Propolis Samples Based on PCA from Different Regions of Anatolia. *Molecules* **2023**, *28*, 1121. https://doi.org/10.3390/molecules28031121.
3. Olędzki, R.; Harasym, J. Assessment of the Effects of Roasting, Contact Grilling, Microwave Processing, and Steaming on the Functional Characteristics of Bell Pepper (*Capsicum annuum* L.). *Molecules* **2024**, *29*, 77. https://doi.org/10.3390/molecules29010077.
4. Sousa, M.M.; Ferreira, D.M.; Machado, S.; Lobo, J.C.; Costa, A.S.G.; Palmeira, J.D.; Nunes, M.A.; Alves, R.C.; Ferreira, H.; Oliveira, M.B.P.P. Effect of Different Time/Temperature Binomials on the Chemical Features, Antioxidant Activity, and Natural Microbial Load of Olive Pomace Paste. *Molecules* **2023**, *28*, 2876. https://doi.org/10.3390/molecules28062876.
5. Ramarao, K.D.R.; Razali, R.; Somasundram, C.; Kunasekaran, W.; Jin, T.L. Effects of Drying Methods on the Antioxidant Properties of *Piper betle* Leaves. *Molecules* **2024**, *29*, *in press*.
6. Núñez-Gómez, V.; San Mateo, M.; González-Barrio, R.; Periago, M.J. Chemical Composition, Functional and Antioxidant Properties of Dietary Fibre Extracted from Lemon Peel after Enzymatic Treatment. *Molecules* **2024**, *29*, 269. https://doi.org/10.3390/molecules29010269.
7. dos Santos, F.A.R.; Xavier, J.A.; da Silva, F.C.; Merlin, J.P.J.; Goulart, M.O.F.; Vasantha Rupasinghe, H.P. Antidiabetic, Antiglycation, and Antioxidant Activities of Ethanolic Seed Extract of *Passiflora edulis* and Piceatannol In Vitro. *Molecules* **2022**, *27*, 4064. https://doi.org/10.3390/molecules27134064.
8. Maliar, T.; Maliarová, M.; Blažková, M.; Kunštek, M.; Uváčková, L.; Viskupičová, J.; Purdešová, A.; Beňovič, P. Simultaneously Determined Antioxidant and Pro-Oxidant Activity of Randomly Selected Plant Secondary Metabolites and Plant Extracts. *Molecules* **2023**, *28*, 6890. https://doi.org/10.3390/molecules28196890.
9. Yimam, M.; Horm, T.; O'Neal, A.; Jiao, P.; Hong, M.; Brownell, L.; Jia, Q.; Lin, M.; Gauthier, A.; Wu, J.; et al. A Standardized Botanical Composition Mitigated Acute Inflammatory Lung Injury and Reduced Mortality through Extracellular HMGB1 Reduction. *Molecules* **2023**, *28*, 6560. https://doi.org/10.3390/molecules28186560.

10. Truong, V.-L.; Rarison, R.H.G.; Jeong, W.-S. Protective Effects of Orange Sweet Pepper Juices Prepared by High-Speed Blender and Low-Speed Masticating Juicer against UVB-induced Skin Damage in SKH-1 Hairless Mice. *Molecules* **2022**, *27*, 6394. https://doi.org/10.3390/molecules27196394.
11. Yang, S.; Pei, T.; Wang, L.; Zeng, Y.; Li, W.; Yan, S.; Xiao, W.; Cheng, W. Salidroside Alleviates Renal Fibrosis in SAMP8 Mice by Inhibiting Ferroptosis. *Molecules* **2022**, *27*, 8039. https://doi.org/10.3390/molecules27228039.
12. Mattioli, S.; Moretti, E.; Castellini, C.; Signorini, C.; Corsaro, R.; Angelucci, E.; Collodel, G. Can Dietary *n*-3 Polyunsaturated Fatty Acids Affect Apelin and Resolvin in Testis and Sperm of Male Rabbits? *Molecules* **2023**, *28*, 6188. https://doi.org/10.3390/molecules28176188.

References

1. Olufunmilayo, E.O.; Gerke-Duncan, M.B.; Holsinger, R.M.D. Oxidative Stress and Antioxidants in Neurodegenerative Disorders. *Antioxidants* **2023**, *12*, 517. [CrossRef]
2. Othón-Díaz, E.D.; Fimbres-García, J.O.; Flores-Sauceda, M.; Silva-Espinoza, B.A.; López-Martínez, L.X.; Bernal-Mercado, A.T.; Ayala-Zavala, J.F. Antioxidants in Oak (*Quercus* sp.): Potential Application to Reduce Oxidative Rancidity in Foods. *Antioxidants* **2023**, *12*, 861. [CrossRef] [PubMed]
3. Signorini, C.; Collodel, G.; Pannuzzo, G.; Graziano, A.C.E.; Moretti, E.; Noto, D.; Belmonte, G.; Cardile, V. Decreased Resolvin D1 and Increased Fatty Acid Oxidation Contribute to Severity Score of Krabbe Disease in Twitcher Mice. *J. Biol. Regul. Homeost. Agents.* **2024**, *38*, 913–924. [CrossRef]

molecules

MDPI

Review

Antiradical and Antioxidant Activity of Compounds Containing 1,3-Dicarbonyl Moiety: An Overview

Laima Bērziņa and Inese Mieriņa *

Institute of Technology of Organic Chemistry, Faculty of Materials Science and Applied Chemistry, Riga Technical University, LV-1048 Riga, Latvia; laima.berzina@inbox.lv
* Correspondence: inese.mierina@rtu.lv

Abstract: Free radicals and oxidants may cause various damages both to the lifeworld and different products. A typical solution for the prophylaxis of oxidation-caused conditions is the usage of various antioxidants. Among them, various classes are found—polyphenols, conjugated polyalkenes, and some sulfur and nitrogen derivatives. Regarding the active site in the molecules, a widely discussed group of compounds are 1,3-dicarbonyl compounds. Among them are natural (e.g., curcumin and pulvinic acids) and synthetic (e.g., 4-hydroxy coumarins, substituted Meldrum's acids) compounds. Herein, information about various compounds containing the 1,3-dicarbonyl moiety is covered, and their antiradical and antioxidant activity, depending on the structure, is discussed.

Keywords: antiradical activity; antioxidant activity; 1,3-dicarbonyl compound; β-dicarbonyl compound; 1,3-diketone; keto–enol tautomerism

Citation: Bērziņa, L.; Mieriņa, I. Antiradical and Antioxidant Activity of Compounds Containing 1,3-Dicarbonyl Moiety: An Overview. *Molecules* **2023**, *28*, 6203. https://doi.org/10.3390/molecules28176203

Academic Editors: José Virgílio Santulhão Pinela, Maria Inês Moreira Figueiredo Dias, Carla Susana Correia Pereira and José Ignacio Alonso-Esteban

Received: 20 July 2023
Revised: 12 August 2023
Accepted: 21 August 2023
Published: 23 August 2023

1. Introduction

Free radicals and oxidants are an integral part of an aerobic world like planet Earth. Our bodies naturally produce reactive oxygen and nitrogen species in various endogenous systems [1]. However, their roles are like two sides of a coin—they may be helpful and harmful to our bodies [2]. Reactive oxygen species are involved in various physiological processes as signal mediators [3]. On the other hand, free radicals cause damage to organs, cells, mitochondria, and bio-molecules. These processes result in oxidative stress, and aging occurs [4]. Oxidative stress may cause various chronic and degenerative illnesses [2], e.g., cardiovascular conditions [2,3,5], cancer, autoimmune disorders [2], etc. Furthermore, oxidized lipids, proteins, and nucleic acids have been detected in the brain tissue of people who have Parkinson's disease [6]. Overproduction of reactive oxygen species (ROS) alternates the mammalian target of rapamycin (mTOR) pathway, which is associated with the development of Alzheimer's disease [7]. Oxidative stress has demonstrated a dual role in cancer cells: the formation of ROS stimulates cellular damage and mutations, thus promoting the development of a tumor; simultaneously, the overexposure of cancer cells to ROS may cause their apoptosis [8].

In addition to affecting complex processes in the human body, it is well established that various products, from food and medicine to technical applications, containing (poly)unsaturated fatty acid derivatives are suspected of autooxidation. Some recent reviews have covered lipid peroxidation kinetic studies [9], lipid oxidation in emulsions [10], autooxidation of both triglycerides and minor components in extra-virgin olive oil during cooking [11], and the oxidative stability of biodiesel [12] issues. Briefly, lipid autooxidation is a free radical reaction cascade involving initiation (the formation of lipid radicals) and propagation, leading to various peroxides. Then, these peroxides, via cyclization, rearrangement, scission, and condensation, terminate the reaction chain, leading to secondary oxidation products (Figure 1). Heat- [13] and sunlight- [14] induced collapse of various polymer materials results from rapid autoxidation. In contrast, oxidation-response polymers have found application in medicine [15]. The oxidation of polyunsaturated fats in food products results in oxidative rancidity [16]. Typically, for minimizing autooxidation processes and increasing the shelf-life, various antioxidants are added before the oxidation process develops uncontrollably. Polyphenols are well-known tools for controlling such undesired changes in the molecular structure of foods [17]. It is worth mentioning that free radicals have turned into valuable soldiers for the degradation of microplastics [18] and wastewater treatment [19,20].

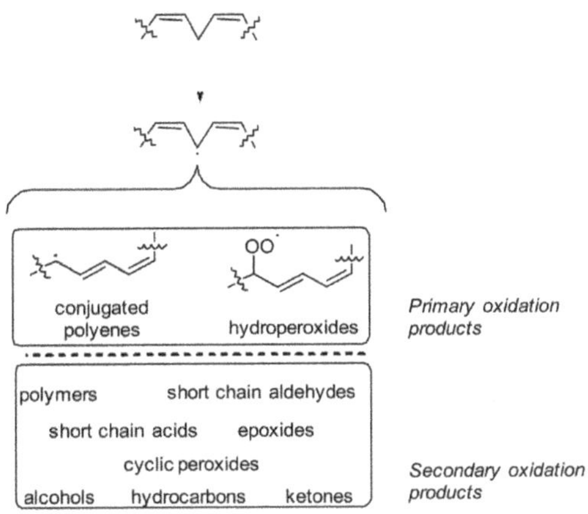

Figure 1. Autooxidation of polyunsaturated lipids. Autooxidation is initiated in the position with the lowest bond dissociation energy—usually the autooxidation in fatty acid triglycerides starts at the *bis*-allyl position, followed by the allyl position. The generated carbon radical is suspected to rearrange, leading to an allyl-type radical. A later reaction with oxygen leads to a peroxyl radical. The isomerization, scission, rearrangement, and polymerization of primary oxidation products generates secondary oxidation products.

Different antioxidants may act as influential players in the prophylaxis of disorders caused by oxidative stress. Anthocyanins have been reported to be protective against atherosclerosis and cardiovascular disease [21]. Dietary flavonoids have demonstrated inhibition of low-density lipoprotein (LDL) oxidation and platelet aggregation. Khan et al. have even speculated that moderate consumption of red wine, due to its flavonoids, may reduce the risk of atherosclerosis and thrombosis [22].

Several mechanisms have been provided for explaining the reactions between free radicals and antioxidants. The role of phenol-type antioxidants in hindering the Fenton reaction by complexing both Fe^{3+} and Fe^{2+} ions is well established [23]. On the other hand, carnosic acid effectively neutralizes peroxyl radicals, leading to quinoid structures [24]. Experimental and theoretical studies have demonstrated that disulfides scavenge $HO^{\bullet}$ via cleavage of the -S-S- bond [25]. The mechanism strongly depends on the media; e.g., it is postulated that phenylpropanoids react with free radicals via electron transfer and sequential proton transfer in solvents with a low ionizing ability [26]. It is worth mentioning that it is not always possible to distinguish which mechanism is preferable for explaining antioxidant and/or antiradical activity. E.g., potentiometric studies of various flavonoids and coumarins have highlighted that their activity is due to synergism among electron transfer, hydrogen atom transfer, and metal chelation mechanisms [27]. The density functional theory calculations for various ingredients of essential oils have turned to the hydrogen atom transfer (HAT) mechanism as the most favorable in nonpolar solvents, whilst the sequential proton loss and electron transfer (SPLET) mechanism is dominant in polar solutions [28]. If the antioxidant contains only one structural moiety responsible for antiradical or antioxidant activity, the deduction of the most plausible mechanism is relatively easy. The situation is complex when the structure contains several units, which may affect the activity. E.g., the natural polyphenol silybin contains several hydroxy groups, and each of them has a different role in providing antiradical and/or antioxidant activity [29]. When new antioxidants are constructed, it is usually considered if the antioxidant can scavenge several free radicals: e.g., coumarin–chalcone hybrids are effective for two SPLET and two HAT cycles [30].

Recent studies on antioxidants, mainly covering antioxidative vitamins, polyphenols, and enzymatic antioxidants, have been reviewed by Parcheta et al. [31]. Various natural antioxidants are established as useful for preventing UV radiation-induced ROS, thus turning them into promising agents for cosmetical applications, for example, in sunscreens [32]. A few classes of common antioxidants—carotenoids [33,34], different polyphenols (flavonoids [34,35], chalcones [36], and anthocyanins [34,37]), and coumarins [38,39]—have been well-reviewed in the last several years. Another structurally differential class of compounds—spiro-cyclic systems [40]—have been highlighted as potential antioxidants and free radical scavengers. Nano-science has revealed various nanoparticles, like CeO_2 [41], nano-selenium, nano-zinc [42], and nitroxides [43], as promising antioxidative scaffolds. Nanostructured forms of organic materials are not an exception: e.g., lignin seems to be a promising antioxidant and UV shield in sunscreen [44]. Polymeric materials have turned into a promising solution for delivering antioxidants [45].

To date, there has been no review covering various antioxidants containing the 1,3-dicarbonyl moiety. Thus, this review covers another exciting class of antioxidants—CH acids containing β-dicarbonyl moieties. The structure–activity relationship is analyzed, and the role of dicarbonyl moiety is discussed.

One of the most widely explored 1,3-dicarbonyl compounds is curcumin. It naturally occurs (up to 2%) mainly in turmeric (*Curcuma longa*) [46]. Since ancient times, the benefits of phytochemical curcumin have been applied in traditional medicine [47]. In recent years, several reviews highlighting various biological activities of curcumin have seen the light of day, indicating that curcumin is a green anticancer weapon [48–54]. The wide application of curcumin in cancer therapy has forced Kong et al. [55] to ask whether "curcumin is the answer to future chemotherapy cocktail?". This compound has been found to be helpful for the treatment of various diseases caused by free radicals: e.g., metabolic-associated fatty liver disease [56], Huntington's disease [57], diabetes mellitus [58] and diabetic complications [59], neurodegenerative diseases [60,61], and dermatological disorders [62]. In addition, both structural modifications of curcumin [49,63] and the synthesis of metal complexes [64] have led to compounds with promising pharmacological effects. It should be noted that the outstanding properties of curcumin have encouraged the development of nanoformulations for cancer treatment [65,66], therapy for aging-related diseases [67], and antimicrobial purposes [68]. Guo et al. analyzed curcumin's plausible antioxidant activity mechanism [69]. However, no attention has been turned to structure–activity analysis. Hunyadi covered the metabolic processes of curcumin in the body [70]. The special biological activities of curcumin have encouraged effective isolation and purification [71–74], conventional synthetic [74], and biosynthetic [75] routes.

2. An Overview of the Method Used for Testing Antioxidant and Antiradical Activity

A general overview of the various methods available for testing antioxidant and antiradical activity was published by Carlos et al. [76]. Briefly, two types of methods are applied: (1) reactions between antioxidant molecules and the free radicals, and (2) processes where a real system is suspected to rapidly oxidize and the decrease in the oxidation of the substrate in the presence and absence of the antioxidant is monitored. The term "antiradical activity" should be attributed to the methods where the test compound scavenges the free radicals. The term "antioxidant activity" is provided for methods where the test compound is used for inhibiting the oxidation of the substrate [77]. The main advantages of the first group are as follows: usually, the reactions are rather fast, standard laboratory equipment (UV-Vis spectrometer) is required, the solubility of the antioxidants can be achieved, and the reaction products may be isolated and characterized. On the other hand, often, the used radicals are not naturally occurring, and the results may not correlate with the activity in real systems. The main advantages of the second group of methods are the following: the antioxidant is tested in the presence of an oxidative unstable substrate and the obtained results may be characteristic for predicting antioxidant effectivity in similar real products. Some of the drawbacks of these methods may include rather long experiment times, depending on the substrate solubility limitations of plausible antioxidants, and in some cases, rather specific equipment may be required.

2.1. *Tests for Determining Antiradical Activity*

Widely used antiradical assays involve the reaction with 2,2-diphenyl-1-picrylhydrazyl (DPPH), galvinoxyl (GO), 2,2′-azino-*bis*(3-ethylbenzothiazoline-6-sulfonic acid) radical cation (ABTS), and hydroxyl radical. DPPH and GO are commercial stable free radicals. In cases of other test systems, the free radical or radical cation should be generated prior to the experiment.

The DPPH test (Scheme 1) is widely used to evaluate the antiradical activity of both individual compounds and plant extracts. Typically, it is a rapid, spectrophotometric method used for solutions. An overview of this assay was published by Kedare et al. [78]. The quenching of DPPH **1** is monitored at $\lambda = 515$ nm. The reaction is run in any solvent; however, widely used solvents are ethanol and methanol. The reaction rate strongly depends on the solvents; it was established that the reaction is faster in polar solvents than in unpolar ones [79]. Some modern DPPH assays involve electrochemical principles [80], chromatographic routes like thin layer chromatography (TLC) on silica plates [81] or high-performance liquid chromatography (HPLC) [82], as well as the application of analyte (solid sample [83] or the sample doped on silica and modified silica plates [84]) or DPPH (immobilized in 96-well plates [85], pharmaceutical blisters [86], or paper [87]) in solid form.

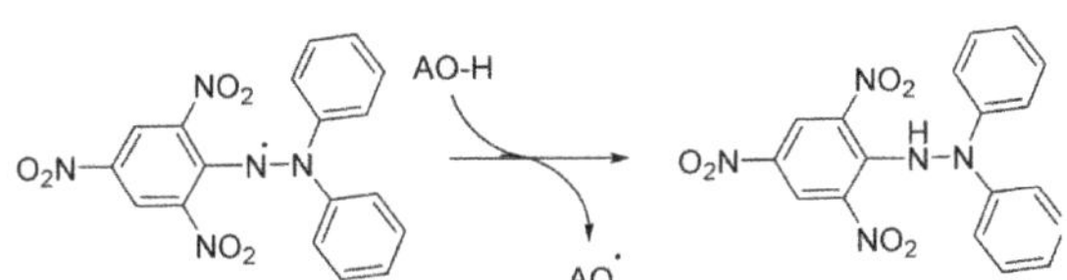

1 2

Scheme 1. Representative reaction between DPPH and an antioxidant.

A less used free radical for the rapid testing of antiradical activity is galvinoxyl **3** (GO) (Scheme 2). Similarly to DPPH assays, the test is implemented in a solution, and the absorbance ($\lambda = 428$ nm [88]) after a certain time is registered [89]. Other routes for monitoring GO inhibition efficiency involve electron spin resonance [90], HPLC with a post-column injection of GO [91], and electrochemical assay with galvinoxyl immobilized on electrodes [92].

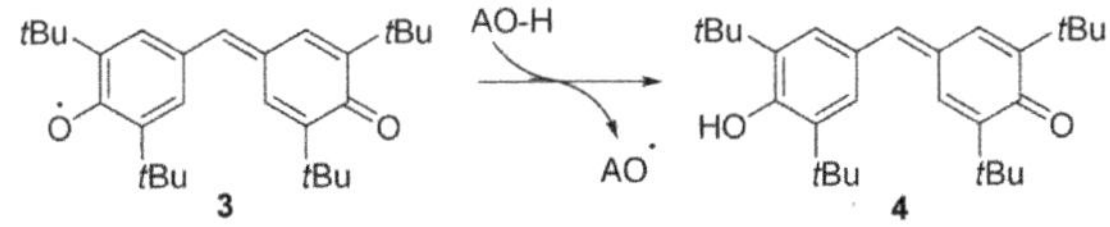

3 4

Scheme 2. Representative reaction between GO and an antioxidant.

Contrary to the DPPH and GO tests, ABTS radical cation **6** (Scheme 3) should be generated from the corresponding acid **5** prior to the analysis. A typical reagent for this transformation is potassium persulfate [93], but faster electrolysis may be useful too [94]. The amount of the radical cation is monitored by a spectrophotometer ($\lambda = 730\ldots750$ nm) [93]. Similarly to GO and DPPH, an HPLC post-column treatment with a radical cation is elaborated for ABTS assay [95].

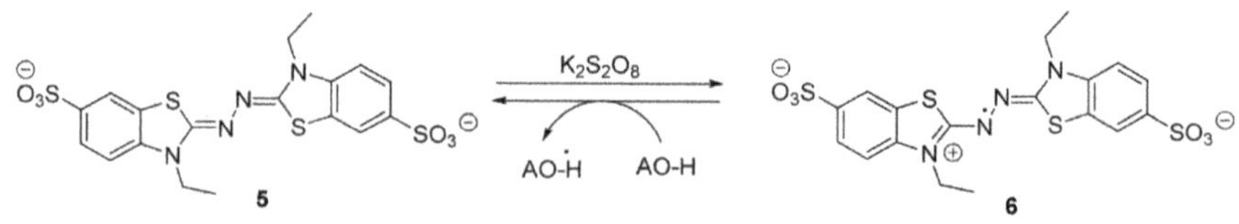

Scheme 3. Generation of ABTS radical cation and its reaction with an antioxidant.

Widely used free radicals for assessing the effectiveness of antioxidants are hydroxyl radicals. These radicals are usually generated through the Fenton reaction (Scheme 4) using hydrogen peroxide. Then, the amount of hydroxyl radicals is detected via the reaction of thiobarbituric acid **7** and malondialdehyde **8** (Scheme 5), which leads to chromophore **9** (λ = 532 nm) [96]. The presence of antioxidants reduces the oxidation of 2-deoxy-D-ribose **10**. The principles of this method are well summarized by Apak et al. [97].

A:

$Fe^{3+} + O_2^{\bullet -} \longrightarrow {}^{2+} + O_2$

B:

$Fe^{2+} + H_2O_2 \longrightarrow Fe^{3+} + HO^- + HO^\bullet$

$Fe^{3+} + HO^- \longrightarrow Fe^{2+} + HO^\bullet$

Scheme 4. Fe^{2+} ions are produced in the Haber–Weiss reaction (**A**), and then, in their reaction with hydrogen peroxide, hydroxyl radicals are formed (**B**). Hydroxyl radical formation from hydroxide ions may be catalyzed by Fe^{3+}.

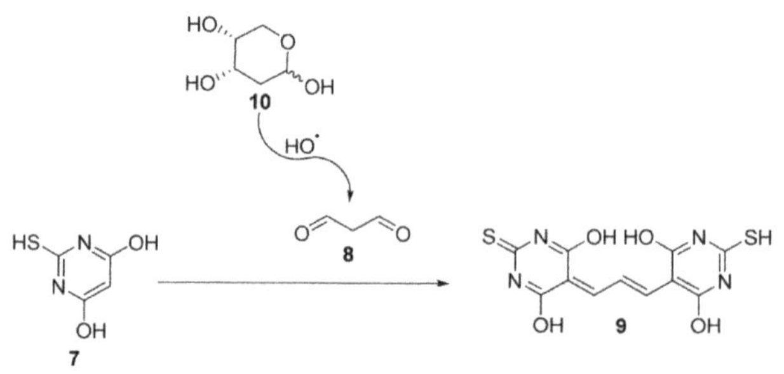

Scheme 5. The principle for testing antioxidants for their ability to scavenge hydroxyl radicals: the hydroxyl radicals convert 2-deoxy-d-ribose into malonaldehyde. The latter is trapped by thiobarbituric acid.

The systems described above were elaborated based on hydrogen transfer (or sequential proton loss and electron transfer). Some tests involve electron transfer processes. A widely used method is the ferric-reducing antioxidant power (FRAP). This procedure involves antioxidant-induced reduction in Fe(III) to Fe(II), leading to blue-colored complex **12** (Scheme 6) with an absorption maximum of λ = 570 nm [98].

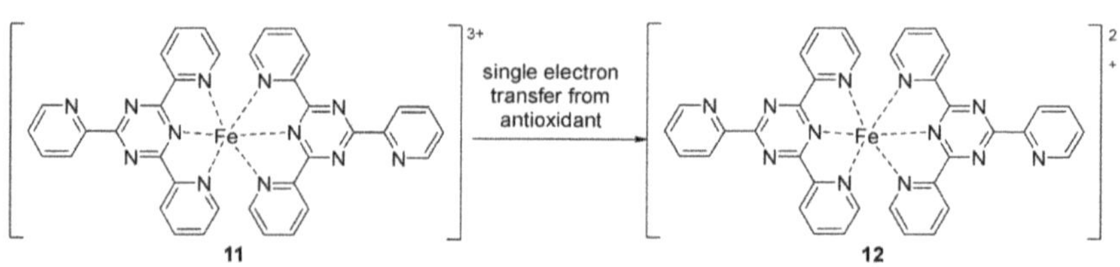

Scheme 6. Reduction of Fe(III)–tripyridyltriazine complex **11** in the presence of antioxidants to deep blue-colored Fe(II)–tripyridyltriazine complex **12**.

Another group of test methods are those for the evaluation of the antioxidant ability to complex with various metals. This property is of particular interest in the context of the Fenton reaction, which may occur in real systems. A typical assay involves the complexation of Fe(II) with antioxidants in the presence of ferrozine. When the analyte leads to a strong complex with Fe(II), no characteristic ferrozine complex with absorption at 562 nm is observed [99].

2.2. *Determining Antioxidant Activity*

Number of methods used for testing antioxidant activity in various lipid systems are available. Usually these methods are based on the analysis of the primary and secondary oxidation products. An overview of such methods is given by Abeyrathne et al. [100].

One such method is the β-carotene bleaching test. Herein, the oxidation processes of linoleic acid **13** in the presence of air are studied, and the rate of degradation of carotene **15** at $\lambda = 470$ nm is monitored (Scheme 7) [101].

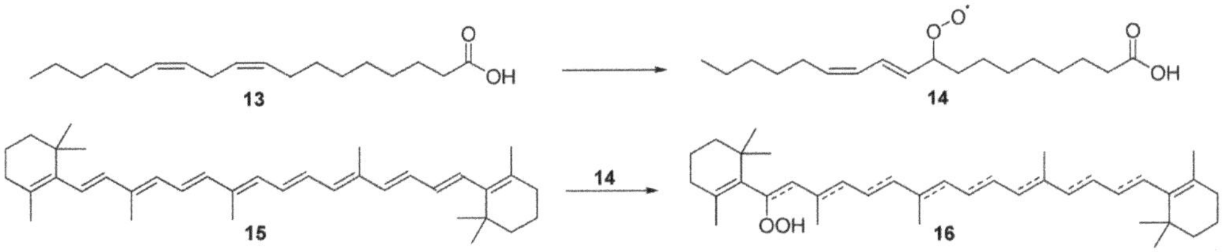

Scheme 7. Linoleic acid **13** undergoes autooxidation, leading to linoleic acid peroxyl radical **14**. The latter induces the oxidation of β-carotene **15**, leading to colorless peroxide **16**.

In addition to air-induced methods, assays involving the controlled release of free radicals are elaborated. Such an example is the AIBN **17**-induced peroxidation of linoleates (Scheme 8) [102].

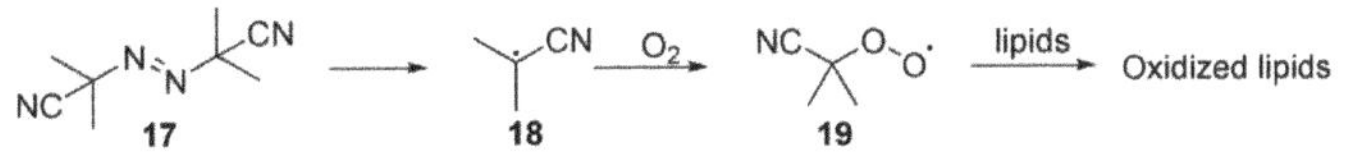

Scheme 8. Generation of C-centered free radical **18** from AIBN **17** and sequential formation of peroxyl radical **19**.

Similarly, antioxidants are tested not only in model systems but also in real samples. The substrates used for testing antioxidant efficiency are various vegetable oils and emulsions. The tests are run under different accelerated oxidation conditions. These methods are well summarized in some recent reviews [103–105].

3. Tautomerism of 1,3-Dicarbonyl Compounds

It is well known that 1,3-dicarbonyl compounds exist in a tautomeric equilibrium between the keto- **20** and enol **21** forms (Scheme 9). Tautomerization is crucial from a medicinal chemistry viewpoint. The binding between small molecules with biological targets may strongly be affected by the keto–enol tautomerism of the compound [106]. Thus, understanding the keto–enol equilibrium might also be helpful in constructing antioxidants and/or free radical scavengers containing the 1,3-dicarbonyl moiety.

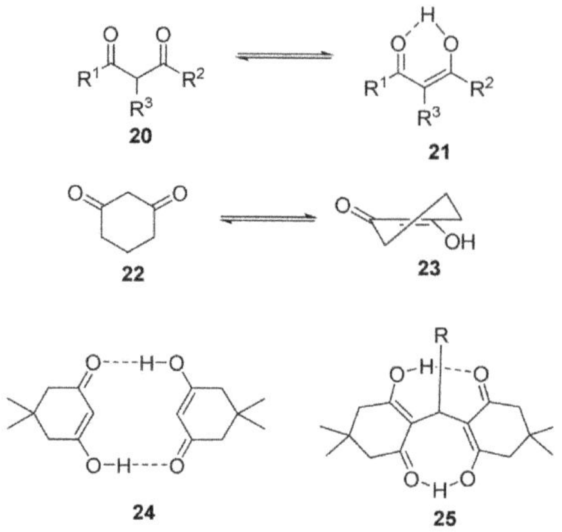

Scheme 9. Keto–enol tautomerism: keto form of acyclic 1,3-dicarbonyl compound **20**, enol form of acyclic 1,3-dicarbonyl compound **21** representing intramolecular hydrogen bonds, cyclic 1,3-diketone **22**, twist form of cyclic 1,3-dicarbonyl compound **23**, intermolecular hydrogen bonds between two cyclic 1,3-diketones **24** and Vorländer adduct **25**.

The scientific community has studied these equilibria in β-dicarbonyl compounds for more than a century—one of the first papers was published by Knorr in 1896 [107]. The general trends of the keto–enol equilibrium, both in 1,3-diketones and β-ketoesters, are well summarized by Iglesias [108]. The equilibrium between diketone and keto–enol forms is shifted by the temperature, solvents, various additives, and the structure of the 1,3-dicarbonyl compound.

In general, the enol form is facilitated due to the formation of both intra- and intermolecular hydrogen bonds. A cyclic six-member transition state, caused by an intramolecular hydrogen bond, favors the enol form in acyclic systems [109]. The formation of intramolecular hydrogen bonds is not sterically possible in cyclic 1,3-dicarbonyl compounds **22**. Such keto–enol tautomers exist as twisted structures **23** [110]. On the other hand, it is known that such enols in dimedone are stabilized by a network of intermolecular hydrogen bonds, leading to complex **24**. However, when two dimedone units are linked via a methylene bridge (so-called Vorländer adducts **25**), the structure is stabilized through intramolecular hydrogen bonds [111]. The tautomerization in cyclic 1,3-dicarbonyl compounds is affected by the size of the cycle—tautomerization does not occur in four-member cycles, while for other 1,3-dicarbonyl compounds, the activation energy is reduced by increasing the size of the cycle [112]. The enol form is dominant in diketones, while the keto form is characteristic of β-ketoesters. It has been observed that the introduction of an additional substituent in the α-position favors the enol form [108]. However, the diketone form is favored by α-fluorosubstituted diketones (compound **20**, R^3 = F) [113]. Modifications of the substituents R^1–R^3 (structures **20** and **21**) in the diketone influence the ratio between the keto–enol and diketone forms; however, the former is still dominant [114]. The tautomeric equilibrium has also been studied for various curcumin derivatives: usually, the enol form is the dominant one; however, some exceptions are known. Both forms are present in similar amounts when an electron acceptor group is introduced to the benzene ring or a methyl group is introduced in the α-position [115].

Regarding the polarity of the media, the enol form in "inert" solvents is favored due lower polarity of the enol [116], whilst polar solvents facilitate the keto–tautomer [108]. Basic solvents (or additives of bases) for ketoesters shift the equilibrium to the keto tautomer, while for diketones—to the enol tautomer [116]. Usually, the addition of water mediates the tautomerization process [112].

The equilibrium of tautomeric forms is also affected by concentration [116]. An increase in temperature reduces the proportion of the enol form in favor of the diketone form [114]. The amount of keto form may be facilitated by UV irradiation [113], which may be significant for products exposed to light.

4. Acyclic 1,3-Dicarbonyl Compounds

4.1. Curcumin and Its Derivatives

Although curcumin demonstrates various beneficial properties, its application is limited due to its rather low stability. Its decrease in stability is attributed to two main aspects. One of them is the degradation caused by certain pH values, exposure to light, and temperature. These degradation products and general relationships are documented in a recent review article [117]. Briefly, the alkaline conditions, sunlight, and autooxidation lead to (a) scission of the C-C bond in 1,3-dicarbonyl moiety, leading to dehydrozingerone, ferulic acid, vanillin, and other low-molecular phenolic compounds, and (b) free radical-caused dimerization. Experimental data demonstrate that during storage, the amount of curcumin is reduced. However, no clear correlation between the degree of curcumin reduction and the antioxidant and/or antiradical activity is described.

In vivo applications of curcumin-containing compositions are subjected to enzymatic degradation, leading to different dehydrocurcumin derivatives [118,119]. The impact of a fully conjugated system on the antiradical activity of curcumin is analyzed here below. It should be noted that tetrahydrocurcumin, due to its lower autooxidation, exhibits better antioxidant properties in comparison to curcumin [119].

Due to the unique and wide biological activity of curcumin, it is essential to provide effective delivery systems that allow for transport through the gastrointestinal system and storing curcumin-containing products for their effective application. The latest trends in different delivery systems of curcumin are summarized by Chang et al. [120].

The antiradical activity of curcumin (**26**) is provided by both the aromatic rings (as a phenolic antioxidant) and the 1,3-carbonyl moiety (Figure 2).

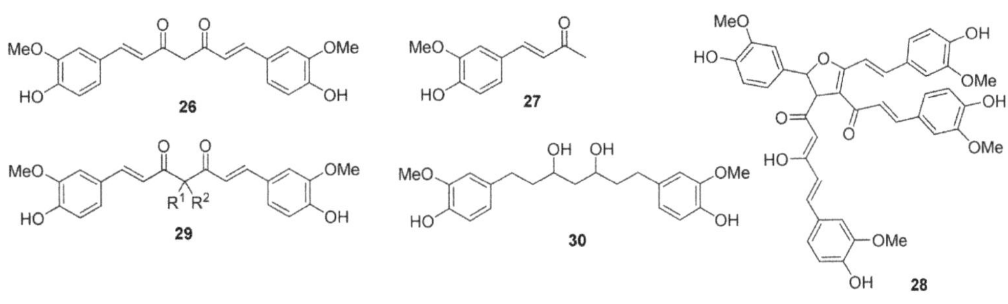

Figure 2. Structures of curcumin and its derivatives: curcumin (**26**), dehydrozingerone (**27**), radical dimerization product **28** of curcumin, α,α-disubstituted curcumin **29**, reduced form ("diol") **30** of curcumin.

For more than a decade, the reasons for providing curcumin's antiradical and antioxidant activity have been discussed. Jovanovic et al. strongly support the role of the methylene moiety in reactions with free radicals. The reaction with the methyl radical justifies this: curcumin rapidly reacted with the radicals generated by pulse radiolysis, while dehydrozingerone **27** (half-curcumin) did not enter the reaction [121]. However, later studies with other radicals do not support this statement. Structure–activity studies of curcumin and its various derivatives have shown that hydroxy groups are crucial for donating a hydrogen atom to *N*- or *O*-centered free radicals. Curcumin inhibited 2,2′-azo*bis*(2-amidinopropane) (AAPH)-induced oxidation of deoxyribonucleic acid (DNA) and erythrocytes, and hemin-induced hemolysis of erythrocytes [122].

Meanwhile, the hydroxyl groups are not essential for reducing the 2,2′-azino-*bis*(3-ethylbenzothiazoline-6-sulfonic acid) (ABTS) cation radical: curcumin and *O*-benzyl-protected curcumin demonstrated similar behavior [122]. The azobisisobutyronitrile

(AIBN)-initiated peroxidation of methyl linoleate and styrene was suppressed by curcumin, while derivatives without free hydroxyl groups did not reduce oxygen uptake. Thus, the unique role of hydroxyl groups over the enolate moiety for quenching peroxyl radicals was demonstrated [123]. Although a hard discussion on the radical formed from curcumin was published, the existence of a C-centered radical is firmly proved by the isolation of oxidation products. In addition to decomposition products (vanillin and ferulic acid), radical dimerization product **28** has been isolated [124]. Litwinienko and Ingold have well established the solvent effect on the reactivity of curcumin with 2,2-diphenyl-1-picrylhydrazyl (DPPH)—the reaction rate remarkably increased in solvents, which facilitated ionization of the antioxidant molecule, thus supporting the SPLET mechanism. These observations postulate that the first step is the formation of a mono-ion in the most acidic enol moiety, followed by electron transfer (Scheme 10) [125].

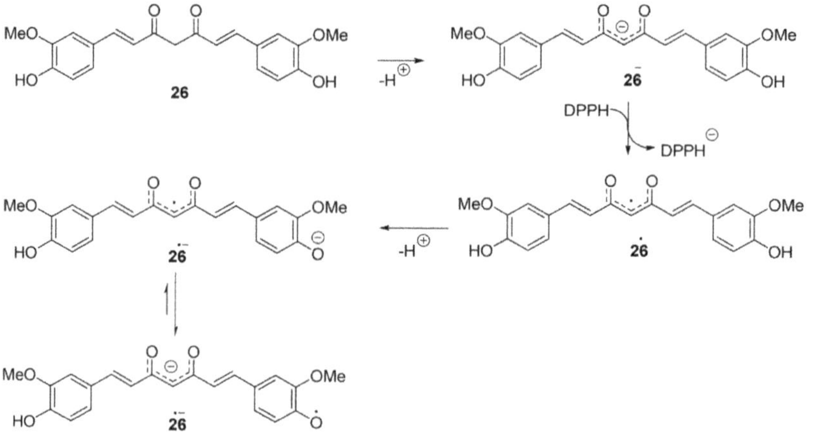

Scheme 10. The plausible mechanism for the reaction between curcumin and DPPH. It is postulated that the first step is dissociation of curcumin **26** in the ions, leading to curcumin anion **26**$^-$. Then, it is expected that an electron is transferred to the DPPH, leading to curcumin radical **26**$^\bullet$. The formed curcumin radical may undergo another proton loss, leading to either O- or C-centered radical anion **26**$^{\bullet-}$.

The role of the dicarbonyl moiety has been demonstrated via compounds **29** containing one or two methyl groups at the α-carbon. The monosubstituted compound **29** was 3- and 24-fold more reactive against DPPH than curcumin **26** and α,α-dimethyl curcumin. The increased reactivity of the monosubstituted derivative compared to curcumin might be due to the lower ionization potential of the former [126]. However, structure–activity correlations for curcumin derivatives give contradictory statements: according to the data provided by Somparn et al., not only the dicarbonyl moiety but also its reduced form (the corresponding diol **30**) may enhance the activity of these compounds in DPPH and linoleic acid oxidation assays [127]. Thus, the role of dicarbonyl fragments is diminished.

Curcumin has demonstrated both antiradical activity and antioxidant activity in sunflower oil in various test systems. The oxidative stability of sunflower oil triglycerides increased by nearly 14-fold in the presence of a curcumin additive (1 mM) [128]. Curcumin effectively reduced the thermal and photo-induced oxidation of β-carotene in emulsions [129]. Aftab and Vieira demonstrated that curcumin is an effective antioxidant for the inhibition of heme-enhanced oxidation: curcumin was more than twice as active as resveratrol [130]. Curcumin has also been studied in various binary mixtures with other antioxidants. A weak synergistic effect (ABTS test) was observed for a mixture of curcumin **26** and (−)-epicatechin **31** or (−)-epicatechin-rich extract of green tea. The synergistic effect is explained by the regeneration of curcumin's phenoxyl radical by an antioxidant with a lower redox potential (0.277 V and ~0.8 V for (−)-epicatechin and curcumin, respectively) (Scheme 11).

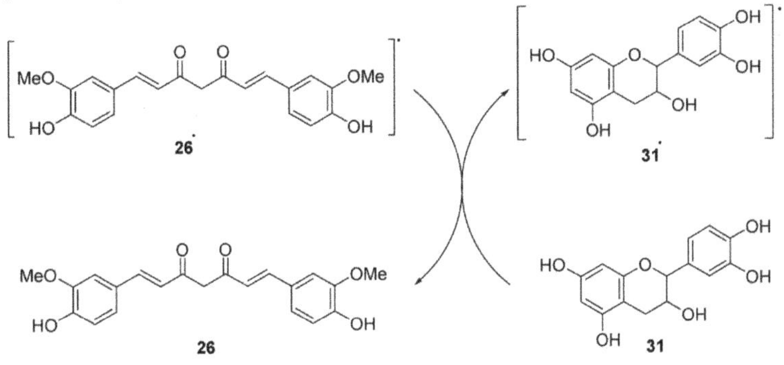

Scheme 11. Plausible synergistic effect between curcumin **26** and (−)-epicatechin **31**. Curcumin **26** is postulated as the primary antioxidants, which quenches free radicals, leading to curcumin radical **26•**. Then, the curcumin may be regenerated through the reaction of curcumin radical **26•** and an antioxidant with a lower redox potential. Thus, (−)-epicatechin **31** regenerate curcumin **26** and the corresponding (−)-epicatechin radical **31•** by-product is formed.

Binary antioxidant mixtures were tested to inhibit the autoxidation of sunflower oil. The results were unpredicted: the mixture containing (−)-epicatechin demonstrated a remarkable synergistic effect, while the mixture with the green tea extract even showed an antagonistic effect [128]. A green tea–curcumin drink helped lower the levels of serum redox, iron, and lipid peroxidation [131]. A synergistic effect was found for an equimolar mixture of curcumin and resveratrol in heme-enhanced oxidation of *N,N,N′,N′*-tetramethyl-1,4-phenylenediamine. This synergistic effect may be caused by the regeneration of curcumin by another antioxidant and the chelating of Fe(III) ions [130]. In addition to phenol-type antioxidants, ascorbate is also helpful in regenerating the parent molecule from the curcumin radical [132].

Hydroxy groups in the phenol rings significantly increase the antiradical activity of curcumin (Figure 3). Curcumin derivatives **32** with three or more hydroxy groups in the phenol rings (at least three of the substituent R^1–R^8 are hydroxy groups) show remarkable antiradical activity (inhibition concentration IC_{50} = 4.6–8.0 μM in the DPPH test) which is up to five times higher than for ascorbic acid (IC_{50} = 25.1 μM) and curcumin itself (IC_{50} = 21 μM). A hydroxy group in the *para*-position (in compound **32**, the R^3 and/or R^7 is OH) is crucial for high antiradical activity. In contrast, compounds with *ortho*-hydroxy-substituted (in compound **32**, the R^1 and/or R^5 are OH) and *para*-methoxy-substituted (in compound **32**, the R^3 and/or R^7 is OMe) aromatic rings do not significantly inhibit DPPH (IC_{50} > 500 μM). Electron-withdrawing groups, such as carboxylic acids, and their esters (in compound **32**, one of the substituents R^1–R^8 is COOH or COOAlk, IC_{50} > 100 μM) or a bromine atom (in compound **32**, one of the substituents R^1–R^8 is Br IC_{50} = 43 μM) tend to reduce the activity of the compound [133,134].

The presence of protons in the α-CH_2 group is essential for the antiradical activity of 1,3-dicarbonyl compounds. Replacing these protons with an arylidene group significantly reduces the antiradical activity of curcumin—DPPH inhibition at a 100 μM concentration is 61% (in compound **33**, R = Ph) and 72% (in compound **33**, R = 4-HO-3-MeO-C_6H_3-). When replaced by a butylidene group (in compound **33**, R = C_3H_7), thus creating a less conjugated system than the arylidene substituent, the antiradical activity is even lower (43%). Curcumin inhibits 91% of DPPH at the same concentration [135]. It can be concluded that the overall degree of conjugation in the molecule is also crucial for the antiradical activity of curcumin.

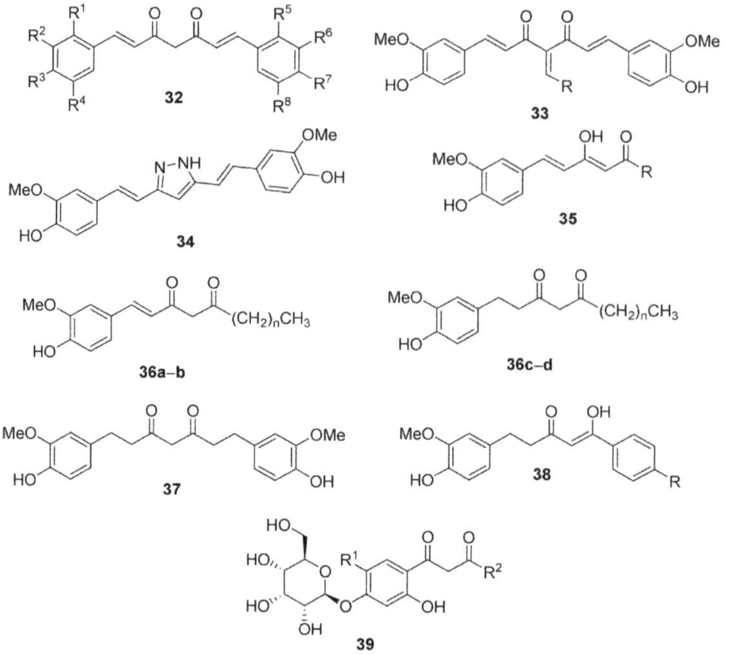

Figure 3. Synthetic analogs of curcumin: curcumin analogs **32** with various substituents in benzene rings, α-arylidene curcumin **33**, pyrazole moiety bearing curcumin derivative **34**, feruloyl acetone derivatives **35**, asymmetric derivatives of curcumin **36** found in ginger, tetrahydrocurcumin **37**, curcumin analogs with shortened conjugate chain **38**, β-D-glucopyranosyl modified acyclic 1,3-diketone derivatives **39**.

Fully conjugated systems also play a vital role in ensuring antioxidant activity. For example, keto–enol tautomerization (thus creating a continuous conjugated chain between the phenol rings) is not possible in compound **34**. The antioxidant activity of this compound is higher when compared to curcumin, as evidenced by the electrochemical oxidation potential of these compounds (for curcumin, E^0 = 0.66 V; for compound **34,** E^0 = 0.62 V) [135].

Shortening of the conjugated system decreases antioxidant activity. For example, in compound **35**, where R = Ph or CF_3, the IC_{50} is about two-fold lower than that of curcumin [134].

Asymmetrical derivatives of curcumin **36** are found in ginger. They show weaker antioxidant activity than curcumin in DPPH and lipid peroxidation tests. However, compared to the reference compounds, butylhydroxyanisole (BHA) and α-tocopherol, their activity against peroxides is relatively high. In this case, compounds with a higher degree of conjugation and a shorter alkyl chain are more effective. Metal ions in the solution influence the inhibition of peroxy radicals. In a test system where Fe^{2+} ions induce lipid peroxides, the antiradical activity of curcumin is much higher than for its asymmetrical analogs **36**. However, when using a test system without the presence of Fe^{2+} ions, this difference is much smaller. The activity of the most potent asymmetrical analog **36** is about 80% of the activity of curcumin. Considering that there are twice as many hydroxy groups in the aromatic system of curcumin, which may provide antiradical activity, the activity of compound **36** is very high. The lipophilicity of the alkyl chain can explain this in the ginger compound. It has also been observed that the activity of compound **36** decreases with the length of the alkyl chain [136].

Tetrahydrocurcumin **37** is a curcumin analog where conjugation is not possible. Still, it has also shown antioxidant activity: against certain types of radicals (e.g., peroxy and alkoxy radicals), it is even higher than that of curcumin (inhibition of lipid peroxidation at a 150 μM concentration for curcumin is 71%, for its derivative 18–83%). The hydrogenated form **37** has better bioavailability, and therefore, it is believed that a large part of dietary curcumin is converted and used in the form of tetrahydrocurcumin [137].

Less conjugated curcumin derivatives **38** (R is H, Me or *tert*-Bu) show weaker antioxidant properties than curcumin—the time to achieve the same concentration of lipid peroxidation products (peroxide value (PV) = 100 meq/kg) for compound **38** is two times shorter than that of curcumin. However, it has been observed that these compounds have a synergic effect with α-tocopherol. Furthermore, this effect is more significant than for the antioxidant mixture of ascorbic acid–tocopherol, which is currently used industrially. The induction period for lipid peroxidation for mixtures of compound **38** and tocopherol is more than 21 h, while for an ascorbic acid–tocopherol mixture at the same concentration, it is about 16 h. Other weak antioxidants could show similar properties in binary mixtures with more potent antioxidants [138].

Although many 1,3-dicarbonyl compounds have high antioxidant activity, they are mostly insoluble in water and have limited bioavailability. To solve this problem, β-D-glucopyranosyl derivatives **39** have been synthesized. Their antioxidant activity (DPPH radical inhibition of 74–85%) is weaker than that of ascorbic acid (98%), but the difference is not significant. Thus, water-soluble 1,3-dicarbonyl-type antioxidants with an activity similar to ascorbic acid could be obtained [139]. To increase the bioavailability, conjugates with pectin are also helpful. These macromolecules tend to self-assemble, leading to nanomicelles with improved stability compared to curcumin [140].

Various systems for delivering curcumin, in addition to the synthesis of water-soluble derivatives, have been developed to increase solubility. The bioavailability of curcumin was increased by introducing it to lecithin-based inverse hexagonal liquid crystals. The antiradical activity of such nanoformulations increased by about 100-fold and 10-fold compared to the free curcumin according to ABTS and DPPH tests, respectively [141]. The increased antiradical activity might be explained by lecithin and castor oil's composition and minor ingredients. Chen et al. loaded curcumin onto CeO_2/SiO_2-PEG nanoparticles for its efficient delivery. Such systems showed antioxidant activity against H_2O_2-induced oxidative damage in Hep G2 cells [142]. Curcumin loaded onto mesoporous silica nanoparticles protected against ROS-induced cell damage [143]. Curcumin-loaded iron oxide nanoparticles reduced oxidative stress parameters in depressed rats [144]. A halloysite–cyclodextrin hybrid material was effective for the encapsulation of curcumin. The proposed material was influential in delivering curcumin and scavenging DPPH [145]. The loading of curcumin on both sodium caseinate (NaCas) and sodium caseinate–laponite (NaCas-LAP) nanoparticles leads to a more soluble form of curcumin. The carrier facilitated the antiradical activity of curcumin against ABTS 3–4-fold higher compared to curcumin. The results were more challenging in the DPPH test: the prepared nanoparticles demonstrated weaker antiradical activity than free curcumin, but a laponite additive to NaCas improved the free radical scavenging activity [146]. Curcumin has also been introduced in nano-emulsions, and its stability through the gastrointestinal tract has been established: it was observed that such emulsions preserve curcumin from degradation in the gastrointestinal tract, and their antiradical activity is greater than that of pure curcumin [147]. Lecithin-encapsulated curcumin demonstrated plasma antioxidant activity [148]. Another delivery system for curcumin with increased antiradical activity is a xanthan–starch matrix [149]. Chitosan-encapsulated nano-curcumin demonstrated in vivo reduction of arsenic-induced oxidative stress [150]. Curcumin loaded onto invitrogen cyanine3-labeled *N*-palmitoyl chitosan (Cy3-NPCS), and poly(1,4-phenyleneacetone dimethylene thioketal) (PPADT) nanoparticles effectively delivered medicine to the inflammation site and reduced oxidative stress [151].

The antiradical activity of the curcumin niobium complex **40** (Figure 4) is comparable to that of ascorbic acid. In the DPPH test, the IC_{50} of the compound is 120 μg/mL (IC_{50} of ascorbic acid—101 μg/mL). Similar results are seen in the ABTS test, where the IC_{50} of the compound is 110 μg/mL (21% of the IC_{50} of ascorbic acid, when comparing molar concentrations) [152]. On the other hand, curcumin and calcium or magnesium complexes (two molecules of curcumin per cation) were synthesized. In these complexes, the counter ion did not influence the free radical scavenging activity (DPPH test) in comparison to curcumin [153].

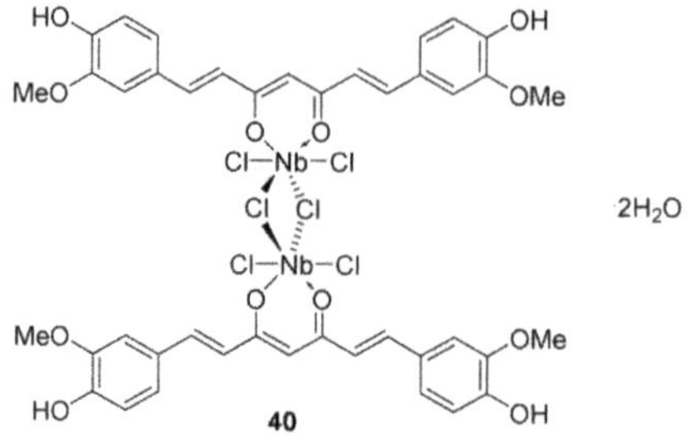

Figure 4. Niobium complex.

4.2. Other Aliphatic 1,3-Dicarbonyl Compounds

Eucalyptus leaves contain various aliphatic 1,3-diketones **41** (Figure 5). These compounds do not have a phenol group, so their antioxidant activity is provided solely by the 1,3-diketone moiety. The most active of them is n-tritriacontane-16,18-dione ($R^1 = R^2 = C_{15}H_{31}$), which shows higher antioxidant properties against peroxy radicals than both butylated hydroxytoluene (BHT) and tocopherol in water/ethanol systems. However, it did not show any inhibition of autooxidation in oil systems. This means that the presence of water is essential for the antioxidant activity of these compounds and keto–enol tautomerization may be essential, wherein the enol form mainly provides the activity. Tautomerization is only one of the factors determining the antioxidant activity of a compound. The alkyl chain length is essential too; e.g., the simplest β-diketone acetylacetone (compound **41** $R^1 = R^2 = CH_3$) has the same keto–enol form ratio as for n-trithriacontan-16,18-dione (1:6), but this compound shows antioxidant properties neither in the oil nor the water/ethanol system [154]. Both R^1 and R^2 in diketone **41** should be long alkyl chains to ensure antioxidant activity. This is demonstrated by the asymmetric compound **41** ($R^1 = CH_3$ and $R^2 = C_{11}H_{23}$), which, similarly to acetylacetone, does not show antioxidant properties.

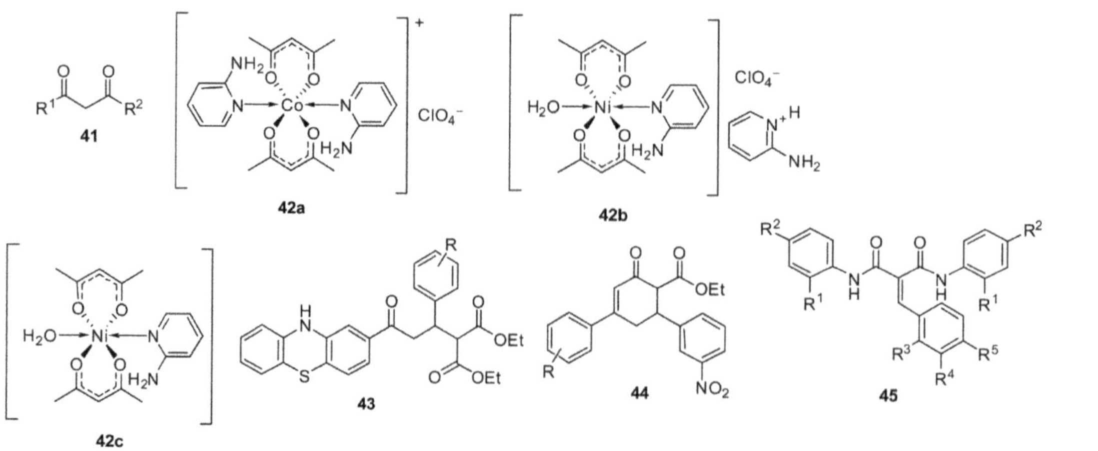

Figure 5. Other acyclic 1,3-dicarbonyl-type antioxidants: general structure of acyclic diketone **41**, acetylacetone complexes with transition metals **42**, α-monosubstituted malonate derivative **43**, curcumin–malonate hybrid **44** and arylidene malonamides **45**.

A β-diketone, which contains an additional hydroxy group (in the compound **41** $R^1 = C_{15}H_{31}$, $R^2 = (CH_2)_{11}CH(OH)CH_2CH_2CH_3$) shows similar activity to its analog without a hydroxy group. This means that hydroxy groups (and the presence of an "isopropanol" unit—a scaffold that is well-known due to its high affinity to oxygen) in the alkyl chains do not affect activity but may potentially improve water solubility, thus making the compound applicable in various water–oil mixtures, such as foods [155].

In the ferric ion-reducing antioxidant power (FRAP) test, both Co(III) and Ni(II) complexes with acetylacetone have higher antiradical activity than ascorbic acid. The anions of the complex play a role in ensuring the overall antioxidant activity of the resulting chelates. For example, the antioxidant activity of complex **42c** is about twice as high as that of complex **42b**. The Co(III) complex **42a** has a lower activity than the Ni(II) complex with the same ligand, which means that the nature of the metal also influences the activity—oxidation state, size, and other factors [156].

Malonate derivative **43** contains a benzyl group at the α-carbon. The benzyl group's type and position of substituents affect the antiradical properties. The unsubstituted compound **43** (R = H) has good antioxidant activity in the ABTS test (IC_{50} = 40.0 µg/mL), which is comparable to the well-known natural antioxidant quercetin (IC_{50} = 44.5 µg/mL). Antiradical activity increases if an electron-donating moiety is present in the benzene ring in the *para*-position (e.g., compound **43** R = 4-Me, IC_{50} = 36.0 µg/mL). If there is an electron-withdrawing group in the *para*-position (e.g., compound **43** R = 4-CHO, IC_{50} = 43.0 µg/mL), it decreases. If the electron-withdrawing group is in the *meta*-position (e.g., compound **43** R = 3-NO_2, IC_{50} = 41.2 µg/mL), the antiradical activity is similar to the unsubstituted analog [157].

Compounds **44** could be considered curcumin and malonate hybrids. However, the antiradical activity of these compounds is fragile (in the DPPH test, the IC_{50} is 4–10-fold higher than that of ascorbic acid). The reason could be a steric hindrance and/or electronic effects, but the exact reason is unknown [158].

The antiradical activity of other malonate derivatives **45** is weak: DPPH radical inhibition at a 1:1 molar ratio (100 µM concentration) is around 0.5–7.0%. The only case where slightly better activity was achieved was when R^4 = MeO and R^5 = OH (inhibition of 51%, IC_{50} = 110 µM). However, the compound most likely acts as a phenol-type antioxidant in this case. Like the curcumin compounds, the lack of α-protons appears to suppress compounds **26** from acting as 1,3-dicarbonyl-type antioxidants [159].

5. Cyclic 1,3-Dicarbonyl Compounds

5.1. Carbocyclic 1,3-Dicarbonyl Compounds

One of the simplest carbocyclic 1,3-dicarbonyl compounds is 1,3-cyclohexadione (Figure 6). Arylmethyl dimedone **46** shows good antiradical activity against DPPH (IC_{50} = 23.0 µM), which is comparable to ascorbic acid (IC_{50} = 25.7 µM) and *tert*-butylhydroq uinone (IC_{50} = 19.5 µM). In the galvinoxyl test, its activity was even greater when compared to the reference compounds: the IC_{50} for compound **46** was 20.3 µM, which is comparable only to *tert*-butylhydroquinone (IC_{50} = 22.6 µM). The activity of other commonly used antioxidants, such as ascorbic acid (IC_{50} = 83.0 µM) or α-tocopherol (IC_{50} = 39.2 µM), under the same conditions, was considerably lower [160].

Additionally, compounds containing two 1,3-diketone moieties were elaborated. Tetraketone **47** shows antioxidant properties. However, no clear relationship has been observed between the position of the substituents in the benzene ring (R = Ar) and the antiradical activity of the compounds. For example, in the DPPH test, one of the most active compounds was *meta*-bromophenyl-substituted tetraketone (IC_{50} = 52.4 µM), but an analog containing chlorine showed no antiradical activity. A cinnamic aldehyde derivative where R is a styryl group exhibited the highest antioxidant activity (IC_{50} = 33.6 µM), which is higher than that of the reference antioxidant *tert*-butyl-4-hydroxyanisole (IC_{50} = 44.7 µM). Compound **47**, where R is BrC = CHPh, also has relatively high activity (IC_{50} = 58.4 µM), so an extended conjugated system may positively affect the antiradical activity of the compound [161].

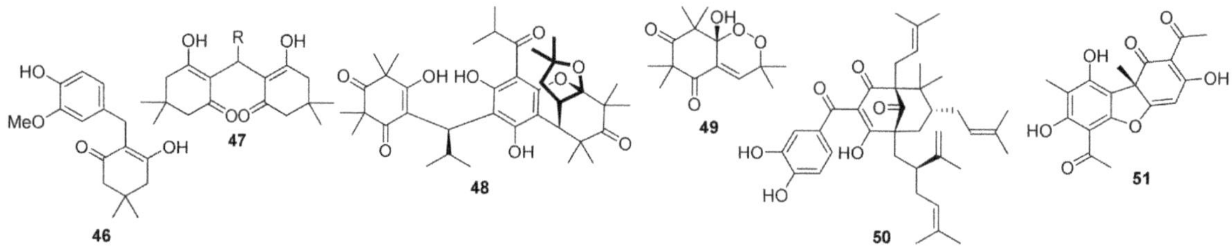

Figure 6. Various cyclic 1,3-diketone-type antioxidants: arylmethyl substituted dimedone **46**, *bis*-dimedone derivatives **47**, myrtucommuacetalone **48**, G3-factor **49**, garcinol **50**, and usnic acid **51**.

The activity of tetraketone derivative **47**, where R is a 3,5-dimethylphenyl group, is comparable to ascorbic acid. In the DPPH test, the *bis*-dimedone derivative inhibited 82% of free radicals (concentration of 100 μg/mL), while ascorbic acid inhibited 80%. When expressed in molar concentrations, approximately twice as much ascorbic acid as *bis*-dimedone is needed to inhibit an equal amount of DPPH radicals [162].

Several compounds containing a 1,3-dicarbonyl group have been isolated from *Myrtus communis* plants. Compounds **48** and **49** have demonstrated a good inhibition capacity against NO· radicals. Compound **49** inhibits 59% NO· at a 25 μg/mL concentration, which is comparable to the reference compound, NG-methyl-L-arginine acetate, which inhibits 66%. Compound **48** is an even more effective inhibitor (inhibiting 82% of NO·) [163].

Some compounds contain several 1,3-dicarbonyl units. One such antioxidant is garcinol. Similarly to compounds **48** and **49**, garcinol **50** showed moderate antioxidant activity in the DPPH test (approximately 85% of the inhibition of ascorbic acid). However, its NO·radical scavenging activity was more significant than for ascorbic acid. It is also a somewhat effective superoxide anion scavenger compared to gallic acid [164,165].

Usnic acid **51** has shown moderate antioxidant activity. In the DPPH test, its IC_{50} value (49.5 μg/mL) was more than five times that of α-tocopherol (IC_{50} = 9.8 μg/mL). However, the O_2^- scavenging ability of usnic acid and α-tocopherol is comparable (IC_{50} = 20.4 and 21.0 μg/mL, respectively) [166]. It has been reported that the interaction of usnic acid with a human blood cell culture promotes mild cell apoptosis and ROS levels and stimulates DNA synthesis. The activity of ursnic acid strongly depends on the concentration; at higher concentrations, it shows cytotoxicity, oxidative stress, and other harmful effects [167].

Carbocycles containing 1,3-dicarbonyl moieties are also found in hops (Figure 7). The antiradical activity (DPPH test) of lupulones **52** and humulones **53** is similar to that exhibited by α-tocopherol and ascorbic acid. It is worth mentioning that acylation of the hydroxy group and the introduction of an additional methyl group (compounds **54** and **55**, respectively) increases the activity of the compounds by up to 10 times. These compounds effectively reduced rat brain lipid peroxidation [168]. The presence of bitter acids is essential for providing antioxidant and/or antiradical activity in hop extracts. The inhibition of hydrogen peroxide increases with the amount of humulone in the extract. More humulone is a more effective scavenger than the phenol-type antioxidant rutin [169]. Wietstock et al. analyzed the antioxidant behavior of bitter acids during wort boiling. A separate analysis of various ingredients showed similar free radical trapping for both α- and β-acids, while iso-α-acid **56**—a heat-induced isomerization product of α-acid—even showed some prooxidant effects. The free radical-scavenging activity of the α-acids was attributed to the presence of three β-keto units, which facilitated the formation of stabilized phenoxyl radicals [170]. In addition, the Fenton and the Haber–Weiss reactions may be limited by the complexation of Fe(II) and Cu(I) ions [171]. Due to the complexation of iron ions, α- and iso-α-acids protected 2-deoxyribose from oxidative degradation. It should be highlighted that α-acids were more effective [170]. Karabin et al. indirectly reported on the antioxidant activity of iso-α-acids. The analysis of iso-α-acids during the storage of lager beer showed a reduction in both the compounds and the antioxidative activity [172]. In contrast, other research indicated that beers with a lower amount of iso-α-acids demonstrate higher antioxidant activity according to an oxygen radical absorbance capacity (ORAC) test [173]. Humulone- and lupulone-rich hop extract demonstrated antioxidant activity against reactive oxygen species in cells under simulated solar irradiation [174]. β-Acids have increased free radical-scavenging properties of chitosan films: a 0.1–0.3% additive of β-acids to the film increases the inhibition of DPPH to 55–75%. Furthermore, these films have been found to be valuable covers for soybean oil packages, thus reducing the autoxidation processes in the oil (the peroxide value of the non-filmed oil after 25 days was >25 meq O_2/kg, while the peroxide value of the samples covered with β-acid-improved chitosan film was around 10 meq O_2/kg) [175].

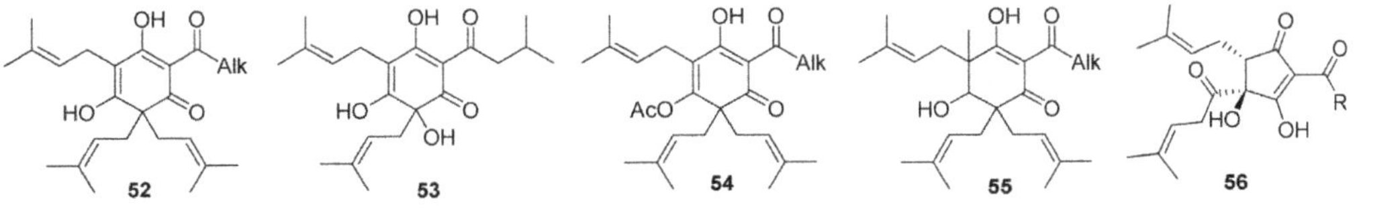

Figure 7. Hop acids: lupulones **52**, humulones **53**, acylated lupulones **54**, methyl-substituted lupulones **55**, and iso-α-acids **56**.

5.2. O-Heterocyclic 1,3-Dicarbonyl Compounds

5.2.1. Dihydropyran-2,4-Diones

Overall, the DPPH radical scavenging activity of compound **57** is relatively weak (Figure 8). However, it can be observed that in cases where R = *para*-$MeOC_6H_4$, the activity of this compound was even weaker. This means that the substituent in the fifth position of the dihydropyran ring affects the activity of the compound [176].

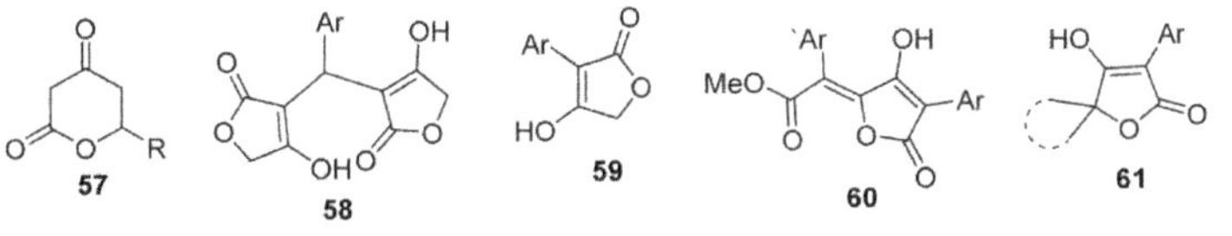

Figure 8. Antioxidants with dihydropyran-2,4-diones moieties.

5-Membered lactones **58–61** have been described, too. In silico and in vitro DPPH assays have turned pulvinic acid derivatives into promising antioxidants. A slight increase in antiradical activity was observed when several moieties of tetronic acid **58** were introduced into the structure [177]. Compounds with the general structure **60** demonstrated good thymidine and plasmid DNA protection against reactive oxygen species and Fenton-type oxidative systems [178].

5.2.2. Meldrum's Acid Derivatives

Various arylmethyl Meldrum's acids have shown good antioxidant properties (Figure 9). It has been observed that a hydroxy group at the *para*-position of the aryl group (in compound **62**, R^3 = OH) improves the activity of the compound (e.g., for the compound, where R^3 = OH and R^2 = OMe, IC_{50} = 20.3 μM, but if R^3 = OH and R^2 = R^4 = Ome, then IC_{50} = 14.5 μM). A fluorine atom (compound **62**, R^3 = F, IC_{50} = 18.2 μM) and an acetoxy group (compound **62**, R^3 = OAc, IC_{50} = 16.8 μM) in this position gives a similar effect. Conversely, a methoxy (compound **62**, R^3 = OMe, IC_{50} = 26.7 μM) or nitro (compound **62**, R^3 = NO_2, IC_{50} = 35.0 μM) *para*-substituted arylmethyl Meldrum's acids have relatively low activity [179]. Some dendritic architectures **63** containing arylmethyl Meldrum's acid moieties as surface groups have also been studied. Among them, the derivatives with a flexible glycerol core were determined to be a promising scaffold for elaborating powerful antioxidants [180]. The role of *C*-centered radicals was proven via trapping the radicals with DPPH with di-substituted Meldrum's acid **65** [179].

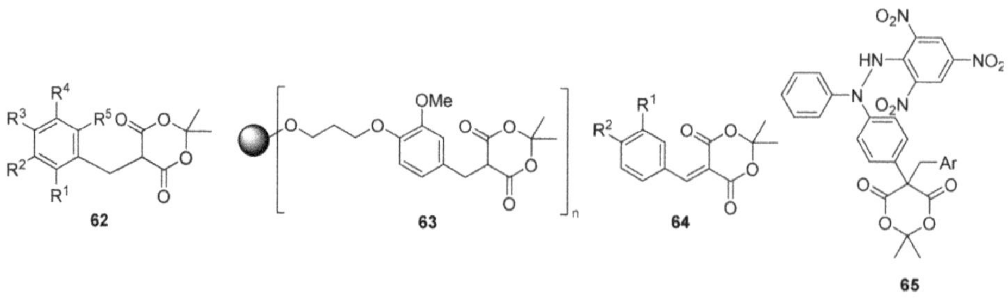

Figure 9. Antioxidants with Meldrum's acid residue: arylmethyl Meldrum's acids **62**, dendrimers decorated with arylmethyl Meldrum's acid units **63**, arylidene Meldrum's acids **64**, and arylmethyl Meldrum's acid–DPPH conjugate **65**.

In contrast, arylidene Meldrum's acids **64**, compared to their reduced analogs **62**, have low antioxidant activity. However, by varying the type and position of the substituents in the aromatic ring, it is possible to improve the activity, e.g., in compound **64**, where R^1 = OMe and R^2= $CH_2CH_2CH_2Br$, IC_{50} = 55.6 μg/mL in the DPPH test. However, it was still significantly less than the reference compound—ascorbic acid (IC_{50} = 6.4 μg/mL) [181]. Compounds with a hydroxy group in the *para*-position and a methoxy (compound **64**, R^2 = OH and R^1 = OMe, IC_{50} = 24 μg/mL) or ethoxy group (compound **64**, R^2 = OH and R^1 = OEt, IC_{50} = 39 μg/mL) in the *meta*-position showed good antiradical activity in the DPPH test in comparison to ascorbic acid (IC_{50} = 22 μg/mL). In these cases, activity is most likely due to the phenol fragment, and the compound acts as a phenol-type antioxidant [182].

5.2.3. Coumarin and Its Derivatives

Unsubstituted 4-hydroxycoumarin **66** (Figure 10) has moderate antiradical activity in the DPPH test (EC_{50} = 3.8 mM) [183–185]. Unsubstituted coumarin also effectively scavenges $HO^{\bullet}$. Electron paramagnetic resonance (EPR) spectra analysis has demonstrated the presence of carbon-centered radicals. It has been suggested that 4-hydroxycoumarin reacts with $HO^{\bullet}$ through radical adduct formation, followed by hydrogen atom abstraction. Compound **70** can be formed through sequential hydrogen atom transfer and radical-radical coupling (Scheme 12) [185].

8-Substituted compounds **74** containing a 4-dihydropyrimidin-2(1*H*)-one or 4-dihydropyrimidin-2(1*H*)-thione unit can be considered 3-unsubstituted 4-hydroxy coumarins. The activity of these compounds varied from 9 to 78% (DPPH test), although a strong correlation between the substituents and the antiradical activity was not observed [186].

The activity of 3-phenyl-4-hydroxycoumarin is slightly lower (EC_{50} = 4.2 mM) than for the unsubstituted 4-hydroxycoumarin. Compounds **75**, where R^2 = H and R^1 = electron-withdrawing group (EWG), showed similar activity to 3-phenyl-4-hydroxycoumarin, for example, chlorine-substituted compounds (in compound **75**, R^1 = *ortho*-Cl, EC_{50} = 4.14 mM or R^1 = *para*-Cl, EC_{50} = 4.15 mM). The antioxidant activity of a naphthyl-substituted 4-hydroxycoumarin **75** (4-hydroxy-3-naphtyl coumarin) was even weaker.

If there is an electron-donating group (EDG) in the benzene ring at the *meta*- or *para*-position (compound **75**, R^1 = *ortho*- or *para*-EDG), the antiradical activity increases [183,184]. The results of the ORAC test show that a *para*-methyl group (compound **75**, R^1 = *para*-CH_3, ORAC = 6.5) or a *meta*-hydroxy group (compound **75**, R^1 = *meta*-OH, ORAC = 4.9) containing 3-phenyl-4-hydroxycoumarins have a higher antioxidant activity than the unsubstituted 4-hydroxycoumarin (**66**, ORAC = 4.2).

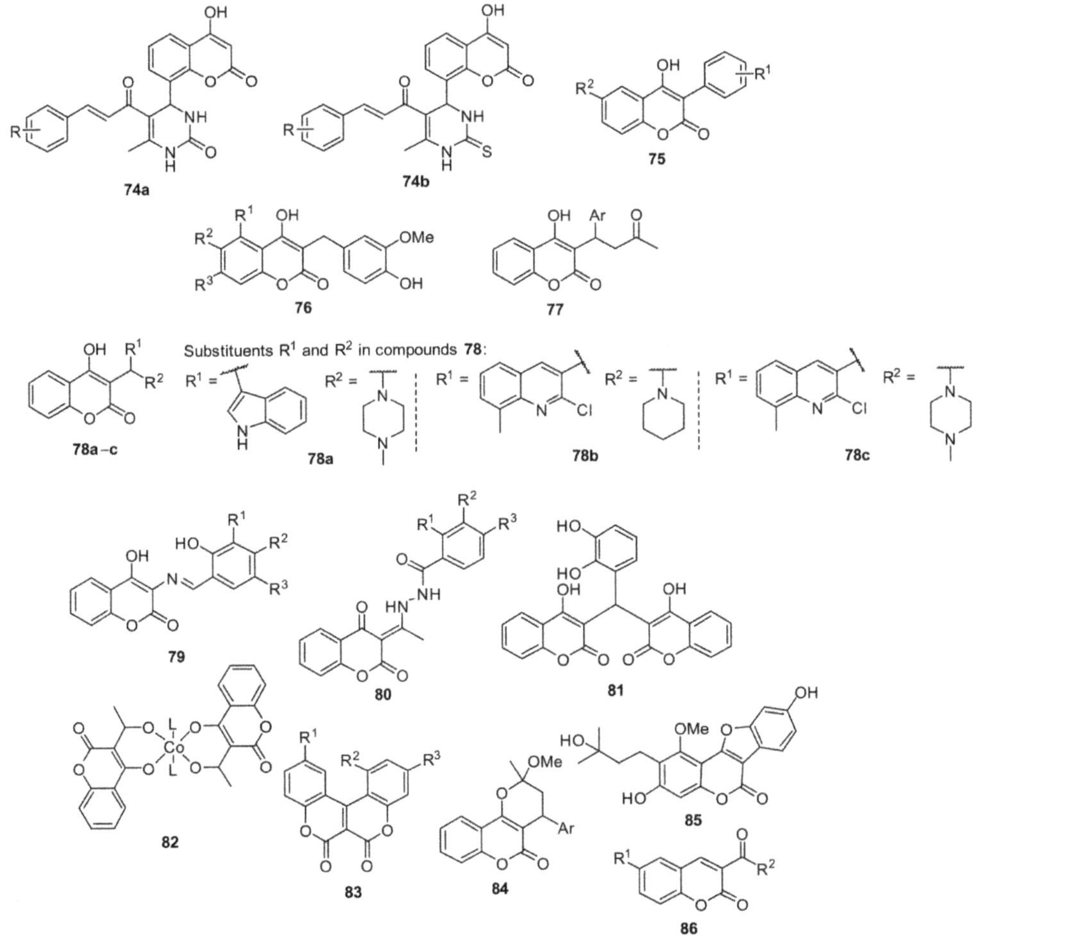

Figure 10. 4-Hydroxycoumarin derivatives: 8-substituted 4-hydroxycoumarins **74**, 3-aryl-4-hydoxycoumarins **75**, 3-arylmethyl-4-hydroxycoumarins **76** and **77**, amino group containing 3-hetarylmethyl 4-hydroxycoumarins **78**, 3-benzylideneamino-4-hydroxycoumarins **79**, 3-aminomethylene-4-hydroxycoumarins **80**, *bis*-(4-hydroxycoumarins) **81**, 3-acetyl-4-hydroxycoumarin–Co complex **82**, coumarin derivatives with rigid **83** or masked **84** and **85** 1,3-dicarbonyl units, and 3-acylcoumarin **86**.

Among compounds **75**, 6-chloro-4-hydroxy-3-(3-hydroxyphenyl) coumarin (R^1 = 3-OH, R^2 = Cl) shows the highest activity. The activity of this compound (ORAC = 7.7) is comparable to the reference compound quercetin (ORAC = 7.3). The presence of an EWG group in the coumarin aromatic ring (compound **75** R^2 = Cl) also improves the antioxidant activity of 4-hydroxycoumarin [187]. The activity of compounds may be affected not only by the electronic characteristics of the substituents but also by steric hindrance. Smaller molecules with better steric availability to the DPPH radical tend to have higher antiradical activity [183].

3-Arylmethyl substituted coumarins **76** also possess considerable antioxidant activity. The vanillin derivatives **76** are more active (IC_{50} = 10.0–36.4 μM) against DPPH radicals than their parent molecules **66** (inhibition at a 1:1 ratio with DPPH is below 15%). Furthermore, the 6-methyl substituted compound **76** (R^2 = Me, IC_{50} = 36.4 μM) shows weaker antioxidant properties than its unsubstituted analog (R^2 = H, IC_{50} = 10.0 μM). The activity of 6-unsubstituted arylmethyl coumarin **76** is even greater than for such antioxidants as ascorbic acid (IC_{50} = 25.7 μM) and *tert*-butylhydroquinone (IC_{50} = 19.5 μM) [183]. It appears to be a tendency for electron-withdrawing groups in the coumarin aromatic ring to increase the antioxidant activity of coumarin derivatives [187], while EWGs decrease it [188]. Several 3-benzyl 4-hydroxy coumarins **77** were analyzed in silico. The results correlate with the experimental data—they are promising structures for constructing powerful antioxidants. According to the preliminary theoretical calculation, the activity does not strongly depend on the substituents [189].

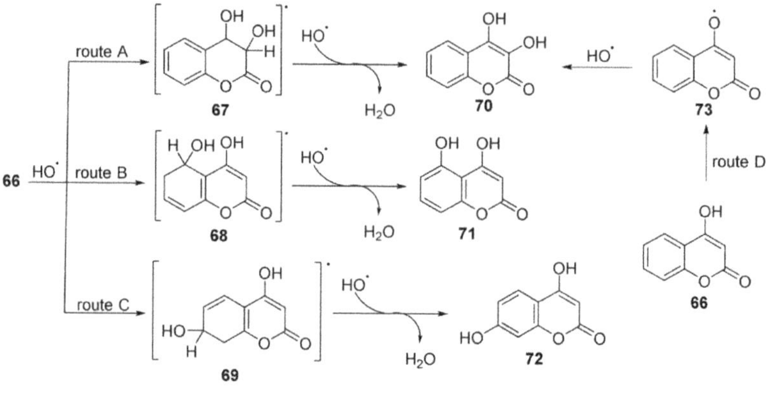

Scheme 12. The plausible mechanism for the reaction between coumarin **66** and hydroxyl radical. Several pathways for the reaction between 4-hydroxycoumarin and hydroxyl radical are provided. Route A foresees the reaction of the hydroxyl radical with the 1,3-dicarbonyl moiety, followed by abstraction of the hydrogen atom, leading to 3-hydroxyl coumarin **70**. The same compound can be obtained through route D: the first step would be homolytic dissociation of the OH bond and a following reaction with hydroxyl radical. Routes B and C foresee the addition of hydroxyl radical to the benzene ring, leading to intermediates **68** and **69**, and sequential re-aromatization of the system leads to 5- or 7-hydroxy coumarins **71** and **72**.

The substitution of 4-hydroxycoumarin in the third position with various nitrogen-containing heterocycles through a methylene group produces compounds with good antioxidant properties. The antiradical activity of compounds **78a–c** in the DPPH test (IC_{50} = 10.5–11 μM) is higher than that of ascorbic acid (IC_{50} = 12.5 μM). Perhaps in these compounds, the methyl groups provide additional radical stabilization in 4-methylpiperazine and the eighth position of quinoline. Other compounds of this type also show good antioxidant properties. Most of them possess higher activity than BHT [190].

The antiradical activity of 4-hydroxycoumarin **79** with an imine moiety in the third position is generally weaker than curcumin (IC_{50} = 39.6 μM). Only in some cases, e.g., compounds **79**, where R^1 = OH (IC_{50} = 48.5 μM), R^1 = R^2 = OH (IC_{50} = 45.8 μM), and especially, R^3 = NO_2 (IC_{50} = 38.6 μM), the IC_{50} value is comparable to the reference compound. This shows that the position of the OH groups in the benzene ring is important. The presence of a nitro group may also positively affect the antioxidant activity of the compound. Conversely, replacing OH groups with OMe, Me groups, or halogens reduces the activity of compounds **79** [191]. Antonijevic et al. have studied some similar compounds **80**, containing hydrazide residue. Compounds that could be considered the derivatives of vanillic acid (R^1 = H, R^2 = OMe, R^3 = OH) demonstrated superior DPPH inhibition compared to other tested compounds. Thus, the role of phenol-type antioxidants in these structures was highlighted [192].

Dicoumarol **81** shows weak antioxidant activity. After 20 min inhibition of ABTS was only 40% at a 2 mM concentration. However, the intramolecular hydrogen bonds within the molecule may hold it in a suitable position to bind to enzymes in biological systems, helping them interact with other antioxidants [193].

4-Hydroxycoumarin chelates **82** show excellent antioxidant properties as well. For compounds **82** (L = H_2O or L = OEt), the IC_{50} in the DPPH test was 0.022 μmol/L and 0.021 μmol/L, respectively, while for the 4-hydroxycoumarin ligand itself, IC_{50} = 0.059 μmol/L [194].

Wang et al. also studied coumarin-fused coumarins **83**. Although the compounds could be considered 1,3-dicarbonyl compounds, they act as phenol-type antioxidants. Theoretical calculations have revealed compound **83** (R^1–R^3 = OH) as the most active one—the compound may be involved even in two HAT or SPLET steps. The compounds in nonpolar media react via the HAT mechanism, but in polar, the SPLET mechanism is preferable [195]. Theoretical calculations for a set of compounds **84** containing a fused cycle, thus restricting keto–enol tautomerism in the 4-hydroxycoumarin system, have been determined to be promising agents for DPPH inhibition [189]. The plausible explanation for the activity of these compounds might be the presence of a benzylic position. A hidden enolate moiety can be seen in glycyrurol **85** isolated from *Glycyrrhiza uralensis*. This coumarin derivative effectively reduced MPP^+-induced reactive oxygen species in PC12D cells. However, it did not demonstrate any significant effect in the *β*-carotene bleaching test and the DPPH system [196].

Few compounds **86** containing a 3-carboxy group, and thus bearing an exocyclic fragment of the 1,3-dicarbonyl compound, have been studied. However, their antiradical activity against DPPH was provided due to 6-hydroxy groups (the inhibition of DPPH reached 60%) [197]. Similar systems were studied by Vazquez-Rodriguez et al. The hydroxyl radical scavenging activity of these compounds is comparable to well-known antioxidants, like quercetin and catechin. However, the activity of these compounds should be attributed to the free hydroxyl groups [198].

5.3. N-Heterocyclic 1,3-Dicarbonyl Compounds

Herein, two groups of *N*-heterocycles are studied (Figure 11): derivatives of barbituric acid and 4-hydroxyquinolinones. The non-substituted compound **87** (R = H) possesses antiradical activity against DPPH (76% inhibition at a 100 μg/mL concentration), which is comparable to ascorbic acid (inhibition is 80%). The presence of halogens in the benzene ring (in compound **87** R = Hal) reduces activity, while a methoxy group (in compound **68** R = OMe) improves activity [162]. In silico and in vitro data have revealed arylidene barbiturates **88** to be promising free radical scavengers [177].

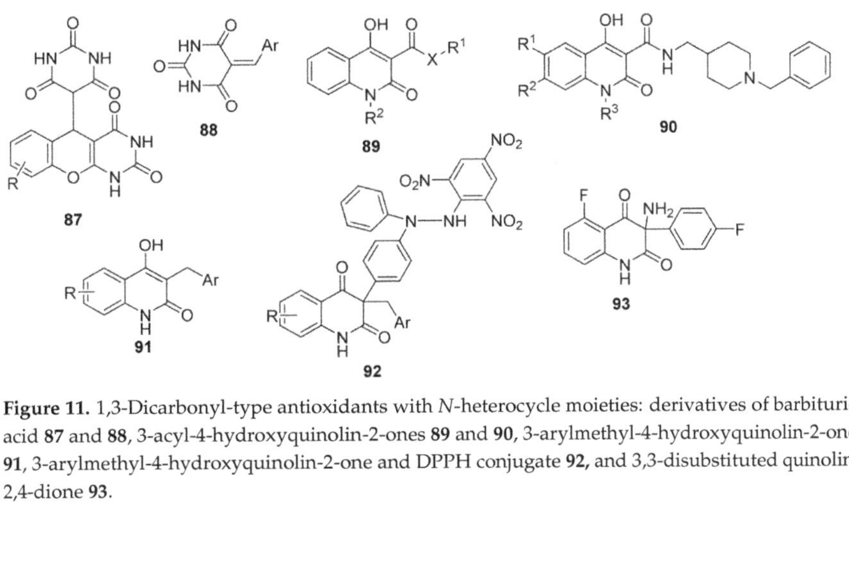

Figure 11. 1,3-Dicarbonyl-type antioxidants with *N*-heterocycle moieties: derivatives of barbituric acid **87** and **88**, 3-acyl-4-hydroxyquinolin-2-ones **89** and **90**, 3-arylmethyl-4-hydroxyquinolin-2-one **91**, 3-arylmethyl-4-hydroxyquinolin-2-one and DPPH conjugate **92**, and 3,3-disubstituted quinolin-2,4-dione **93**.

DPPH test results have shown that compounds **89**, where X = NH and R^1 = 4-$NH_2C_6H_4$, have high antiradical activity. After 20 min at a concentration of 0.1 mM, the DPPH inhibition of these compounds (R^2 = C_6H_5 or CH_3) was 90%, which was higher than for the reference compound nordihydroguaiaretic acid (NDGA) (81%). These compounds have the position of amine in the benzene ring in common. Compared to other compounds of this type, the increased activity is ensured by the absence of intramolecular forces between the primary amine and the amide carbonyl group. Although the antiradical activity in the DPPH test of other derivatives **89** is negligible, compounds where X = O, R^1 = CH_3, R^2 = C_6H_5 or X = NH, R^1 = 2-$NH_2C_6H_4$, R^2 = C_6H_5 showed excellent results in a $HO^\bullet$ quenching test. Their inhibitions were 95% and 98%, respectively. For comparison, Trolox inhibited 88% under the same conditions. The previously mentioned *para*-amino phenyl compounds **89** showed weak activity in the $OH^\bullet$ test (21% and 10%) [199].

The antiradical activity of quinoline derivatives **90** can be improved by introducing hydroxy groups into positions 6- and 7- of the quinoline ring. Compound **90** without free hydroxy groups (R^1 = R^2 = OMe and R^3 = H) did not show antiradical activity in the DPPH test, while the activity of compound **90** (R^1 = R^2 = OH and R^3 = H, effective concentration, EC_{50} = 12.2 μM) is comparable to that of quercetin (EC_{50} = 10.3 μM). Compounds **90** with R^3 = Me (EC_{50} = 21.7 μM) or R^3 = Bn (EC_{50} = 20.8 μM) also showed high activity (higher than curcumin, EC_{50} = 26.6 μM), but the activity of these compounds was weaker than that of compound **90**, where R^3 = H. The *N*-substituent in the quinoline ring also affects the activity of compound **90**: *N*-unsubstituted quinoline is a better antioxidant than the *N*-methyl- or *N*-benzyl substituted analogs **90** [200].

3-Arylmethyl 4-hydroxyquinoliones **91** are effective DPPH and GO quenchers. Although their activity strongly depends on the substituents in the arylmethyl moiety, the presence of 1,3-dicarbonyl moiety is reasonable—the corresponding product **92** was isolated [201].

Antioxidants that perform a specific function are also important—inhibiting certain radicals from treating specific medical disorders. For example, when *cis*-platinum is used in cancer treatment, patients often experience hearing impairment caused by additional ROS generation. Compound **93** has shown an ability to inhibit these ROS, thus preventing hearing damage [202].

6. Conclusions

The 1,3-dicarbonyl unit is found in many compounds with antiradical and antioxidant properties. Evidence for the crucial role of the 1,3-dicarbonyl moiety for its antiradical activity has been demonstrated for various compounds in different test systems. On the other hand, competing studies are suggesting the dominant role of the substituents in the benzene ring, thus revealing these compounds as antioxidants—mainly as classical phenol-type antioxidants. However, the current information has revealed 1,3-dicarbonyl compounds as promising scaffolds for promising antioxidants. Further investigations into this series of compounds should result in excellent antioxidants due to the presence of acidic C-H bonds, thus leading to compounds that may be easily adjusted both for lipophilic and polar media. The general correlation between the structural elements and antiradical/antioxidant activity are as follows: (a) both acyclic and cyclic α-monosubstituted 1,3-dicarbonyl compounds demonstrate increased activity in comparison both with unsubstituted and α,α-disubstituted compounds; (b) the acyclic structures are more sensitive to various structural modifications; (c) the antiradical activity of cyclic 1,3-dicarbonyl compounds are less affected by the structure of the heterocycle. Thus, cyclic 1,3-dicarbonyl compounds could be more effective for constructing powerful antioxidants.

Author Contributions: L.B. and I.M. contributed equally to the preparation of the presented manuscript: L.B. prepared the original draft version; I.M. reviewed and edited the manuscript. All authors have read and agreed to the published version of the manuscript.

Funding: This research was funded by the Latvian Council of Science, grant LZP-2020/2-0165.

Institutional Review Board Statement: Not applicable.

Informed Consent Statement: Not applicable.

Data Availability Statement: Not applicable.

Conflicts of Interest: The authors declare no conflict of interest.

Sample Availability: Not applicable.

References

1. Lobo, V.; Patil, A.; Phatak, A.; Chandra, N. Free radicals, antioxidants and functional foods: Impact on human health. *Pharmacogn. Rev.* **2010**, *4*, 118–126. [CrossRef]
2. Pham-Huy, L.A.; He, H.; Pham-Huy, C. Free radicals, antioxidants in disease and health. *Int. J. Biomed. Sci.* **2008**, *4*, 89–96. [PubMed]
3. Akhigbe, R.; Ajayi, A. The impact of reactive oxygen species in the development of cardiometabolic disorders: A review. *Lipids Health Dis.* **2021**, *20*, 23. [CrossRef] [PubMed]
4. Mehdi, M.M.; Solanki, P.; Singh, P. Oxidative stress, antioxidants, hormesis and calorie restriction: The current perspective in the biology of aging. *Arch. Gerontol. Geriatr.* **2021**, *95*, 104413. [CrossRef] [PubMed]
5. Shah, A.K.; Bhullar, S.K.; Elimban, V.; Dhalla, N.S. Oxidative stress as a mechanism for functional alterations in cardiac hypertrophy and heart failure. *Antioxidants* **2021**, *10*, 931. [CrossRef] [PubMed]
6. Dorszewska, J.; Kowalska, M.; Prendecki, M.; Piekut, T.; Kozłowska, J.; Kozubski, W. Oxidative stress factors in Parkinson's disease. *Neural Regen. Res.* **2021**, *16*, 1383–1391. [CrossRef]
7. Fang, C.; Gu, L.; Smerin, D.; Mao, S.; Xiong, X. The Interrelation between Reactive Oxygen Species and Autophagy in Neurological Disorders. *Oxid. Med. Cell. Longev.* **2017**, *2017*, 8495160. [CrossRef]
8. Ramundo, V.; Giribaldi, G.; Aldieri, E. Transforming growth factor-β and oxidative stress in cancer: A crosstalk in driving tumor transformation. *Cancers* **2021**, *13*, 3093. [CrossRef]
9. Farhoosh, R. New insights into the kinetic and thermodynamic evaluations of lipid peroxidation. *Food Chem.* **2022**, *375*, 131659. [CrossRef]
10. Feng, J.; Berton-Carabin, C.C.; Fogliano, V.; Schroen, K. Maillard reaction products as functional components in oil-in-water emulsions: A review highlighting interfacial and antioxidant properties. *Trends Food Sci. Technol.* **2022**, *121*, 129–141. [CrossRef]
11. Lozano-Castellon, J.; Rinaldi de Alvarenga, J.F.; Vallverdu-Queralt, A.; Lamuela-Raventos, R.M. Cooking with extra-virgin olive oil: A mixture of food components to prevent oxidation and degradation. *Trends Food Sci. Technol.* **2022**, *123*, 28–36. [CrossRef]
12. Longanesi, L.; Pereira, A.P.; Johnston, N.; Chuck, C.J. Oxidative stability of biodiesel: Recent insights. *Biofuels Bioprod. Biorefining* **2022**, *16*, 265–289. [CrossRef]
13. Celina, M.C. Review of polymer oxidation and its relationship with materials performance and lifetime prediction. *Polym. Degrad. Stab.* **2013**, *98*, 2419–2429. [CrossRef]
14. Yousif, E.; Haddad, R. Photodegradation and photostabilization of polymers, especially polystyrene: Review. *SpringerPlus* **2013**, *2*, 398. [CrossRef] [PubMed]
15. Song, C.C.; Du, F.S.; Li, Z.C. Oxidation-responsive polymers for biomedical applications. *J. Mater. Chem. B* **2014**, *2*, 3413–3426. [CrossRef]
16. Amaral, A.B.; Da Solva, M.V.; Lannes, S.C.D.S. Lipid oxidation in meat: Mechanisms and protective factors—A review. *Food Sci. Technol.* **2018**, *38*, 1–15. [CrossRef]
17. Lund, M.N. Reactions of plant polyphenols in foods: Impact of molecular structure. *Trends Food Sci. Technol.* **2021**, *112*, 241–251. [CrossRef]
18. Ricardo, I.A.; Alberto, E.A.; Silva Júnior, A.H.; Macuvele, D.L.P.; Padoin, N.; Soares, C.; Gracher Riella, H.; Starling, M.C.V.M.; Trovó, A.G. A critical review on microplastics, interaction with organic and inorganic pollutants, impacts and effectiveness of advanced oxidation processes applied for their removal from aqueous matrices. *Chem. Eng. J.* **2021**, *424*, 130282. [CrossRef]
19. Bolujoko, N.B.; Unuabonah, E.I.; Alfred, M.O.; Ogunlaja, A.; Ogunlaja, O.O.; Omorogie, M.O.; Olukanni, O.D. Toxicity and removal of parabens from water: A critical review. *Sci. Total Environ.* **2021**, *792*, 148092. [CrossRef]
20. Maria, M.C.; de Mendonça Neto, R.P.; Pires, G.F.F.; Vilela, P.B.; Amorim, C.C. Combat of antimicrobial resistance in municipal wastewater treatment plant effluent via solar advanced oxidation processes: Achievements and perspectives. *Sci. Total Environ.* **2021**, *786*, 147448. [CrossRef]
21. Garcia, C.; Blesso, C.N. Antioxidant properties of anthocyanins and their mechanism of action in atherosclerosis. *Free Radic. Biol. Med.* **2021**, *172*, 152–166. [CrossRef] [PubMed]
22. Khan, J.; Deb, P.K.; Priya, S.; Medina, K.D.; Devi, R.; Walode, S.G.; Rudrapal, M. Dietary flavonoids: Cardioprotective potential with antioxidant effects and their pharmacokinetic, toxicological and therapeutic concerns. *Molecules* **2021**, *26*, 4021. [CrossRef] [PubMed]
23. Perron, N.R.; Brumaghim, J.L. A review of the antioxidant mechanisms of polyphenol compounds related to iron binding. *Cell Biochem. Biophys.* **2009**, *53*, 75–100. [CrossRef] [PubMed]

24. Masuda, T.; Inaba, Y.; Takeda, Y. Antioxidant Mechanism of Carnosic Acid: Structural Identification of Two Oxidation Products. *J. Agric. Food Chem.* **2001**, *49*, 5560–5565. [CrossRef]
25. Adhikari, S.; Crehuet, R.; Anglada, J.M.; Francisco, J.S.; Xia, Y. Two-step reaction mechanism reveals new antioxidant capability of cysteine disulfides against hydroxyl radical attack. *Proc. Natl. Acad. Sci. USA* **2020**, *117*, 18216–18223. [CrossRef]
26. Belaya, N.I.; Belyi, A.V.; Davydova, A.A. Mechanism of the Antiradical Action of Natural Phenylpropanoids in Nonionizing Polar Media. *Kinet. Catal.* **2020**, *61*, 839–845. [CrossRef]
27. Gerasimova, E.; Gazizullina, E.; Radosteva, E.; Ivanova, A. Antioxidant and antiradical properties of some examples of flavonoids and coumarins—Potentiometric studies. *Chemosensors* **2021**, *9*, 112. [CrossRef]
28. Boulebd, H. DFT study of the antiradical properties of some aromatic compounds derived from antioxidant essential oils: C–H bond vs. O–H bond. *Free Radic. Res.* **2019**, *53*, 1125–1134. [CrossRef]
29. Gažák, R.; Sedmera, P.; Vrbacký, M.; Vostálová, J.; Drahota, Z.; Marhol, P.; Walterová, D.; Křen, V. Molecular mechanisms of silybin and 2,3-dehydrosilybin antiradical activity-role of individual hydroxyl groups. *Free Radic. Biol. Med.* **2009**, *46*, 745–758. [CrossRef]
30. Xue, Y.; Liu, Y.; Luo, Q.; Wang, H.; Chen, R.; Liu, Y.; Li, Y. Antiradical Activity and Mechanism of Coumarin-Chalcone Hybrids: Theoretical Insights. *J. Phys. Chem. A* **2018**, *122*, 8520–8529. [CrossRef]
31. Parcheta, M.; Świsłocka, R.; Orzechowska, S.; Akimowicz, M.; Choińska, R.; Lewandowski, W. Recent developments in effective antioxidants: The structure and antioxidant properties. *Materials* **2021**, *14*, 1984. [CrossRef] [PubMed]
32. Dunaway, S.; Odin, R.; Zhou, L.; Ji, L.; Zhang, Y.; Kadekaro, A.L. Natural Antioxidants: Multiple Mechanisms to Protect Skin from Solar Radiation. *Front. Pharmacol.* **2018**, *9*, 392. [CrossRef] [PubMed]
33. Dadali, V.A.; Tutelyan, V.A.; Dadali, Y.V.; Kravchenko, L.V. Carotenoids. Bioavailability, biotransformation, antioxidant properties. *Vopr. Pitan.* **2010**, *79*, 4–18. [PubMed]
34. Lu, W.; Shi, Y.; Wang, R.; Su, D.; Tang, M.; Liu, Y.; Li, Z. Antioxidant activity and healthy benefits of natural pigments in fruits: A review. *Int. J. Mol. Sci.* **2021**, *22*, 4945. [CrossRef] [PubMed]
35. Burda, S.; Oleszek, W. Antioxidant and Antiradical Activities of Flavonoids. *J. Agric. Food Chem.* **2001**, *49*, 2774–2779. [CrossRef] [PubMed]
36. Basappa, V.C.; Ramaiah, S.; Penubolu, S.; Kariyappa, A.K. Recent developments on the synthetic and biological applications of chalcones—A review. *Biointerface Res. Appl. Chem.* **2022**, *12*, 180–195. [CrossRef]
37. Gonçalves, A.C.; Nunes, A.R.; Falcão, A.; Alves, G.; Silva, L.R. Dietary Effects of Anthocyanins in Human Health: A Comprehensive Review. *Pharmaceuticals* **2021**, *14*, 690. [CrossRef]
38. Kostova, I.; Bhatia, S.; Grigorov, P.; Balkansky, S.; Parmar, V.S.; Prasad, A.K.; Saso, L. Coumarins as Antioxidants. *Curr. Med. Chem.* **2012**, *18*, 3929–3951. [CrossRef]
39. Kostova, I. Synthetic and Natural Coumarins as Antioxidants. *Mini-Rev. Med. Chem.* **2006**, *6*, 365–374. [CrossRef]
40. Acosta-Quiroga, K.; Rojas-Peña, C.; Nerio, L.S.; Gutiérrez, M.; Polo-Cuadrado, E. Spirocyclic derivatives as antioxidants: A review. *RSC Adv.* **2021**, *11*, 21926–21954. [CrossRef]
41. Naidi, S.N.; Harunsani, M.H.; Tan, A.L.; Khan, M.M. Green-synthesized CeO_2 nanoparticles for photocatalytic, antimicrobial, antioxidant and cytotoxicity activities. *J. Mater. Chem. B* **2021**, *9*, 5599–5620. [CrossRef] [PubMed]
42. Abdelnour, S.A.; Alagawany, M.; Hashem, N.M.; Farag, M.R.; Alghamdi, E.S.; Ul Hassan, F.; Bila, R.M.; Elnesr, S.S.; Dawood, M.A.O.; Nagadi, S.A.; et al. Nanominerals: Fabrication methods, benefits and hazards, and their applications in ruminants with special reference to selenium and zinc nanoparticles. *Animals* **2021**, *11*, 1916. [CrossRef] [PubMed]
43. Genovese, D.; Baschieri, A.; Vona, D.; Baboi, R.E.; Mollica, F.; Prodi, L.; Amorati, R.; Zaccheroni, N. Nitroxides as Building Blocks for Nanoantioxidants. *ACS Appl. Mater. Interfaces* **2021**, *13*, 31996–32004. [CrossRef]
44. Piccinino, D.; Capecchi, E.; Tomaino, E.; Gabellone, S.; Gigli, V.; Avitabile, D.; Saladino, R. Nano-structured lignin as green antioxidant and uv shielding ingredient for sunscreen applications. *Antioxidants* **2021**, *10*, 274. [CrossRef] [PubMed]
45. Yeo, J.; Lee, J.; Lee, S.; Kim, W.J. Polymeric Antioxidant Materials for Treatment of Inflammatory Disorders. *Adv. Ther.* **2021**, *4*, 2000270. [CrossRef]
46. Anderson, A.M.; Mitchell, M.S.; Mohan, R.S. Isolation of Curcumin from Turmeric. *J. Chem. Educ.* **2000**, *77*, 359–360. [CrossRef]
47. Hatcher, H.; Planalp, R.; Cho, J.; Torti, F.M.; Torti, S.V. Curcumin: From ancient medicine to current clinical trials. *Cell. Mol. Life Sci.* **2008**, *65*, 1631–1652. [CrossRef]
48. Wigner, P.; Bijak, M.; Saluk-bijak, J. The green anti-cancer weapon. The role of natural compounds in bladder cancer treatment. *Int. J. Mol. Sci.* **2021**, *22*, 7787. [CrossRef]
49. Adeluola, A.; Zulfiker, A.H.M.; Brazeau, D.; Amin, A.R.M.R. Perspectives for synthetic curcumins in chemoprevention and treatment of cancer: An update with promising analogues. *Eur. J. Pharmacol.* **2021**, *906*, 174266. [CrossRef]
50. Wang, H.; Zhang, K.; Liu, J.; Yang, J.; Tian, Y.; Yang, C.; Li, Y.; Shao, M.; Su, W.; Song, N. Curcumin Regulates Cancer Progression: Focus on ncRNAs and Molecular Signaling Pathways. *Front. Oncol.* **2021**, *11*, 1202. [CrossRef]
51. Jahanbakhshi, F.; Maleki Dana, P.; Badehnoosh, B.; Yousefi, B.; Mansournia, M.A.; Jahanshahi, M.; Asemi, Z.; Halajzadeh, J. Curcumin anti-tumor effects on endometrial cancer with focus on its molecular targets. *Cancer Cell Int.* **2021**, *21*, 120. [CrossRef] [PubMed]
52. Yao, Z.; Le, T.H.; Du, Q.; Mu, H.; Liu, C.; Zhu, Y. The Potential Clinical Value of Curcumin and its Derivatives in Colorectal Cancer. *Anti-Cancer Agents Med. Chem.* **2020**, *21*, 1626–1637. [CrossRef] [PubMed]

53. Javed, Z.; Khan, K.; Rasheed, A.; Sadia, H.; Shahwani, M.N.; Irshad, A.; Raza, S.; Salehi, B.; Sharifi-Rad, J.; Suleria, H.A.R.; et al. Targeting androgen receptor signaling with MicroRNAs and Curcumin: A promising therapeutic approach for Prostate Cancer Prevention and intervention. *Cancer Cell Int.* **2021**, *21*, 77. [CrossRef] [PubMed]
54. Akbari, A.; Sedaghat, M.; Heshmati, J.; Tabaeian, S.P.; Dehghani, S.; Pizarro, A.B.; Rostami, Z.; Agah, S. Molecular mechanisms underlying curcumin-mediated microRNA regulation in carcinogenesis; Focused on gastrointestinal cancers. *Biomed. Pharmacother.* **2021**, *141*, 111849. [CrossRef]
55. Kong, W.-Y.; Ngai, S.C.; Goh, B.-H.; Lee, L.-H.; Htar, T.-T.; Chuah, L.-H. Is Curcumin the Answer to Future Chemotherapy Cocktail? *Molecules* **2021**, *26*, 4329. [CrossRef]
56. Różański, G.; Kujawski, S.; Newton, J.L.; Zalewski, P.; Słomko, J. Curcumin and biochemical parameters in metabolic-associated fatty liver disease (MAFLD)—A review. *Nutrients* **2021**, *13*, 2654. [CrossRef]
57. Labanca, F.; Ullah, H.; Khan, H.; Milella, L.; Xiao, J.; Dajic-Stevanovic, Z.; Jeandet, P. Therapeutic and Mechanistic Effects of Curcumin in Huntington's Disease. *Curr. Neuropharmacol.* **2020**, *19*, 1007–1018. [CrossRef]
58. Marton, L.T.; Pescinini-e-Salzedas, L.M.; Camargo, M.E.C.; Barbalho, S.M.; Haber, J.F.D.S.; Sinatora, R.V.; Detregiachi, C.R.P.; Girio, R.J.S.; Buchaim, D.V.; Cincotto dos Santos Bueno, P. The Effects of Curcumin on Diabetes Mellitus: A Systematic Review. *Front. Endocrinol.* **2021**, *12*, 669448. [CrossRef]
59. Yang, J.; Miao, X.; Yang, F.J.; Cao, J.F.; Liu, X.; Fu, J.L.; Su, G.F. Therapeutic potential of curcumin in diabetic retinopathy (Review). *Int. J. Mol. Med.* **2021**, *47*, 75. [CrossRef]
60. Silvestro, S.; Sindona, C.; Bramanti, P.; Mazzon, E. A state of the art of antioxidant properties of curcuminoids in neurodegenerative diseases. *Int. J. Mol. Sci.* **2021**, *22*, 3168. [CrossRef]
61. Lee, W.-H.; Loo, C.-Y.; Bebawy, M.; Luk, F.; Mason, R.; Rohanizadeh, R. Curcumin and its Derivatives: Their Application in Neuropharmacology and Neuroscience in the 21st Century. *Curr. Neuropharmacol.* **2013**, *11*, 338–378. [CrossRef]
62. Barbalho, S.M.; de Sousa Gonzaga, H.F.; de Souza, G.A.; de Alvares Goulart, R.; de Sousa Gonzaga, M.L.; de Alvarez Rezende, B. Dermatological effects of Curcuma species: A systematic review. *Clin. Exp. Dermatol.* **2021**, *46*, 825–833. [CrossRef]
63. Rodrigues, F.C.; Kumar, N.A.; Thakur, G. The potency of heterocyclic curcumin analogues: An evidence-based review. *Pharmacol. Res.* **2021**, *166*, 105489. [CrossRef]
64. Prasad, S.; Dubourdieu, D.; Srivastava, A.; Kumar, P.; Lall, R. Metal–curcumin complexes in therapeutics: An approach to enhance pharmacological effects of curcumin. *Int. J. Mol. Sci.* **2021**, *22*, 7094. [CrossRef]
65. Kabir, M.T.; Rahman, M.H.; Akter, R.; Behl, T.; Kaushik, D.; Mittal, V.; Pandey, P.; Akhtar, M.F.; Saleem, A.; Albadrani, G.M.; et al. Potential role of curcumin and its nanoformulations to treat various types of cancers. *Biomolecules* **2021**, *11*, 392. [CrossRef] [PubMed]
66. D'Angelo, N.A.; Noronha, M.A.; Kurnik, I.S.; Câmara, M.C.C.; Vieira, J.M.; Abrunhosa, L.; Martins, J.T.; Alves, T.F.R.; Tundisi, L.L.; Ataide, J.A.; et al. Curcumin encapsulation in nanostructures for cancer therapy: A 10-year overview. *Int. J. Pharm.* **2021**, *604*, 120534. [CrossRef]
67. Mahjoob, M.; Stochaj, U. Curcumin nanoformulations to combat aging-related diseases. *Ageing Res. Rev.* **2021**, *69*, 101364. [CrossRef] [PubMed]
68. Trigo-gutierrez, J.K.; Vega-chacón, Y.; Soares, A.B.; Mima, E.G.d.O. Antimicrobial activity of curcumin in nanoformulations: A comprehensive review. *Int. J. Mol. Sci.* **2021**, *22*, 7130. [CrossRef] [PubMed]
69. Guo, Y.L.; Li, X.Z.; Kuang, C.T. Antioxidant pathways and chemical mechanism of curcumin. *Adv. Mater. Res.* **2011**, *236–238*, 2311–2314. [CrossRef]
70. Hunyadi, A. The mechanism(s) of action of antioxidants: From scavenging reactive oxygen/nitrogen species to redox signaling and the generation of bioactive secondary metabolites. *Med. Res. Rev.* **2019**, *39*, 2505–2533. [CrossRef]
71. Heffernan, C.; Ukrainczyk, M.; Gamidi, R.K.; Hodnett, B.K.; Rasmuson, Å.C. Extraction and Purification of Curcuminoids from Crude Curcumin by a Combination of Crystallization and Chromatography. *Org. Process Res. Dev.* **2017**, *21*, 821–826. [CrossRef]
72. Pawar, H.A.; Gavasane, A.J.; Choudhary, P.D. A Novel and Simple Approach for Extraction and Isolation of Curcuminoids from Turmeric Rhizomes. *Nat. Prod. Chem. Res.* **2018**, *6*, 1000300. [CrossRef]
73. Verghese, J. Isolation of curcumin from *Curcuma longa* L. rhizome. *Flavour Fragr. J.* **1993**, *8*, 315–319. [CrossRef]
74. Priyadarsini, K.I. The chemistry of curcumin: From extraction to therapeutic agent. *Molecules* **2014**, *19*, 20091–20112. [CrossRef] [PubMed]
75. Wang, L.; Han, X.; Wang, F.; Sun, L.; Xin, F. Research progresses in the biosynthesis of curcuminoids. *Shengwu Gongcheng Xuebao/Chin. J. Biotechnol.* **2021**, *37*, 404–417. [CrossRef]
76. Marchi, R.C.; Campos, I.A.S.; Santana, V.T.; Carlos, R.M. Chemical implications and considerations on techniques used to assess the in vitro antioxidant activity of coordination compounds. *Coordin. Chem. Rev.* **2022**, *451*, 214275. [CrossRef]
77. Tirzitis, G.; Bartosz, G. Determination of antiradical and antioxidant activity: Basic principles and new insights. *Acta Biochim. Pol.* **2010**, *57*, 139–142. [CrossRef]
78. Kedare, S.B.; Singh, R.P. Genesis and development of DPPH method of antioxidant assay. *J. Food Sci. Technol.* **2011**, *48*, 412–422. [CrossRef]
79. Litwinienko, G.; Ingold, K.U. Abnormal solvent effects on hydrogen atom abstractions. 1. The reactions of phenols with 2,2-diphenyl-1-picrylhydrazyl (dpph•) in alcohols. *J. Org. Chem.* **2003**, *68*, 3433–3438. [CrossRef]

80. Gerasimova, E.; Gazizullina, E.; Kolbaczkaya, S.; Ivanova, A. The novel potentiometric approach to antioxidant capacity assay based on the reaction with stable radical 2,2′-diphenyl-1-picrylhydrazyl. *Antioxidants* **2022**, *11*, 1974. [CrossRef]
81. Sethiya, N.K.; Raja, M.K.M.M.; Mishra, S.H. Antioxidant markers based TLC-DPPH differentiation on four commercialized botanical sources of *Shankhpushpi* (A Medhya Rasayana): A preliminary assessment. *J. Adv. Pharm. Technol. Res.* **2013**, *4*, 25–30. [CrossRef] [PubMed]
82. Wu, J.-H.; Huang, C.-Y.; Tung, Y.-T.; Chang, S.-T. Online RP-HPLC-DPPH screening method for detection of radical-scavenging phytochemicals from flowers of *Acacia confusa*. *J. Agric. Food. Chem.* **2008**, *56*, 328–332. [CrossRef] [PubMed]
83. Condezo-Hoyos, L.; Abderrahim, F.; Arriba, S.M.; Gonzalez, M.C. A novel, micro, rapid and direct assay to assess total antioxidant capacity of solid foods. *Talanta* **2015**, *138*, 108–116. [CrossRef] [PubMed]
84. Ciesla, L.; Kryszen, J.; Stochmal, A.; Oleszek, W.; Waksmundzka-Hajnos, M. Approach to develop a standardized TLC-DPPH• radical dot test for assessing free radical scavenging properties of selected phenolic compounds. *J. Pharm. Biomed. Anal.* **2012**, *70*, 126–135. [CrossRef] [PubMed]
85. Musa, K.H.; Abdullah, A.; Kuswandi, B.; Hidayant, H.A. A novel high throughput method based on the DPPH dry reagent array for determination of antioxidant activity. *Food Chem.* **2013**, *141*, 4102–4106. [CrossRef] [PubMed]
86. Hidayat, M.A.; Sari, P.; Kuswandi, B. Simple scanometric assay based on DPPH immobilized on pharmaceutical blister for determination of antioxidant capacity in the herbal extracts. *Marmara Pharm. J.* **2018**, *22*, 450–459. [CrossRef]
87. Sirivibulkovit, K.; Nounthavong, S.; Sameenoi, Y. Paper-based DPPH assay for antioxidant activity analysis. *Anal. Sci.* **2018**, *34*, 795–800. [CrossRef]
88. Chen, Z.; Liu, Q.; Zhao, Z.; Bai, B.; Sun, Z.; Cai, L.; Fu, Y.; Ma, Y.; Wang, Q.; Xi, G. Effect of hydroxyl on antioxidant properties of 2,3-dihydro-3,5-dihydroxy-6-methyl-4H-pyran-4-one to scavenge free radicals. *RSC Adv.* **2021**, *11*, 34456. [CrossRef]
89. Bendjedid, S.; Lekmine, S.; Tadjine, A.; Djelloul, R.; Bensouici, C. Analysis of phytochemical constituents, antibacterial, antioxidant, photoprotective activities and cytotoxic effect of leaves extracts and fractions of *Aloe vera*. *Biocatal. Agric. Biotechnol.* **2021**, *33*, 101991. [CrossRef]
90. Watanabe, A.; Noguchi, N.; Fujisawa, A.; Kodama, T.; Tamura, K.; Cynshi, O.; Niki, E. Stability and reactivity of aryloxyl radicals derived from a novel antioxidant BO-653 and related compounds. Effects of substituent and side chain in solution and membranes. *J. Am. Chem. Soc.* **2000**, *122*, 5438–5442. [CrossRef]
91. Zhang, Q.; van der Klift, E.J.C.; Janssen, H.-G.; van Beek, T.A. An on-line normal-phase high performance liquid chromatography method for the rapid detection of radical scavengers in non-polar food matrixes. *J. Chromat. A* **2009**, *1216*, 7268–7274. [CrossRef] [PubMed]
92. Ziyatdinova, G.; Zelenova, Y.; Budnikov, H. Novel modified electrode with immobilized galvinoxyl radical for the voltammetric determination of antioxidant activity. *J. Electroanal. Chem.* **2020**, *856*, 113677. [CrossRef]
93. Dong, J.-W.; Cai, L.; Xing, Y.; Yu, J.; Ding, Z.-T. Re-evaluation of $ABTS^{\bullet+}$ assay for total antioxidant capacity of natural products. *Nat. Prod. Commun.* **2015**, *10*, 2169–2172. [CrossRef] [PubMed]
94. Konan, K.V.; Tien, C.L.; Mateescu, M.A. Electrolysis-induced fast activation of the ABTS reagent for an antioxidant capacity assay. *Anal. Methods* **2016**, *8*, 5638–5644. [CrossRef]
95. Kalili, K.M.; De Smet, S.; van Hoeylandt, T.; Lynen, F.; de Villiers, A. Comprehensive two-dimensional liquid chromatography coupled to the ABTS radical scavenging assay: A powerful method for the analysis of phenolic antioxidants. *Anal. Bioanal. Chem.* **2014**, *406*, 4233–4242. [CrossRef] [PubMed]
96. Hazra, B.; Biswas, S.; Mandal, N. Antioxidant and free radical scavenging activity of *Spondias pinnata*. *BMC Complement. Altern. Med.* **2008**, *8*, 63. [CrossRef]
97. Apak, R.; Calokerinos, A.; Gorinstein, S.; Segundo, M.A.; Hibbert, D.B.; Gulcin, I.; Cekic, S.D.; Guclu, K.; Ozyurek, M.; Celik, S.E.; et al. Methods to evaluate the scavenging activity of antioxidants toward reactive oxygen and nitrogen species (IUPAC Technical Report). *Pure Appl. Chem.* **2022**, *94*, 87–144. [CrossRef]
98. Payne, A.C.; Mazzer, A.; Clarkson, G.J.J.; Taylor, G. Antioxidant assays—Consistent findings from FRAP and ORAC reveal a negative impact of organic cultivation on antioxidant potential in spinach but not watercress or rocket leaves. *Food Sci. Nutr.* **2013**, *1*, 439–444. [CrossRef]
99. González-Palma, I.; Escalona-Buendía, H.B.; Ponce-Alquicira, E.; Téllez-Téllez, M.; Gupta, V.K.; Díaz-Godínez, G.; Soriano-Santos, J. Evaluation of the Antioxidant Activity of Aqueous and Methanol Extracts of *Pleurotus ostreatus* in Different Growth Stages. *Front. Microbiol.* **2016**, *7*, 1099. [CrossRef]
100. Abeyrathne, E.D.N.S.; Nam, K.; Ahn, D.U. Analytical methods for lipid oxidation and antioxidant capacity in food systems. *Antioxidants* **2021**, *10*, 1587. [CrossRef]
101. Prieto, M.A.; Rodríguez-Amado, I.; Vázquez, J.A.; Murado, M.A. β-Carotene assay revisited. application to characterize and quantify antioxidant and prooxidant activities in a microplate. *J. Agric. Food Chem.* **2012**, *60*, 8983–8993. [CrossRef] [PubMed]
102. Niki, E.; Saito, T.; Kawakami, A.; Kamiya, Y. Inhibition of oxidation of methyl linoleate in solution by vitamin E and vitamin C. *J. Biol. Chem.* **1984**, *259*, 4177–4182. [CrossRef] [PubMed]
103. Shahidi, F.; Zhong, Y. Measurement of antioxidant activity. *J. Funct. Foods* **2015**, *18*, 757–781. [CrossRef]
104. Musakhanian, J.; Rodier, J.D.; Dave, M. Oxidative stability in lipid formulations: A review of the mechanisms, drivers, and inhibitors of oxidation. *AAPS PharmSciTech* **2022**, *23*, 151. [CrossRef] [PubMed]

105. Laguerre, M.; Lecomte, J.; Villeneuve, P. Evaluation of the ability of antioxidants to counteract lipid oxidation: Existing methods, new trends and challenges. *Prog. Lipid Res.* **2007**, *46*, 244–282. [CrossRef]
106. Katritzky, A.R.; Dennis Hall, C.; El-Gendy, B.E.D.M.; Draghici, B. Tautomerism in drug discovery. *J. Comput. Aided Mol. Des.* **2010**, *24*, 475–484. [CrossRef]
107. Knorr, L. Studien über Tautomerie. *Justus Liebigs Ann. Chem.* **1896**, *293*, 70–72. [CrossRef]
108. Iglesias, E. Application of Organized Microstructures to Study Keto-Enol Equilibrium of β-Dicarbonyl Compounds. *Curr. Org. Chem.* **2005**, *8*, 1–24. [CrossRef]
109. Karabulut, S.; Namli, H.; Leszczynski, J. Detection of tautomer proportions of dimedone in solution: A new approach based on theoretical and FT-IR viewpoint. *J. Comput. Aided Mol. Des.* **2013**, *27*, 681–688. [CrossRef]
110. Lacerda, V.; Constantino, M.G.; da Silva, G.V.J.; Neto, Á.C.; Tormena, C.F. NMR and theoretical investigation of the keto-enol tautomerism in cyclohexane-1,3-diones. *J. Mol. Struct.* **2007**, *828*, 54–58. [CrossRef]
111. Sigalov, M.; Shainyan, B.; Krief, P.; Ushakov, I.; Chipanina, N.; Oznobikhina, L. Intramolecular interactions in dimedone and phenalen-1,3-dione adducts of 2(4)-pyridinecarboxaldehyde: Enol-enol and ring-chain tautomerism, strong hydrogen bonding, zwitterions. *J. Mol. Struct.* **2011**, *1006*, 234–246. [CrossRef]
112. Jana, K.; Ganguly, B. DFT Study to Explore the Importance of Ring Size and Effect of Solvents on the Keto-Enol Tautomerization Process of α- And β-Cyclodiones. *ACS Omega* **2018**, *3*, 8429–8439. [CrossRef] [PubMed]
113. Rozatian, N.; Beeby, A.; Ashworth, I.W.; Sandford, G.; Hodgson, D.R.W. Enolization rates control mono- versus di-fluorination of 1,3-dicarbonyl derivatives. *Chem. Sci.* **2019**, *10*, 10318–10330. [CrossRef] [PubMed]
114. Bassetti, M.; Cerichelli, G.; Floris, B. Substituent effects in keto-enol tautomerism. Part 3.1 influence of substitution on the equilibrium composition of of β-dicarbonyl compounds. *Tetrahedron* **1988**, *44*, 2997–3004. [CrossRef]
115. Cornago, P.; Claramunt, R.M.; Bouissane, L.; Alkorta, I.; Elguero, J. A study of the tautomerism of β-dicarbonyl compounds with special emphasis on curcuminoids. *Tetrahedron* **2008**, *64*, 8089–8094. [CrossRef]
116. Rogers, M.T.; Burdett, J.L. Keto–enol tautomerism in β-dicarbonyls studied by nuclear magnetic resonance spectroscopy: II. Solvent effects on proton chemical shifts and on equilibrium constants. *Can. J. Chem.* **1965**, *43*, 1516–1526. [CrossRef]
117. Zheng, B.; McClements, D.J. Formulation of more efficacious curcumin delivery systems using colloid science: Enhanced solubility, stability and, bioavailability. *Molecules* **2020**, *25*, 2791. [CrossRef]
118. Jongjitphisut, N.; Thitikornpong, W.; Wichitnithad, W.; Thanusuwannasak, T.; Vajragupta, O.; Rojsitthisak, P. A stability-indicating assay for tetrahydrocurcumin-diglutaric acid and its applications to evaluate bioaccessability in an in vitro digestive model. *Molecules* **2023**, *28*, 1678. [CrossRef]
119. Zhu, L.; Xue, Y.; Feng, J.; Wang, Y.; Lu, Y.; Chen, X. Tetrahydrocurcumin as a stable and highly active curcumin derivative: A review of synthesis, bioconversion, detection and application. *Food Biosci.* **2023**, *53*, 102591. [CrossRef]
120. Chang, R.; Chen, L.; Qamar, M.; Wen, Y.; Li, L.; Zhang, J.; Li, X.; Assadpour, E.; Esatbeyoglu, T.; Kharazmi, M.S.; et al. The bioavailability, metabolism and microbial modulation of curcumin-loaded nanodelivery systems. *Adv. Colloid Interface Sci.* **2023**, *318*, 102933. [CrossRef]
121. Jovanovic, S.V.; Steenken, S.; Boone, C.W.; Simic, M.G. H-atom transfer is a preferred antioxidant mechanism of curcumin. *J. Am. Chem. Soc.* **1999**, *121*, 9677–9681. [CrossRef]
122. Feng, J.Y.; Liu, Z.Q. Phenolic and enolic hydroxyl groups in curcumin: Which plays the major role in scavenging radicals? *J. Agric. Food Chem.* **2009**, *57*, 11041–11046. [CrossRef] [PubMed]
123. Barclay, L.R.C.; Vinqvist, M.R.; Mukai, K.; Goto, H.; Hashimoto, Y.; Tokunaga, A.; Uno, H. On the antioxidant mechanism of curcumin: Classical methods are needed to determine antioxidant mechanism and activity. *Org. Lett.* **2000**, *2*, 2841–2843. [CrossRef] [PubMed]
124. Masuda, T.; Hidaka, K.; Shinohara, A.; Maekawa, T.; Takeda, Y.; Yamaguchi, H. Chemical studies on antioxidant mechanism of curcuminoid: Analysis of radical reaction products from curcumin. *J. Agric. Food Chem.* **1999**, *47*, 71–77. [CrossRef] [PubMed]
125. Litwinienko, G.; Ingold, K.U. Abnormal solvent effects on hydrogen atom abstraction. 2. Resolution of the curcumin antioxidant controversy. The role of sequential proton loss electron transfer. *J. Org. Chem.* **2004**, *69*, 5888–5896. [CrossRef]
126. Foti, M.C.; Slavova-Kazakova, A.; Rocco, C.; Kancheva, V.D. Kinetics of curcumin oxidation by 2,2-diphenyl-1-picrylhydrazyl (DPPH): An interesting case of separated coupled proton-electron transfer. *Org. Biomol. Chem.* **2016**, *14*, 8331–8337. [CrossRef]
127. Somparn, P.; Phisalaphong, C.; Nakornchai, S.; Unchern, S.; Morales, N.P. Comparative antioxidant activities of curcumin and its demethoxy and hydrogenated derivatives. *Biol. Pharm. Bull.* **2007**, *30*, 74–78. [CrossRef]
128. Slavova-Kazakova, A.; Janiak, M.A.; Sulewska, K.; Kancheva, V.D.; Karamać, M. Synergistic, additive, and antagonistic antioxidant effects in the mixtures of curcumin with (−)-epicatechin and with a green tea fraction containing (−)-epicatechin. *Food Chem.* **2021**, *360*, 129994. [CrossRef]
129. Guo, Q.; Bayram, I.; Shu, X.; Su, J.; Liao, W.; Wang, Y.; Gao, Y. Improvement of stability and bioaccessibility of β-carotene by curcumin in pea protein isolate-based complexes-stabilized emulsions: Effect of protein complexation by pectin and small molecular surfactants. *Food Chem.* **2022**, *367*, 130726. [CrossRef]
130. Aftab, N.; Vieira, A. Antioxidant activities of curcumin and combinations of this curcuminoid with other phytochemicals. *Phyther. Res.* **2010**, *24*, 500–502. [CrossRef]

131. Koonyosying, P.; Tantiworawit, A.; Hantrakool, S.; Utama-Ang, N.; Cresswell, M.; Fucharoen, S.; Porter, J.B.; Srichairatanakool, S. Consumption of a green tea extract-curcumin drink decreases blood urea nitrogen and redox iron in β-thalassemia patients. *Food Funct.* **2020**, *11*, 932–943. [CrossRef]
132. Jovanovic, S.V.; Boone, C.W.; Steenken, S.; Trinoga, M.; Kaskey, R.B. How curcumin works preferentially with water soluble antioxidants. *J. Am. Chem. Soc.* **2001**, *123*, 3064–3068. [CrossRef] [PubMed]
133. Venkateswarlu, S.; Ramachandra, M.S.; Subbaraju, G.V. Synthesis and biological evaluation of polyhydroxycurcuminoids. *Bioorg. Med. Chem.* **2005**, *13*, 6374–6380. [CrossRef] [PubMed]
134. Nieto, C.I.; Cornago, M.P.; Cabildo, M.P.; Sanz, D.; Claramunt, R.M.; Torralba, M.C.; Torres, M.R.; Casanova, D.M.; Sánchez-Alegre, Y.R.; Escudero, E.; et al. Evaluation of the antioxidant and neuroprotectant activities of new asymmetrical 1,3-diketones. *Molecules* **2018**, *23*, 1837. [CrossRef] [PubMed]
135. Jha, N.S.; Mishra, S.; Jha, S.K.; Surolia, A. Antioxidant activity and electrochemical elucidation of the enigmatic redox behavior of curcumin and its structurally modified analogues. *Electrochim. Acta* **2015**, *151*, 574–583. [CrossRef]
136. Patro, B.S.; Rele, S.; Chintalwar, G.J.; Chattopadhyay, S.; Adhikari, S.; Mukherjee, T. Protective activities of some phenolic 1,3-diketones against lipid peroxidation: Possible involvement of the 1,3-diketone moiety. *ChemBioChem* **2002**, *3*, 364–370. [CrossRef]
137. Sugiyama, Y.; Kawakishi, S.; Osawa, T. Involvement of the β-diketone moiety in the antioxidative mechanism of tetrahydrocurcumin. *Biochem. Pharmacol.* **1996**, *52*, 519–525. [CrossRef]
138. Slavova-Kazakova, A.; Angelova, S.; Fabbri, D.; Antonietta Dettori, M.; Kancheva, V.D.; Delogu, G. Antioxidant properties of novel curcumin analogues: A combined experimental and computational study. *J. Food Biochem.* **2021**, *45*, 13584. [CrossRef]
139. Sheikh, J.; Ben Hadda, T. Antibacterial, antifungal and antioxidant activity of some new water-soluble β-diketones. *Med. Chem. Res.* **2013**, *22*, 964–975. [CrossRef]
140. Bai, F.; Diao, J.; Wang, Y.; Sun, S.; Zhang, H.; Liu, Y.; Wang, Y.; Cao, J. A New Water-Soluble Nanomicelle Formed through Self-Assembly of Pectin-Curcumin Conjugates: Preparation, Characterization, and Anticancer Activity Evaluation. *J. Agric. Food Chem.* **2017**, *65*, 6840–6847. [CrossRef]
141. Wei, L.; Li, X.; Guo, F.; Liu, X.; Wang, Z. Structural properties, in vitro release and radical scavenging activity of lecithin based curcumin-encapsulated inverse hexagonal (HII) liquid crystals. *Colloids Surf. A Physicochem. Eng. Asp.* **2018**, *539*, 124–131. [CrossRef]
142. Chen, Z.; Xu, L.; Gao, X.; Wang, C.; Li, R.; Xu, J.; Zhang, M.; Panichayupakaranant, P.; Chen, H. A multifunctional CeO_2@SiO_2-PEG nanoparticle carrier for delivery of food derived proanthocyanidin and curcumin as effective antioxidant, neuroprotective and anticancer agent. *Food Res. Int.* **2020**, *137*, 109674. [CrossRef] [PubMed]
143. Cheng, C.S.; Liu, T.P.; Chien, F.C.; Mou, C.Y.; Wu, S.H.; Chen, Y.P. Codelivery of Plasmid and Curcumin with Mesoporous Silica Nanoparticles for Promoting Neurite Outgrowth. *ACS Appl. Mater. Interfaces* **2019**, *11*, 15322–15331. [CrossRef] [PubMed]
144. Khadrawy, Y.A.; Hosny, E.N.; Magdy, M.; Mohammed, H.S. Antidepressant effects of curcumin-coated iron oxide nanoparticles in a rat model of depression. *Eur. J. Pharmacol.* **2021**, *908*, 174384. [CrossRef]
145. Massaro, M.; Riela, S. Organo-clay nanomaterials based on halloysite and cyclodextrin as carriers for polyphenolic compounds. *J. Funct. Biomater.* **2018**, *9*, 61. [CrossRef]
146. Qu, B.; Xue, J.; Luo, Y. Self-assembled caseinate-laponite® nanocomposites for curcumin delivery. *Food Chem.* **2021**, *363*, 130338. [CrossRef]
147. Shah, B.R.; Zhang, C.; Li, Y.; Li, B. Bioaccessibility and antioxidant activity of curcumin after encapsulated by nano and Pickering emulsion based on chitosan-tripolyphosphate nanoparticles. *Food Res. Int.* **2016**, *89*, 399–407. [CrossRef]
148. Takahashi, M.; Uechi, S.; Takara, K.; Asikin, Y.; Wada, K. Evaluation of an Oral Carrier System in Rats: Bioavailability and Antioxidant Properties of Liposome-Encapsulated Curcumin. *J. Agric. Food Chem.* **2009**, *57*, 9141–9146. [CrossRef]
149. Jeon, W.Y.; Yu, J.Y.; Kim, H.W.; Park, H.J. Production of customized food through the insertion of a formulated nanoemulsion using coaxial 3D food printing. *J. Food Eng.* **2021**, *311*, 110689. [CrossRef]
150. Kushwaha, P.; Yadav, A.; Samim, M.; Flora, S.J.S. Combinatorial drug delivery strategy employing nano-curcumin and nano-MiADMSA for the treatment of arsenic intoxication in mouse. *Chem. Biol. Interact.* **2018**, *286*, 78–87. [CrossRef]
151. Pu, H.-L.; Chiang, W.-L.; Maiti, B.; Liao, Z.-X.; Ho, Y.-C.; Shim, M.S.; Chuang, E.-Y.; Xia, Y.; Sung, H.-W. Nanoparticles with Dual Responses to Oxidative Stress and Reduced pH for Drug Release and Anti-inflammatory Applications. *ACS Nano* **2014**, *8*, 1213–1221. [CrossRef] [PubMed]
152. Nowzari, Z.; Khorshidi, A. Synthesis, characterization, and antibacterial, antioxidant, and anticancer activity of di-μ-chlorobis[dichlorocurcuminatoniobium(V)] dehydrate. *Res. Chem. Intermed.* **2018**, *44*, 6339–6349. [CrossRef]
153. Altundağ, E.M.; Özbilenler, C.; Ustürk, S.; Kerküklü, N.R.; Afshani, M.; Yilmaz, E. Metal-based curcumin and quercetin complexes: Cell viability, ROS production and antioxidant activity. *J. Mol. Struct.* **2021**, *1245*, 131107. [CrossRef]
154. Osawa, T.; Namiki, M. A novel type of antioxidant isolated from leaf wax of eucalyptus leaves. *Agric. Biol. Chem.* **1981**, *45*, 735–739. [CrossRef]
155. Osawa, T.; Namiki, M. Natural Antioxidants Isolated from Eucalyptus Leaf Waxes. *J. Agric. Food Chem.* **1985**, *33*, 777–780. [CrossRef]

156. Salehi, M.; Galini, M.; Kubicki, M.; Khaleghian, A. Synthesis and Characterization of New Cobalt(III) and Nickel(II) Complexes Derived from Acetylacetone and 2-Aminopyridine: A New Precursor for Preparation of NiO Nanoparticles. *Russ. J. Inorg. Chem.* **2019**, *64*, 18–27. [CrossRef]
157. Saranya, A.V.; Ravi, S.; Venkatachalapathi, S. In-vitro Antioxidant activity of Diethyl malonate adducts of Phenothiazine. *Res. J. Chem. Sci.* **2013**, *3*, 82–85.
158. Mazimba, O.; Wale, K.; Loeto, D.; Kwape, T. Antioxidant and antimicrobial studies on fused-ring pyrazolones and isoxazolones. *Bioorg. Med. Chem.* **2014**, *22*, 6564–6569. [CrossRef]
159. Mieriņa, I.; Kostjuņina, D.; Skrastiņa, D.Z.; Jure, M. Synthesis and Antiradical Activity of 2-Arylidenemalonic Acid Dianilides. *Key Eng. Mater.* **2020**, *850*, 230–235. [CrossRef]
160. Stikute, A.; Skestere, K.; Mierina, I.; Mishnev, A.; Jure, M. Crystal structure of 3-hydroxy-2-(4-hydroxy-3-methoxyphenylmethyl)-5,5-dimethylcyclohex-2-enone. *Acta Crystallogr. Sect. E Crystallogr. Commun.* **2018**, *74*, 796–798. [CrossRef]
161. Maharvi, G.M.; Ali, S.; Riaz, N.; Afza, N.; Malik, A.; Ashraf, M.; Iqbal, L.; Lateef, M. Mild and efficient synthesis of new tetraketones as lipoxygenase inhibitors and antioxidants. *J. Enzym. Inhib. Med. Chem.* **2008**, *23*, 62–69. [CrossRef] [PubMed]
162. Naidu Kalla, R.M.; Karunakaran, R.S.; Balaji, M.; Kim, I. Catalyst-Free Synthesis of Xanthene and Pyrimidine-Fused Heterocyclic Derivatives at Water-Ethanol Medium and Their Antioxidant Properties. *ChemistrySelect* **2019**, *4*, 644–649. [CrossRef]
163. Choudhary, M.I.; Khan, N.; Ahmad, M.; Yousuf, S.; Fun, H.K.; Soomro, S.; Asif, M.; Mesaik, M.A.; Shaheen, F. New inhibitors of ROS generation and T-cell proliferation from myrtus communis. *Org. Lett.* **2013**, *15*, 1862–1865. [CrossRef] [PubMed]
164. Yamaguchi, F.; Saito, M.; Ariga, T.; Yoshimura, Y.; Nakazawa, H. Free Radical Scavenging Activity and Antiulcer Activity of Garcinol from Garcinia indica Fruit Rind. *J. Agric. Food Chem.* **2000**, *48*, 2320–2325. [CrossRef]
165. Yamaguchi, F.; Ariga, T.; Yoshimura, Y.; Nakazawa, H. Antioxidative and Anti-Glycation Activity of Garcinol from Garcinia indica Fruit Rind. *J. Agric. Food Chem.* **2000**, *48*, 180–185. [CrossRef]
166. Cetin Cakmak, K.; Gülçin, İ. Anticholinergic and antioxidant activities of usnic acid-an activity-structure insight. *Toxicol. Rep.* **2019**, *6*, 1273–1280. [CrossRef]
167. Popovici, V.; Matei, E.; Cozaru, G.C.; Aschie, M.; Bucur, L.; Rambu, D.; Costache, T.; Cucolea, I.E.; Vochita, G.; Gherghel, D.; et al. Usnic acid and usnea barbata (L.) F.H. wigg. dry extracts promote apoptosis and DNA damage in human blood cells through enhancing ROS levels. *Antioxidants* **2021**, *10*, 1171. [CrossRef]
168. Tagashira, M.; Watanabe, M.; Uemitsu, N. Antioxidative Activity of Hop Bitter Acids and Their Analogues. *Biosci. Biotechnol. Biochem.* **1995**, *59*, 740–742. [CrossRef]
169. Gorjanović, S.; Pastor, F.T.; Vasić, R.; Novaković, M.; Simonović, M.; Milić, S.; Sužnjević, D. Electrochemical versus spectrophotometric assessment of antioxidant activity of hop (*Humulus lupulus* L.) products and individual compounds. *J. Agric. Food Chem.* **2013**, *61*, 9089–9096. [CrossRef]
170. Wietstock, P.C.; Shellhammer, T.H. Chelating properties and hydroxyl-scavenging activities of hop A- and iso-α-acids. *J. Am. Soc. Brew. Chem.* **2011**, *69*, 133–138. [CrossRef]
171. Liégeois, C.; Lermusieau, G.; Collin, S. Measuring antioxidant efficiency of wort, malt, and hops against the 2,2′-azobis(2-amidinopropane) dihydrochloride-induced oxidation of an aqueous dispersion of linoleic acid. *J. Agric. Food Chem.* **2000**, *48*, 1129–1134. [CrossRef] [PubMed]
172. Karabín, M.; Rýparová, A.; Jelínek, L.; Kunz, T.; Wietstock, P.; Methner, F.J.; Dostálek, P. Relationship of iso-α-acid content and endogenous antioxidative potential during storage of lager beer. *J. Inst. Brew.* **2014**, *120*, 212–219. [CrossRef]
173. Wannenmacher, J.; Cotterchio, C.; Schlumberger, M.; Reuber, V.; Gastl, M.; Becker, T. Technological influence on sensory stability and antioxidant activity of beers measured by ORAC and FRAP. *J. Sci. Food Agric.* **2019**, *99*, 6628–6637. [CrossRef] [PubMed]
174. Weber, N.; Biehler, K.; Schwabe, K.; Haarhaus, B.; Quirin, K.W.; Frank, U.; Schempp, C.M.; Wölfle, U. Hop extract acts as an antioxidant with antimicrobial effects against Propionibacterium acnes and Staphylococcus aureus. *Molecules* **2019**, *24*, 223. [CrossRef]
175. Tian, B.; Xu, D.; Cheng, J.; Liu, Y. Chitosan-silica with hops β-acids added films as prospective food packaging materials: Preparation, characterization, and properties. *Carbohydr. Polym.* **2021**, *272*, 118457. [CrossRef]
176. De Souza, L.C.; De Araújo, S.M.S.; De Oliveira Imbroisi, D. Determination of the free radical scavenging activity of dihydropyran-2,4-diones. *Bioorg. Med. Chem. Lett.* **2004**, *14*, 5859–5861. [CrossRef]
177. Martinčič, R.; Mravljak, J.; Švajger, U.; Perdih, A.; Anderluh, M.; Novič, M. In silico discovery of novel potent antioxidants on the basis of pulvinic acid and coumarine derivatives and their experimental evaluation. *PLoS ONE* **2015**, *10*, e0140602. [CrossRef]
178. Habrant, D.; Poigny, S.; Ségur-Derai, M.; Brunei, Y.; Heurtaux, B.; Le Gall, T.; Strehle, A.; Saladin, R.; Meunier, S.; Mioskowski, C.; et al. Evaluation of antioxidant properties of monoaromatic derivatives of pulvinic acids. *J. Med. Chem.* **2009**, *52*, 2454–2464. [CrossRef]
179. Mierina, I.; Jure, M.; Zeberga, S.; Makareviciene, V.; Zicane, D.; Tetere, Z.; Ravina, I. Novel type of carbon-centered antioxidants arylmethyl Meldrum's acids—Inhibit free radicals. *Eur. J. Lipid Sci. Technol.* **2017**, *119*, 1700172. [CrossRef]
180. Mieriņa, I.; Peipiņa, E.; Aišpure, K.; Jure, M. 1st generation dendrimeric antioxidants containing Meldrum's acid moieties as surface groups. *New J. Chem.* **2022**, *46*, 607–620. [CrossRef]
181. Janković, N.; Muškinja, J.; Ratković, Z.; Bugarčić, Z.; Ranković, B.; Kosanić, M.; Stefanović, S. Solvent-free synthesis of novel vanillidene derivatives of Meldrum's acid: Biological evaluation, DNA and BSA binding study. *RSC Adv.* **2016**, *6*, 39452–39459. [CrossRef]

182. Sandhu, H.S.; Sapra, S.; Gupta, M.; Nepali, K.; Gautam, R.; Yadav, S.; Kumar, R.; Jachak, S.M.; Chugh, M.; Gupta, M.K.; et al. Synthesis and biological evaluation of arylidene analogues of Meldrum's acid as a new class of antimalarial and antioxidant agents. *Bioorg. Med. Chem.* **2010**, *18*, 5626–5633. [CrossRef]
183. Rodríguez, S.A.; Nazareno, M.A.; Baumgartner, M.T. Effect of different C3-aryl substituents on the antioxidant activity of 4-hydroxycoumarin derivatives. *Bioorg. Med. Chem.* **2011**, *19*, 6233–6238. [CrossRef] [PubMed]
184. Rouaiguia-Bouakkaz, S.; Benayahoum, A. The antioxidant activity of 4-hydroxycoumarin derivatives and some sulfured analogs. *J. Phys. Org. Chem.* **2015**, *28*, 714–722. [CrossRef]
185. Milenković, D.A.; Dimić, D.S.; Avdović, E.H.; Amić, A.D.; Dimitrić Marković, J.M.; Marković, Z.S. Advanced oxidation process of coumarins by hydroxyl radical: Towards the new mechanism leading to less toxic products. *Chem. Eng. J.* **2020**, *395*, 124971. [CrossRef]
186. Konidala, S.K.; Kotra, V.; Danduga, R.C.S.R.; Kola, P.K.; Bhandare, R.R.; Shaik, A.B. Design, multistep synthesis and in-vitro antimicrobial and antioxidant screening of coumarin clubbed chalcone hybrids through molecular hybridization approach. *Arab. J. Chem.* **2021**, *14*, 103154. [CrossRef]
187. Pérez-Cruz, F.; Serra, S.; Delogu, G.; Lapier, M.; Maya, J.D.; Olea-Azar, C.; Santana, L.; Uriarte, E. Antitrypanosomal and antioxidant properties of 4-hydroxycoumarins derivatives. *Bioorg. Med. Chem. Lett.* **2012**, *22*, 5569–5573. [CrossRef]
188. Verpakovska, I.; Skrastiņa, D.Z.; Mieriņa, I.; Jure, M. 4-Substituted Coumarin Antioxidants. *Key Eng. Mater.* **2019**, *800*, 30–35. [CrossRef]
189. Jorge, E.G.; Rayar, A.M.; Barigye, S.J.; Rodríguez, M.E.J.; Veitía, M.S.I. Development of an in silico model of DPPH free radical scavenging capacity: Prediction of antioxidant activity of coumarin type compounds. *Int. J. Mol. Sci.* **2016**, *17*, 881. [CrossRef]
190. Zaheer, Z.; Kalam Khan, F.A.; Sangshetti, J.N.; Patil, R.H.; Rafiq, Z.; Campus, Y.B. Expeditious synthesis, antileishmanial and antioxidant activities of novel 3-substituted-4-hydroxycoumarin derivatives. *Chin. Chem. Lett.* **2016**, *27*, 287–294. [CrossRef]
191. Wang, Z.M.; Xie, S.S.; Li, X.M.; Wu, J.J.; Wang, X.B.; Kong, L.Y. Multifunctional 3-Schiff base-4-hydroxycoumarin derivatives with monoamine oxidase inhibition, anti-β-amyloid aggregation, metal chelation, antioxidant and neuroprotection properties against Alzheimer's disease. *RSC Adv.* **2015**, *5*, 70395–70409. [CrossRef]
192. Antonijević, M.R.; Simijonović, D.M.; Avdović, E.H.; Ćirić, A.; Petrović, Z.D.; Marković, J.D.; Stepanić, V.; Marković, Z.S. Green one-pot synthesis of coumarin-hydroxybenzohydrazide hybrids and their antioxidant potency. *Antioxidants* **2021**, *10*, 1106. [CrossRef] [PubMed]
193. Hamdi, N.; Puerta, M.C.; Valerga, P. Synthesis, structure, antimicrobial and antioxidant investigations of dicoumarol and related compounds. *Eur. J. Med. Chem.* **2008**, *43*, 2541–2548. [CrossRef] [PubMed]
194. Bejaoui, L.; Rohlicek, J.; Ben Hassen, R. New cobalt (II) complexes of '3-acetyl-4-hydroxy-2H-chromene-2-one': Crystal structure and Hirshfeld surface analysis, fluorescence behaviour and antioxidant activity. *J. Mol. Struct.* **2018**, *1173*, 574–582. [CrossRef]
195. Wang, G.; Liu, Y.; Zhang, L.; An, L.; Chen, R.; Liu, Y.; Luo, Q.; Li, Y.; Wang, H.; Xue, Y. Computational study on the antioxidant property of coumarin-fused coumarins. *Food Chem.* **2020**, *304*, 125446. [CrossRef]
196. Fujimaki, T.; Saiki, S.; Tashiro, E.; Yamada, D.; Kitagawa, M.; Hattori, N.; Imoto, M. Identification of licopyranocoumarin and glycyrurol from herbal medicines as neuroprotective compounds for Parkinson's disease. *PLoS ONE* **2014**, *9*, e100395. [CrossRef]
197. Martínez-Martínez, F.J.; Razo-Hernández, R.S.; Peraza-Campos, A.L.; Villanueva-García, M.; Sumaya-Martínez, M.T.; Cano, D.J.; Gómez-Sandoval, Z. Synthesis and in vitro antioxidant activity evaluation of 3-carboxycoumarin derivatives and qsar study of their dpph radical scavenging activity. *Molecules* **2012**, *17*, 14882–14898. [CrossRef]
198. Vazquez-Rodriguez, S.; Figueroa-Guíñez, R.; Matos, M.J.; Santana, L.; Uriarte, E.; Lapier, M.; Maya, J.D.; Olea-Azar, C. Synthesis of coumarin-chalcone hybrids and evaluation of their antioxidant and trypanocidal properties. *Medchemcomm* **2013**, *4*, 993–1000. [CrossRef]
199. Detsi, A.; Bouloumbasi, D.; Prousis, K.C.; Koufaki, M.; Athanasellis, G.; Melagraki, G.; Afantitis, A.; Igglessi-Markopoulou, O.; Kontogiorgis, C.; Hadjipavlou-Litina, D.J. Design and synthesis of novel quinolinone-3-aminoamides and their α-lipoic acid adducts as antioxidant and anti-inflammatory agents. *J. Med. Chem.* **2007**, *50*, 2450–2458. [CrossRef]
200. Pudlo, M.; Luzet, V.; Ismaïli, L.; Tomassoli, I.; Iutzeler, A.; Refouvelet, B. Quinolone-benzylpiperidine derivatives as novel acetylcholinesterase inhibitor and antioxidant hybrids for Alzheimer Disease. *Bioorg. Med. Chem.* **2014**, *22*, 2496–2507. [CrossRef]
201. Mieriņa, I.; Stikute, A.; Jure, M. A green and effective route leading to antiradical agents with 3-arylmethyl 4-hydroxyquinolin-2(1H)-one moiety. *Tetrahedron Lett.* **2022**, *99*, 153847. [CrossRef]
202. Shin, Y.S.; Song, S.J.; Kang, S.U.; Hwang, H.S.; Choi, J.W.; Lee, B.H.; Jung, Y.S.; Kim, C.H. A novel synthetic compound, 3-amino-3-(4-fluoro-phenyl)-1H-quinoline-2,4-dione, inhibits cisplatin-induced hearing loss by the suppression of reactive oxygen species: In vitro and in vivo study. *Neuroscience* **2013**, *232*, 1–12. [CrossRef] [PubMed]

molecules

Article

Phenolic Constituents, Antioxidant and Antimicrobial Activity and Clustering Analysis of Propolis Samples Based on PCA from Different Regions of Anatolia

Ümit Altuntaş [1,2,*], İsmail Güzel [3] and Beraat Özçelik [1,4]

1 Department of Food Engineering, İstanbul Technical University, 34469 İstanbul, Turkey
2 Department of Food Engineering, Gümüşhane University, 29000 Gümüşhane, Turkey
3 Department of Mathematical Engineering, İstanbul Technical University, 34469 İstanbul, Turkey
4 Bioactive Research & Innovation, Food Manuf. Indust. Trade Ltd., Teknokent ARI-3, 34469 Istanbul, Turkey
* Correspondence: ualtuntas@itu.edu.tr

Abstract: This study aimed to evaluate the biochemical composition and biological activity of propolis samples from different regions of Türkiye to characterize and classify 24 Anatolian propolis samples according to their geographical origin. Chemometric techniques, namely, principal component analysis (PCA) and a hierarchical clustering algorithm (HCA), were applied for the first time to all data, including antioxidant capacity, individual phenolic constituents, and the antimicrobial activity of propolis to reveal the possible clustering of Anatolian propolis samples according to their geographical origin. As a result, the total phenolic content (TPC) of the propolis samples varied from 16.73 to 125.83 mg gallic acid equivalent per gram (GAE/g) sample, while the number of total flavonoids varied from 57.98 to 327.38 mg quercetin equivalent per gram (QE/g) sample. The identified constituents of propolis were phenolic/aromatic acids (chlorogenic acid, caffeic acid, *p*-coumaric acid, ferulic acid, and trans-cinnamic acid), phenolic aldehyde (vanillin), and flavonoids (pinocembrin, kaempferol, pinobanksin, and apigenin). This study has shown that the application of the PCA chemometric method to the biochemical composition and biological activity of propolis allows for the successful clustering of Anatolian propolis samples from different regions of Türkiye, except for samples from the Black Sea region.

Keywords: propolis; Anatolia; antioxidant activity; phenolics; clustering; PCA

Citation: Altuntaş, Ü.; Güzel, İ.; Özçelik, B. Phenolic Constituents, Antioxidant and Antimicrobial Activity and Clustering Analysis of Propolis Samples Based on PCA from Different Regions of Anatolia. *Molecules* **2023**, *28*, 1121. https://doi.org/10.3390/molecules28031121

Academic Editors: Carla Pereira, José Pinela, Maria Inês Dias and José Ignacio Alonso-Esteban

Received: 19 December 2022
Revised: 16 January 2023
Accepted: 19 January 2023
Published: 22 January 2023

1. Introduction

Propolis is a complex substance consisting of natural resinous and sticky materials collected by bees, *Apis mellifera*, collected from living plant buds and exudates [1–7]. Propolis is a wax-like natural product formed by honeybees by combining their own wax with various plant sources [8] and can be considered a non-toxic 'glue or cement' for bees [5,8] that provides many benefits to honeybees, such as sealing cracks and plugging holes in the walls of the hive, flattening the inner surface of the hive to minimize moisture loss, regulating humidity and temperature in their nest, and embalming dead insects [1,2,4,9].

The biochemical composition of propolis is quite variable and complex. It depends on factors such as local flora surrounding the hive accessible to honeybees, collection time, geographical origin, type of honeybee, diversity of trees, and plant species collected by honeybees [1,8]. The chemical patterns of propolis types indicate the geographical distribution of plant species [5]. Thus, the chemical composition of propolis differs according to its phytogeographic origin. The variability of colors (yellow, green, brown, and red) of propolis depends on the resin sources found in the region (botanical source), as well as the preparation period [1,2]. However, the chemical composition of some types of propolis, such as European poplar propolis (from *Popolus* spp.), Brazilian green (from *Baccharis dracunculifolia*), and red propolis (from *Dalbergia ecastophyllum*), has been clarified and standardized [4,8].

Propolis consists of about 50% resin (i.e., the polyphenolic fraction, including flavonoids and phenolic/aromatic acids) and 30% wax (waxes and fatty acids), while the essential and aromatic oils (mainly volatiles) and bee-pollen (free amino acids) are approximately 10% of propolis, and the remaining 5% consists of other organic and mineral substances [1,8]. Propolis contains more than 300 natural compounds. A list of pharmacologically active chemical substances reported in several studies are polyphenols (flavonoids, phenolic acids, and their esters), phenolic aldehydes, flavonoid aglycones, sesquiterpenes, quinones, coumarins, amino acids, fatty acids, steroids, terpenoids, and inorganic compounds [1,3,5,7,8,10]. The bioactivity of propolis is variable and is believed to be related to the variation of its chemical composition [8,10]. Due to the presence of bioactive constituents, propolis produces a wide spectrum of important biological activities, such as being antioxidant (strong scavenger of free radicals), antitumor, antibacterial, antifungal, antiviral, anti-inflammatory, cardioprotective, and hepatoprotective, and it has anticancer and immunomodulatory properties [1,3,5,7,8,10,11]. All this has caused propolis to be known as a natural product that can potentially be used as a therapeutic agent to enhance the immune system and prevent various human diseases [1].

The antioxidant activity of propolis protects the human body from cell damage because of free radicals by lowering the number of oxidative chemical reactions that occur [1,3,6,8,12]. Additionally, propolis was recently assessed for its antimicrobial activity against both Gram-positive (+) and -negative (−) bacteria, and previous research has confirmed the in vitro antibacterial activity of propolis extracts via different assays [1,6]. The antioxidant and antibacterial activities of propolis are reported to be caused by flavonoids and phenolic acids and its esters. The interaction and/or synergism between phenolic and other chemicals in propolis is considered the mechanism of this activity [13]. Flavonoids and phenolic compounds are the main bioactive components of propolis [2,3], and both have proven their ability to neutralize free radicals. The wide spectrum of propolis, i.e., biological properties and multiple applications, has aroused interest in the study of its properties according to its origin [2,6].

From the literature reviewed, propolis collected from different places, even from the same city, may differ substantially in terms of antioxidant capacity, antibacterial activity, individual phenolic compounds, and, thus, its biological activities [2,4,5,7,8,10]. Türkiye has a rich phytogeographical structure, and some researchers have attempted to elucidate the chemical structure of propolis produced there. In one of these studies, Kartal et al. (2002) examined propolis samples from the Ankara and Marmaris regions and reported that the botanical origin of the sample from Marmaris could be *Pinus brutia* L. (pine propolis) [14]. In this context, Popova et al. (2005) studied Turkish propolis, and *Populus nigra* and *P. euphratica* were found to be important sources of propolis [15]. Velikova et al. (2000) examined a Bulgarian and two Turkish propolis samples and found that their chemical compositions were similar; they had the characteristics of poplar propolis, and the samples were particularly rich in caffeic acid and ferulic acid [16]. In a study conducted by Sorkun et al. (2001) on Turkish propolis of different geographical origins, it was also found that the samples from Trabzon and Gümüşhane had similar chemical compositions, and the main components were aromatic and aliphatic acids, esters, and ketones. It was found that flavanones, aromatic acid and its esters, terpenoids, flavones, and ketones were the main compounds in the Bursa sample [17]. However, the biochemical data and biological activities of Anatolian propolis are still scarce.

The present study aims to evaluate the biochemical composition and biological activity of propolis samples from different regions of Türkiye to characterize and classify 24 propolis samples based on their geographical origin. In addition, multivariate analysis, principal component analysis (PCA), and a hierarchical clustering algorithm (HCA) were applied to all data, including moisture (%), total phenolic and flavonoid content, antioxidant capacities (DPPH and CUPRAC), individual phenolic and aromatic constituents, and antimicrobial activity to show, for the first time, a possible clustering of Anatolian propolis samples from different regions of origin according to their geographical origin.

2. Results

2.1. Chemical Properties

The moisture content of the 24 propolis samples belonging to different geographical regions of Türkiye, collected twice in different harvesting seasons, was determined, and the average results are provided in Table 1. According to Table 1, the propolis samples had moisture profiles ranging from 3.83 to 7.13% of total weight. The results of this study are similar to those reported in the literature for other propolis samples, which had moisture content ranging from 3.40 to 9.16% [18–20]. Propolis samples from the Marmara region had higher moisture content than other regions. On the other hand, propolis samples from Central Anatolia had relatively lower moisture content. The variations in moisture content (%) of propolis samples are most likely due to the type of propolis, the geographical region and environmental conditions, and the collection period of the propolis [18,21].

Table 1. Moisture content (%) of Anatolian propolis samples (data are expressed as g water/100 g of the crude sample).

Region	City	Moisture Content (%)
Black Sea	Amasya	5.12 ± 0.1^{g}
Black Sea	Bartın	$5.88 \pm 0.1^{b,c}$
Black Sea	Karabük	$5.05 \pm 0.1^{d,e}$
Black Sea	Kastamonu	$4.52 \pm 0.3^{e,f}$
Black Sea	Ordu	$5.09 \pm 0.1^{d,e}$
Black Sea	Samsun	$5.22 \pm 0.1^{c,d,e}$
Black Sea	Tokat	$6.25 \pm 0.1^{a,b}$
Black Sea	Zonguldak	$5.35 \pm 0.0^{c,d}$
Black Sea	Çorum	$4.01 \pm 0.1^{f,g}$
Central Anatolia	Konya	3.84 ± 0.1^{g}
Central Anatolia	Kütahya	$5.15 \pm 0.1^{c,d,e}$
Central Anatolia	Sivas	3.83 ± 0.1^{g}
Central Anatolia	Yozgat	$4.24 \pm 0.1^{e,f}$
Marmara	Balıkesir	7.13 ± 0.3^{a}
Marmara	Bilecik	$6.07 \pm 0.1^{b,c}$
Marmara	Bursa	$5.72 \pm 0.3^{b,c}$
Marmara	İstanbul	$6.22 \pm 0.1^{a,b}$
Marmara	Kırklareli	$5.26 \pm 0.5^{c,d,e}$
Marmara	Kocaeli	$5.22 \pm 0.1^{c,d,e}$
Marmara	Tekirdağ	$4.17 \pm 0.1^{c,d,e}$
Mediterranean	Adana	$4.61 \pm 0.1^{e,f}$
Mediterranean	Gaziantep	3.70 ± 0.1^{g}
Mediterranean	Hatay	$5.86 \pm 0.1^{b,c}$
Mediterranean	Mersin	3.83 ± 0.1^{g}

Significant differences in the same column are represented by different letters (a–g) ($p < 0.05$).

2.2. Antioxidant Activity

The antioxidant activity of the Turkish propolis samples was determined based on the total phenolic content (TPC) using the Folin–Ciocalteu reagent method, the total flavonoid content (TFC) by indexing the total flavonoids, the cupric ion-reducing antioxidant capacity (CUPRAC) using the copper inducing method, and the scavenging free radical DPPH (2,2-diphenyl-1-picrylhydrazyl) method, since propolis cannot be satisfactorily evaluated by only one method [8,22]. The higher values of the results indicate a high antioxidant capacity for the propolis samples for each method. Table 2 shows the TPC, TFC, and antioxidant capacity (CUPRAC and DPPH) of propolis extract results obtained by analyzing the four different methods mentioned above. Significant differences ($p < 0.05$) in the four different tests between the Anatolian propolis samples are provided in Table 2.

Table 2. Total phenolic and flavonoid contents and antioxidant capacity results (CUPRAC and DPPH) of ethanolic extracts of Anatolian propolis samples (data are expressed as mg/g of the crude propolis sample).

Region	City	CUPRAC (mg TE/g Sample)	DPPH (mg TE/g Sample)	TPC (mg GAE/g Sample)	TFC (mg QE/g Sample)
Black Sea	Amasya	345.60 ± 6.7 ^{a,b,c}	186.84 ± 8.0 ^{a,b}	88.32 ± 1.1 ^{b,c}	325.09 ± 11.3 ^{a}
Black Sea	Bartın	328.23 ± 37.1 ^{a,b,c}	157.58 ± 9.8 ^{b,c,d}	92.51 ± 3.1 ^{b}	251.47 ± 56.3 ^{a,b,c}
Black Sea	Karabük	257.43 ± 28.4 ^{c,d,e}	160.94 ± 30.5 ^{b,c,d}	66.42 ± 8.9 ^{d,e,f}	240.28 ± 38.2 ^{b,c,d}
Black Sea	Kastamonu	297.36 ± 39.0 ^{b,c}	166.81 ± 23.4 ^{a,b,c}	76.51 ± 8.0 ^{c,d}	237.07 ± 32.9 ^{c,d,e}
Black Sea	Ordu	357.73 ± 9.4 ^{a,b}	214.50 ± 22.3 ^{a,b}	70.23 ± 8.2 ^{c,d,e}	301.71 ± 47.8 ^{a,b}
Black Sea	Samsun	196.35 ± 38.7 ^{e,f,g}	128.07 ± 7.3 ^{d,e,f}	70.26 ± 2.2 ^{c,d,e}	222.04 ± 8.8 ^{c,d,e}
Black Sea	Tokat	233.45 ± 20.0 ^{c,d,e}	228.23 ± 30.1 ^{a}	96.00 ± 8.2 ^{b}	289.68 ± 12.3 ^{a,b,c}
Black Sea	Zonguldak	378.93 ± 29.7 ^{a}	175.00 ± 12.0 ^{a,b}	98.89 ± 5.3 ^{b}	323.71 ± 9.4 ^{a}
Black Sea	Çorum	243.99 ± 16.2 ^{c,d,e}	151.82 ± 8.3 ^{c,d}	81.33 ± 3.3 ^{c,d}	217.56 ± 19.8 ^{c,d,e}
Central Anatolia	Konya	127.40 ± 6.5 ^{g,h,i}	164.41 ± 4.0 ^{a,b,c}	110.37 ± 5.0 ^{a,b}	147.04 ± 3.0 ^{fg}
Central Anatolia	Kütahya	360.93 ± 46.8 ^{a,b}	160.82 ± 7.6 ^{b,c,d}	88.44 ± 13.2 ^{b,c}	283.26 ± 37.1 ^{a,b,c}
Central Anatolia	Sivas	173.21 ± 17.2 ^{f,g}	143.26 ± 10,1 ^{c,d}	56.89 ± 8.8 ^{e,f}	167.73 ± 33.9 ^{e,f,g}
Central Anatolia	Yozgat	304.88 ± 12.0 ^{b,c}	195.36 ± 47.4 ^{a,b}	125.83 ± 24.0 ^{a}	241.04 30.9 ^{b,c,d}
Marmara	Balıkesir	225.34 ± 21.6 ^{d,e,f}	167.32 ± 6.2 ^{a,b,c}	68.88 ± 10.3 ^{d,e}	208.50 ± 4.0 ^{d,e}
Marmara	Bilecik	259.75 ± 30.8 ^{c,d,e}	169.74 ± 4.7 ^{a,b,c}	71.14 ± 1.8 ^{c,d,e}	250.87 ± 16.1 ^{a,b,c}
Marmara	Bursa	200.73 ± 22.8 ^{e,f,g}	143.14 ± 38.2 ^{c,d,e}	51.90 ± 9.6 ^{e,f}	150.46 ± 22.4 ^{f,g}
Marmara	İstanbul	227.44 ± 13.7 d^{e,f}	111.48 ± 20.6 ^{d,e,f}	55.69 ± 16.5 ^{e,f}	189.54 ± 35.5 ^{e,f,g}
Marmara	Kırklareli	370.18 ± 31.6 ^{a}	157.61 ± 3.6 ^{b,c,d}	83.20 ± 9.1 ^{c}	327.38 ± 21.6 ^{a}
Marmara	Tekirdağ	101.36 ± 4.0 ^{h,i}	90.46 ± 3.9 ^{e,f}	48.49 ± 5.3 ^{e,f}	91.78 ± 4.8 ^{g}
Marmara	Kocaeli	262.29 ± 28.0 ^{b,c,d}	138.26 ± 1.1 ^{d,e,f}	71.51 ± 6.8 ^{c,d}	182.61 ± 15.0 ^{e,f,g}
Mediterranean	Adana	147.95 ± 40.9 ^{g,h}	80.58 ± 10.9 ^{e,f}	41.91 ± 8.8 ^{e,f}	142.70 ± 27.4 ^{f,g}
Mediterranean	Gaziantep	61.55 ± 3.0 ^{h}	46.72 ± 2.1 ^{f}	16.73 ± 1.0 ^{f}	57.98 ± 1.0 ^{g}
Mediterranean	Hatay	122.04 ± 5.7 ^{g,h,i}	85.72 ± 4.6 ^{e,f}	34.17 ± 2.9 ^{e,f}	109.07 ± 2.9 ^{f,g}
Mediterranean	Mersin	151.79 ± 21.7 ^{f,g,h}	111.74 ± 11.9 ^{d,e,f}	41.38 ± 6.6 ^{e,f}	167.37 ± 6.6 ^{e,f,g}

Significant differences in the same column are represented by different letters (a–h) ($p < 0.05$). TE (Trolox equivalent); GAE (gallic acid equivalent); QE (quercetin equivalent).

The correlation between the method results, DPPH, CUPRAC, TPC, and TFC are shown in Figure 1. A positive, strong correlation is observed between the CUPRAC and TFC (0.92) and between DPPH and TPC (0.80). The lowest correlation was found between the CUPRAC and TPC (0.64).

	CUPRAC	DPPH	TPC	TFC
CUPRAC	1	0.69	0.64	0.92
DPPH	0.69	1	0.8	0.78
TPC	0.64	0.8	1	0.67
TFC	0.92	0.78	0.67	1

Figure 1. Correlations between the DPPH, CUPRAC, TPC, and TFC results.

2.2.1. Total Phenolic Content and DPPH

The Folin–Ciocalteu colorimetric method was used in this study to determine the TPC of Anatolian propolis samples [22]. The TPCs of samples are expressed as gallic acid equivalent (GAE) in 1 g of crude propolis sample. The TPC of the propolis samples from different regions are shown in Table 2. The TPC of the ethanolic extracts of propolis (EEP) samples varied from 16.73 to 125.83 mg GAE/g sample. Based on the region, the highest TPC was found in the Yozgat sample from the Central Anatolia region, whereas the lowest TPC was found in the Gaziantep sample from the Mediterranean region. The lowest TPC values (41.49–16.73 mg GAE/g sample) were found in the samples from the Mediterranean region, while the highest values (125.83–92.51 mg GAE/g sample) were found in the Central Anatolia and Black Sea regions. The TPC values of the samples assayed in this study differed according to the growing region of the cultivars.

The loss of DPPH reagent after the reaction with the samples was measured in the DPPH assay. DPPH is a purple-colored stable radical that turns bright yellow when it combines with free radicals [23]. Propolis's antioxidant ability was measured in Trolox equivalent (TE) per gram of crude propolis sample. The antioxidant abilities of the EEP samples ranged from 46.72 to 228.23 mg TE/g sample according to the results of the DPPH method. The highest DPPH activity result was observed in the Tokat sample from the Black Sea region, whereas the lowest DPPH activity value was observed in the Gaziantep sample from the Mediterranean region. As can be seen in the TFC profile of the propolis samples, the lowest DPPH activity was obtained in samples from the Mediterranean region (the Hatay, Adana, and Gaziantep samples) and some samples from the Marmara region, namely, İstanbul and Tekirdağ.

The TPC and DPPH revealed a substantial correlation coefficient ($r = 0.80$). These results are in agreement with those of Ahn et al. 2007, who found a good correlation between the antioxidant capacity of the TPC and DPPH ($r = 0.76$) for extracts of Chinese propolis collected throughout China [8,24].

2.2.2. Total Flavonoid Content and CUPRAC

Table 2 represents the TFC of the crude propolis samples; flavonoids varied from 57.98 to 327.38 mg QE per gram of propolis. The highest flavonoid content was found in the Kırklareli sample from the Marmara region, while the lowest value was found in the Gaziantep sample from the Mediterranean region. The lowest TFC was mainly found in the samples from the Mediterranean region, the highest TFC was observed in EEP samples from the Black Sea region of Türkiye, and the relatively lowest was found in the samples from the Marmara region. According to the CUPRAC test results, the propolis samples from the Black Sea region (Zonguldak) showed the highest result (378.93 mg TE/g sample), while the radical scavenging capacity of the Kırklareli sample (370.18 mg TE/g sample) from the Marmara region was slightly higher (Table 2). The CUPRAC results of the samples varied between 61.55 and 378.93 mg TE/g sample. The CUPRAC results of the Mediterranean samples showed the lowest values in this test. The TFC and DPPH results also showed a significant correlation coefficient ($r = 0.92$), as shown in Figure 1. This was the highest correlation coefficient obtained in these four different measurement tests.

2.3. *Phenolic Composition of Propolis*

Propolis often has a wide range of chemical components. In general, propolis contains flavonoids, phenolic acids and esters, phenolic aldehydes and ketones, terpenes, amino acids, alcohols, and other acids and derivatives [1,25,26]. Several researchers have used HPLC to determine the polyphenolic components of propolis samples, with most applications involving different propolis samples from Europe (Italy), Argentina, and Brazil [27,28]. The individual compounds identified in Anatolian propolis samples are listed in Table 3. Propolis from mild zones (Asia, Europe, North America, etc.) is mostly composed of phenolic compounds, such as various flavonoids and aromatic acids and their esters, obtained from poplar buds, which seem to be the most common source of propolis [29,30]. In our study of 24 Anatolian propolis samples, the identified substances were flavonoids (pinocembrin, kaempferol, pinobanksin, and apigenin), phenolic/aromatic acids (chlorogenic acid, caffeic acid, *p*-coumaric acid, ferulic acid, and trans-cinnamic acid), and phenolic aldehyde (vanillin). Many of these substances (pinobanksin, kaempferol, ferulic acid, caffeic acid, *p*-coumaric acid, etc.) have already been determined by several authors who studied the polyphenolic composition of Turkish [8,17,29] and European propolis samples [31]. Each sample was analyzed in triplicate to determine the average phenolic compound content in Turkish propolis. In our study, ten compounds were identified using the HPLC-PDA system based on ethanolic extracts from propolis samples from four different regions of Anatolia, Türkiye. The peaks of the polyphenolic compounds were detected using a conventional addition procedure and quantified using calibration formulas based on peak normalization rates.

The concentrations of phenolic compounds were calculated as mg/g of the crude propolis sample. The results summarized in Table 3 show that caffeic acid and ferulic acid were the major phenolic acid constituents, while pinobanksin and pinocembrin were the most abundant flavonoids in all propolis samples. Pinobanksin content was lowest in the sample from Gaziantep (2.95 mg/g), which is from the Mediterranean region, and the highest was found in the Ordu sample (38.76 mg/g), which is from the Black Sea region. The other major flavonoid was pinocembrin, whose content ranged from 1.22 mg/g to 6.80 mg/g, with the lowest value found in the Gaziantep sample and the highest value in the Kocaeli sample.

However, in the samples from the Black Sea region, the pinobanksin content was several times higher than the samples from the Marmara and Central Anatolia regions. There is a positive correlation between the number of individual flavonoids and phenolic compounds and their antioxidant capacities [24,31]. The propolis samples that contained more flavonoids and phenols also had higher antioxidant capacities.

2.4. Antimicrobial Activity

The disc diffusion method was used to determine the inhibition zones of the 24 different propolis extracts for primary screening. A standard Gram-positive and Gram-negative bacterial strain and two fungi (a yeast and a mold) were used. According to the results shown in Table 4, the EEPs of the samples showed antibacterial activity against *S. aureus* and *E. coli* and antifungal activity against *C. albicans* and *A. niger.* The antibacterial activity of the Anatolian propolis samples tested in this study was effective against the pathogens tested. Table 4 shows that statistically significant differences ($p < 0.05$) were found between Anatolian propolis samples for four different microorganisms.

Table 3. Content of individual compounds (phenolic acids, phenolic aldehyde, and flavonoids) determined in Anatolian propolis samples (data are expressed as mg/g of the crude propolis sample).

City	Chlorogenic Acid	Caffeic Acid	Vanillin	*p*-Coumaric Acid	Ferulic Acid	trans-Cinnamic Acid	Pinobanksin	Kaempferol	Apigenin	Pinocembrin
Amasya	-	7.17 ± 2.6^{a}	0.48 ± 0.4^{c}	2.80 ± 0.6^{c}	$3.09 \pm 1.6^{c,d}$	7.84 ± 0.3^{a}	19.55 ± 7.3^{b}	$5.41 \pm 1.4^{a,b}$	$3.78 \pm 0.1^{a,b}$	$2.41 \pm 0.2^{b,c}$
Bartın	-	$4.67 \pm 2.0^{b,c}$	0.83 ± 0.1^{b}	3.95 ± 0.9^{b}	5.17 ± 0.1^{c}	4.65 ± 0.1^{b}	21.53 ± 3.2^{b}	4.06 ± 1.2^{b}	1.73 ± 0.1	$2.24 \pm 0.2^{b,c}$
Karabük	-	$3.42 \pm 1.2^{d,e}$	$0.32 \pm 0.1^{c,d}$	$2.21 \pm 0.9^{c,d}$	0.83 ± 0.2^{e}	1.37 ± 0.8^{e}	10.59 ± 3.1^{c}	4.30 ± 1.4^{b}	2.79 ± 0.1^{b}	3.41 ± 0.8^{b}
Kastamonu	-	4.42 ± 0.8^{c}	0.46 ± 0.3^{c}	3.64 ± 1.6^{b}	$4.21 \pm 3.4^{c,d}$	$0.51 \pm 0.3^{e,f}$	34.66 ± 0.7^{a}	$1.48 \pm 0.1^{c,d}$	$0.43 \pm 0.1^{c,e}$	$2.92 \pm 0.7^{b,c}$
Ordu	-	7.38 ± 3.5^{a}	-	$4.43 \pm 1.9^{a,b}$	$1.78 \pm 1.6^{d,e}$	$2.93 \pm 1.5^{c,d}$	38.76 ± 1.5^{a}	6.95 ± 1.5^{a}	$4.10 \pm 1.0^{a,b}$	3.86 ± 1.0^{b}
Samsun	0.31 ± 0.1^{c}	$1.93 \pm 0.9^{e,f}$	0.18 ± 0.1^{d}	1.11 ± 0.1^{d}	2.05 ± 1.2^{d}	$0.97 \pm 0.5^{e,f}$	$13.75 \pm 0.6^{b,c}$	1.73 ± 0.6^{c}	$1.41 \pm 0.3^{b,c}$	2.09 ± 1.0^{c}
Tokat	-	$3.44 \pm 0.5^{d,e}$	-	$2.25 \pm 0.5^{c,d}$	2.59 ± 0.2^{d}	$0.53 \pm 0.4^{e,f}$	$8.15 \pm 3.2^{c,d}$	2.07 ± 0.2^{c}	5.33 ± 1.6^{a}	1.48 ± 0.2^{c}
Zonguldak	-	$6.62 \pm 2.8^{a,b}$	1.37 ± 0.9^{c}	5.86 ± 3.3^{a}	10.77 ± 1.8^{b}	7.45 ± 0.3^{a}	$26.65 \pm 2.2^{a,b}$	$2.95 \pm 1.2^{b,c}$	0.87 ± 0.1^{c}	6.53 ± 2.4^{a}
Çorum	-	$3.56 \pm 0.7^{d,e}$	0.14 ± 0.1^{d}	1.57 ± 0.1^{d}	$1.54 \pm 0.3^{d,e}$	$0.72 \pm 0.2^{e,f}$	16.18 ± 0.2^{b}	2.12 ± 0.1^{c}	$1.75 \pm 0.4^{b,c}$	$5.25 \pm 0.3^{a,b}$
Konya	-	4.64 ± 0.1^{c}	$0.27 \pm 0.1^{c,d}$	$2.14 \pm 0.1^{c,d}$	0.94 ± 0.1^{e}	$0.86 \pm 0.1^{e,f}$	$12.37 \pm 4.0^{b,c}$	$2.85 \pm 0.1^{b,c}$	$1.84 \pm 0.1^{b,c}$	3.17 ± 1.1^{b}
Kütahya	-	$4.12 \pm 1.3^{c,d}$	-	2.75 ± 0.9^{c}	$1.68 \pm 0.2^{d,e}$	4.02 ± 0.1^{b}	17.30 ± 0.2^{b}	$2.61 \pm 0.1^{b,c}$	$1.55 \pm 0.4^{b,c}$	$2.42 \pm 0.1^{b,c}$
Sivas	-	5.28 ± 0.1^{b}	0.42 ± 0.1^{c}	2.63 ± 0.3^{c}	2.65 ± 0.4^{d}	1.01 ± 0.1^{e}	10.21 ± 1.6^{c}	4.39 ± 0.1^{b}	$1.50 \pm 0.1^{b,c}$	$2.82 \pm 0.1^{b,c}$
Yozgat	-	$3.64 \pm 2.0^{d,e}$	$0.65 \pm 0.2^{b,c}$	2.57 ± 0.3^{c}	5.50 ± 0.5^{c}	$0.94 \pm 0.1^{e,f}$	$13.39 \pm 2.8^{b,c}$	3.90 ± 0.4^{b}	$1.52 \pm 0.1^{b,c}$	$2.31 \pm 0.7^{b,c}$
Balıkesir	-	4.46 ± 1.9^{c}	$0.61 \pm 0.4^{b,c}$	$3.23 \pm 2.3^{b,c}$	2.24 ± 0.5^{d}	2.86 ± 0.7^{d}	18.02 ± 3.1^{b}	-	$1.76 \pm 0.4^{b,c}$	2.03 ± 1.6^{c}
Bilecik	-	$3.98 \pm 0.9^{c,d}$	1.54 ± 1.2^{a}	3.83 ± 2.2^{b}	$17.28 \pm 2,4^{a}$	1.63 ± 0.3^{e}	16.19 ± 2.1^{b}	$3.04 \pm 1.2^{b,c}$	$1.48 \pm 0.2^{b,c}$	6.57 ± 1.6^{a}
Bursa	-	4.36 ± 1.9^{c}	$0.57 \pm 0.1^{b,c}$	2.53 ± 1.5^{c}	$1.27 \pm 0.1^{d,e}$	$0.79 \pm 0.1^{e,f}$	10.79 ± 2.4^{c}	1.72 ± 0.4^{c}	$1.30 \pm 0.3^{b,c}$	$2.55 \pm 0.1^{b,c}$
İstanbul	-	3.24 ± 1.3^{e}	0.74 ± 0.1^{b}	$2.10 \pm 0.1^{c,d}$	5.67 ± 0.3^{c}	$3.49 \pm 0.3^{b,c}$	10.93 ± 6.0^{c}	1.89 ± 0.1^{c}	$1.84 \pm 0.5^{b,c}$	$2.54 \pm 0.1^{b,c}$
Kırklareli	1.56 ± 0.1^{a}	5.78 ± 2.5^{b}	0.97 ± 0.8^{b}	2.51 ± 1.6^{c}	2.75 ± 0.2^{d}	1.60 ± 0.2^{e}	20.74 ± 5.6^{b}	1.75 ± 0.1^{c}	2.31 ± 0.1^{b}	$5.35 \pm 2.1^{a,b}$
Kocaeli	-	$3.69 \pm 1.9^{d,e}$	1.56 ± 1.3^{a}	3.93 ± 2.1^{b}	14.59 ± 2.2^{a}	-	16.82 ± 1.4^{b}	4.00 ± 1.6^{b}	$1.91 \pm 0.1^{b,c}$	6.80 ± 0.3^{a}
Tekirdağ	-	1.09 ± 0.1^{f}	0.74 ± 0.1^{b}	1.11 ± 0.2	5.10 ± 0.7^{c}	0.45 ± 0.1^{f}	10.01 ± 3.3^{c}	1.01 ± 0.1^{d}	$1.12 \pm 0.1^{b,c}$	1.44 ± 0.1^{c}
Adana	-	$2.24 \pm 0.4^{e,f}$	-	1.19 ± 0.5^{d}	0.93 ± 0.1^{e}	1.68 ± 0.3^{e}	$9.11 \pm 3.5^{c,d}$	$2.58 \pm 0.1^{b,c}$	1.80 ± 0.1^{c}	1.68 ± 0.4^{c}
G.antep	-	0.88 ± 0.1^{f}	-	0.20 ± 0.1^{f}	0.45 ± 0.1^{f}	$0.24 \pm 0.1^{f,g}$	2.95 ± 0.1^{d}	1.86 ± 0.1^{c}	$0.41 \pm 0.1^{c,e}$	1.22 ± 0.1^{c}
Hatay	0.69 ± 0.1^{b}	3.75 ± 2.5^{d}	-	0.51 ± 0.2^{e}	0.41 ± 0.1^{f}	$0.56 \pm 0.1^{e,f}$	11.44 ± 1.5^{c}	1.91 ± 0.1^{c}	$1.56 \pm 0.5^{b,c}$	1.99 ± 0.1^{c}
Mersin	-	4.62 ± 2.9^{c}	-	1.17 ± 0.2^{d}	0.83 ± 0.1^{e}	0.32 ± 0.1^{f}	21.70 ± 3.3^{b}	$2.88 \pm 0.1^{b,c}$	$3.82 \pm 0.5^{a,b}$	4.25 ± 0.1^{b}

Significant differences in the same column are represented by different letters (a–f) ($p < 0.05$).

Table 4. Antimicrobial activity results of ethanolic extracts of Anatolian propolis (data are expressed as millimeter (mm) inhibition zone).

Region	City	*S. aureus*	*E. coli*	*C. albicans*	*A. niger*
Black Sea	Amasya	12.00 ± 1.0 [a,b]	10.50 ± 0.5 [a,b]	12.50 ± 1.5 [a,b]	9.50 ± 0.5 [a,b]
Black Sea	Bartın	12.50 ± 1.5 [a,b]	12.50 ± 1.5 [a]	10.00 ± 1.0 [a,b]	11.00 ± 1.0 [a,b]
Black Sea	Karabük	12.00 ± 1.0 [a,b]	9.50 ± 1.5 [a,b]	10.50 ± 0.5 [a,b]	8.75 ± 0.3 [b,c]
Black Sea	Kastamonu	14.00 ± 1.0 [a]	11.50 ± 0.5 [a,b]	10.50 ± 0.5 [a,b]	9.50 ± 0.5 [a,b]
Black Sea	Ordu	10.25 ± 0.2 [a,b]	10.50 ± 0.5 [a,b]	9.00 ± 0.0 [b]	9.00 ± 1.0 [b,c]
Black Sea	Samsun	11.25 ± 1.2 [a,b]	11.50 ± 0.5 [a,b]	10.00 ± 1.0 [a,b]	9.50 ± 0.5 [a,b]
Black Sea	Tokat	9.75 ± 0.7 [a,b]	9.00 ± 0.0 [a,b]	10.50 ± 0.5 [a,b]	10.00 ± 1.0 [a,b]
Black Sea	Zonguldak	13.50 ± 0.5 [a,b]	11.50 ± 0.5 [a,b]	11.00 ± 1.0 [a,b]	11.00 ± 0.0 [a,b]
Black Sea	Çorum	9.25 ± 0.7 [a,b]	9.50 ± 0.5 [a,b]	10.50 ± 0.5 [a,b]	10.00 ± 0.0 [a,b]
Central Anatolia	Konya	9.00 ± 0.2 [a,b]	8.00 ± 0.0 [b]	9.00 ± 1.0 [b]	8.00 ± 0.0 [c]
Central Anatolia	Kütahya	11.25 ± 1.2 [a,b]	9.50 ± 0.5 [a,b]	9.50 ± 0.5 [a,b]	12.00 ± 1.0 [a]
Central Anatolia	Sivas	8.25 ± 0.7 [b]	9.50 ± 0.5 [a,b]	11.50 ± 1.5 [a,b]	9.00 ± 1.0 [b,c]
Central Anatolia	Yozgat	11.00 ± 1.0 [a,b]	9.50 ± 0.5 [a,b]	9.00 ± 0.0 [b]	9.50 ± 0.5 [a,b]
Marmara	Balıkesir	13.50 ± 1.5 [a,b]	12.00 ± 1.0 [a,b]	12.00 ± 1.0 [a,b]	10.50 ± 0.5 [a,b]
Marmara	Bilecik	12.50 ± 1.5 [a,b]	11.00 ± 1.0 [a,b]	10.00 ± 0.0 [a,b]	9.50 ± 0.5 [a,b]
Marmara	Bursa	11.00 ± 1.0 [b]	8.75 ± 0.3 [a,b]	10.00 ± 1.0 [a,b]	8.50 ± 1.0 [b,c]
Marmara	İstanbul	11.00 ± 1.0 [a,b]	11.00 ± 1.0 [a,b]	8.75 ± 0.8	10.00 ± 0.5 [a,b]
Marmara	Kırklareli	13.50 ± 0.5 [a,b]	11.00 ± 1.0 [a,b]	10.00 ± 1.0 [a,b]	12.50 ± 0.5 [a]
Marmara	Kocaeli	12.25 ± 0.2 [a,b]	10.50 ± 0.5 [a,b]	10.75 ± 0.8 [a,b]	9.50 ± 0.5 [a,b]
Marmara	Tekirdağ	11.50 ± 0.5 [a,b]	10.50 ± 0.5 [a,b]	14.00 ± 1.0 [a]	10.50 ± 0.5 [a,b]
Mediterranean	Adana	10.00 ± 1.0 [a,b]	9.50 ± 0.5 [a,b]	12.50 ± 0.5 [a,b]	10.00 ± 0.0 [a,b]
Mediterranean	Gaziantep	11.00 ± 1.0 [a,b]	10.00 ± 0.5 [a,b]	10.00 ± 1.0 [a,b]	9.00 ± 0.0 [b,c]
Mediterranean	Hatay	11.00 ± 1.0 [a,b]	10.50 ± 0.5 [a,b]	10.00 ± 0.0 [a,b]	9.50 ± 0.5 [a,b]
Mediterranean	Mersin	13.50 ± 0.5 [a,b]	10.50 ± 0.5 [a,b]	12.50 ± 0.5 [a,b]	9.50 ± 0.5 [a,b]
	Mean values	11.40	10.28	10.58	9.82
	Negative Control **	0	0	0	0
	Positive Control ***	50	35	30	30

Significant differences in the same column are represented by different letters (a–c) ($p < 0.05$). ** 10% dimethyl sulfoxide (DMSO) solution was used. *** For bacteria, ampicillin (10 µg-Oxoid) commercial discs were used. *** For mold and yeast, vorozanole (11 µg) commercial discs were used.

The antimicrobial activity of propolis has been attributed to both hydrophilic and hydrophobic phenolic chemicals, such as flavonoids and aromatic acids and esters, which can act on the cell walls of bacteria [3]. Caffeic acid and its esters, as well as volatile fractions containing phenols, have antimicrobial properties. However, whether the antibacterial and antifungal activities of the ethanol extracts of propolis depend on the concentration of polyphenolic fractions, such as pinocembrin and caffeic acid derivatives, or on the synergism of these or other compounds is still unknown [29]. Anjum et al. reported that the bioactive compounds of propolis, such as pinocembrin, showed antibacterial activity against *Streptococcus* spp., and artepillin C and *p*-coumaric acid showed antibacterial activity against *Helicobacter pylori* [25].

2.4.1. Antibacterial Activity

Activity against bacterial pathogens was tested for all propolis samples. As can be seen in Table 3, all extracts showed different degrees of antibacterial activity against *S. aureus* and *E. coli*, with the strongest antibacterial activity obtained by the samples from Kastamonu, Bartın, and Kırklareli. Among the 24 samples, Sivas, Konya, and Çorum showed moderate activity against *S. aureus*, with inhibition zone values ranging from 8.25 to 9.75 mm. The samples from Kırklareli and Bartın (from the Marmara and Black Sea regions, respectively) showed high activity against *S. aureus* with a 13.5 mm inhibition zone value and against *E. coli* with a 12.5 mm inhibition zone value, respectively. Consequently, the antimicrobial activity test results clearly showed that propolis samples from the Black Sea region had much stronger antibacterial activity compared with the propolis sample from Central Anatolia (Table 3). Previous studies indicated that caffeic acid esters are the main chemicals responsible for this antimicrobial activity [13]. Considering the composition of propolis from Amasya, it had higher caffeic acid content.

2.4.2. Antifungal Activity

The activity of the propolis samples against fungal strains (*C. albicans* and *A. niger*) was tested. All extracts exhibited significant antifungal activity. However, the antifungal activity was similar for all samples. The zones of inhibition around the disc observed in Petri dishes containing *C. albicans* were slightly larger in the Amasya, Mersin, and Adana samples than in the other samples. The highest antifungal activity was obtained in samples from Amasya and Mersin (12.50 mm) for *C. albicans* and from Kırklareli samples (12.50 mm) for the *A. niger* strain. Flavonoids are known for their antibacterial, antifungal, and antiviral capabilities, and they are assumed to be responsible for the beneficial medicinal properties of propolis [30]. Propolis's antibacterial, antifungal, and antiviral properties have also been observed in the esters of phenolic acids, especially caffeic acids and ferulates [9].

2.5. Principal Component Analysis (PCA)

Principal components (PCs) were determined from the eigenvalues of the correlation matrix of observations. The eigenvalues were found to be 6.75, 3.16, 2.06, 1.49, 1.22, and 1.03 for PC1 to PC6, respectively. As can be seen in Figure 2, the first six PCs explain 82.68% of the total variance. The first two PCs account for 35.51% and 16.66% of the total variance, respectively. Thus, the first two PCs explain the origin of the score plot with a very tight confidence ellipse.

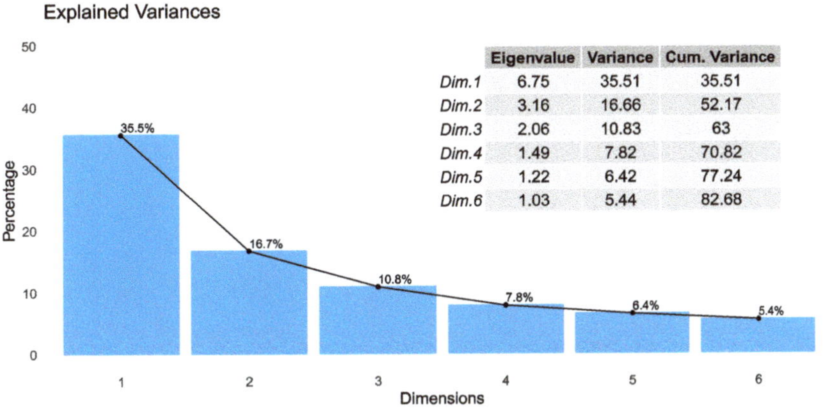

Figure 2. The explained PCs of the total variance and eigenvalues for each of the PCs.

Figure 3 shows the observations and PCs obtained from the analyzed data. The score over PC1 and PC2 separates the groups from the Marmara, Central Anatolia, Mediterranean, and Black Sea regions. The center of the samples from the Marmara region is in the positive parts of both PC1 and PC2, while the samples from Central Anatolia are in the negative parts of both PCs. The Black Sea region samples are grouped along the positive part of PC1, whereas the central point of the samples from the Black Sea region is located on the negative part of PC2. As a result, it can be stated that the DPPH and *E. coli* values as two variables (Figure 4) are useful in clustering the propolis samples from the Marmara and Black Sea regions separately. Mediterranean propolis samples were well separated from the other three groups of samples by being on the negative region of the PC1 due to their significantly higher antifungal activity against *C. albicans*, their relatively low antioxidant capacity determined by four different measurement methods, their low ferulic acid content, and their lack of vanillin content. On the other hand, apart from the *E. coli* and DPPH levels, the *S. aureus*, vanillin, CUPRAC, TFC, and caffeic acid levels played a key role in clustering among the propolis samples from Marmara, the Black Sea, and Central Anatolia.

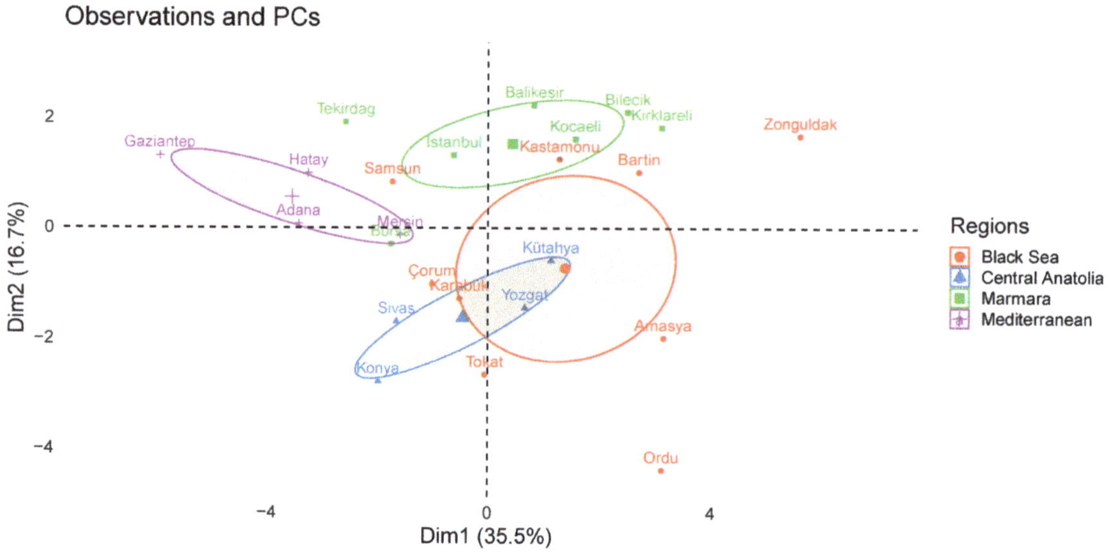

Figure 3. Observations (antimicrobial activity, antioxidant capacity, moisture content (%), and individual phenolic constituents) and PCs belong to Anatolian propolis.

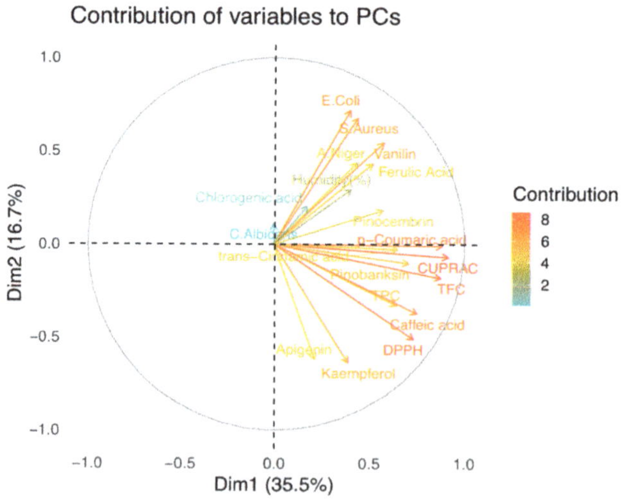

Figure 4. Contribution of variables (TFC, *p*-coumaric acid, DPPH, caffeic acid, etc.) to PCs.

The contribution of the variables to the PCs is shown in Figure 4. The CUPRAC and *p*-coumaric acid values strongly correlate with PC1 in the positive direction, while the values of *E. coli* and kaempferol correlate with PC2 in the positive and negative directions, respectively. When PCA was applied to map the results of the biochemical composition and biological activity of the propolis samples (n = 24), differences in the antioxidant and antimicrobial capacity profiles were observed for samples collected in different parts of Anatolia. This is clearly related to the climatic characteristics and local flora surrounding the hive accessible to honeybees at the collection sites.

A dendrogram of the Turkish propolis samples was generated using a hierarchical clustering algorithm (HCA) with the Ward method using the packages FactoMineR and factoextra in R 4.0.4. This algorithm was applied to the first six PCs since they have eigenvalues greater than 1.0. Figure 5 shows the dendrogram and hierarchical clustering results of the Turkish propolis samples. As can be seen in Figure 5, the propolis samples are divided into four clusters. The Ordu and Amasya samples have a strong relationship with each other and are separated from all the other clusters. Although the Samsun sample belongs to the Black Sea region, it shows a greater similarity with the samples from the Marmara region group. In addition, the samples from the Mediterranean region and two of the samples from the Marmara region (İstanbul and Tekirdağ) are in the same cluster, as shown in Figure 5.

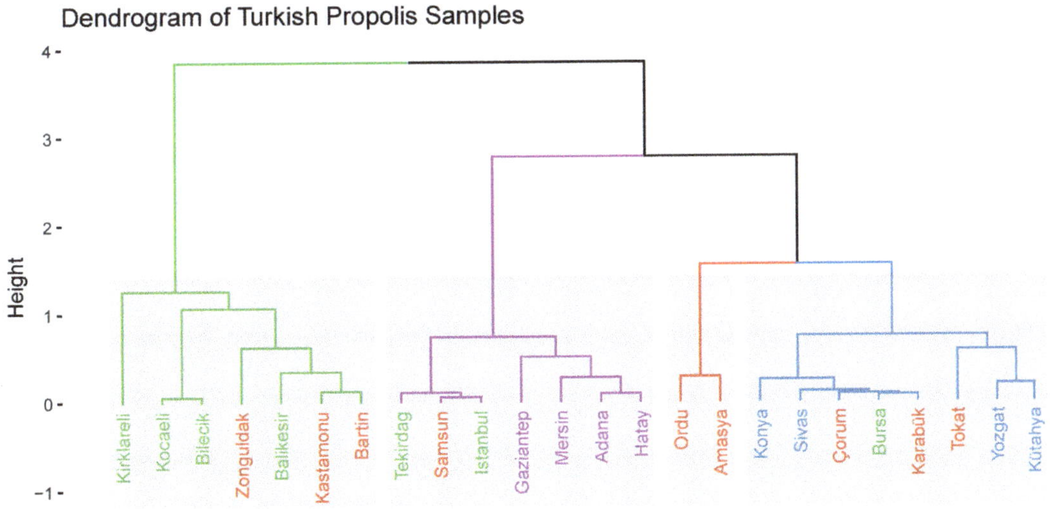

Figure 5. Dendrogram of Turkish propolis samples obtained from the results of HCA by using the factoextra package in R.

3. Discussion

According to the results of the total phenolic content (TPC), total flavonoid content (TFC), and antioxidant capacity measurement methods, phenolic compounds were detected in different amounts and types in the samples from different geographical regions of Türkiye. The antioxidant profile of the propolis samples may be related to the diversity of the geographical areas, plants in the region, and the type of bees [2,32]. Our results for the TPC and DPPH (2,2-diphenyl-1-picrylhydrazyl) methods, remarkably, follow those of Shi et al. (2012) [33], who studied 15 Chinese propolis samples, and Ozdal et al. (2019) [8], who studied 11 Turkish propolis samples. Similarly, the findings of Ahn et al. (2007), who evaluated the TPC of 20 Chinese poplar-type propolis samples, and Yesiltas, who examined 4 Turkish propolis samples, are also quite parallel to our TPC values [24]. However, our TPC results were not as high as those of the samples examined by Yesiltas et al. (2014) [34]. However, the TPC of our samples was relatively higher than the TPC result of Wieczynska et al. (2017) [35], who studied seven Polish propolis samples, and samples of Kubiliene et al. (2015) [36], who obtained the samples under different extraction conditions. Ahn et al. (2007), Moreira et al. (2008) [37], Lagouri et al. (2013) [31], Kumazawa et al. 2004 [33], Ahn et al. (2004) [38], and Shi et al. (2012) reported phenolic compound contents ranging from 42.9 to 329.0 mg/g of the propolis sample in China, Macedonia, Greece, Portugal, and Korea. On the other hand, the results of other studies were reported as IC^{50} units of measurement referring to the prepared solution; this could complicate the comparison of results between different experiments [37]. Alternatively, the results were exressed in mg Trolox or quercetin equivalent per ethanolic propolis extracts with respect to the crude propolis sample [39].

Erdogan et al. (2011) used the DPPH method to assess the antioxidant capabilities of propolis samples from different Anatolian localities, including the cities of Bingol, Rize, Tekirdağ, and Van [40]. According to their results, the highest value for antioxidant activity was obtained in the sample from Rize, with a value of 503.70 mg TE/g sample, which is considerably higher than the DPPH values obtained in our study. The findings of Yesiltas et al. (2014) also showed high DPPH activity values for some of the samples (135–454 TE/g) [34]. Our results for the TFC also follow the TFC results of Shi et al. (2012), Ozdal et al. (2019), and Zarate et al. (2018) [2]. Similarly, the values are also parallel with the results of Ahn et al. (2007) and Yesiltas et al. (2014). However, our findings for the total flavonoid content were higher than the values reported in the studies of Ozdal (2019) and Ahn (2007), respectively. Of the five samples with high TFC values, one sample (Kırklareli) was from the Marmara region, and four of these samples (Amasya, Zonguldak, Ordu, and Tokat) were from the Black Sea region. The surrounding flora and climatic conditions might have contributed to the higher total phenolic and flavonoid content in the propolis samples obtained from these areas. Our results from the cupric ion-reducing antioxidant capacity (CUPRAC) assay are higher than the results (24.00–85.00 TE/g sample) of Ozdal et al. (2019) and slightly lower than the results (575.00 TE/g sample) of Yesiltas et al. (2014).

When the quantitative results of phenols were compared with other publications in the literature (found by Shi et al. (2012), Ahn et al. (2007), Yesiltas et al. (2014), and Pellati et al. (2011)), the number of flavonoids ranged from the lowest value (3.44 mg/g sample) to the highest value (76.50 mg/g sample) for pinobanksin and in a range of 0.43 to 46.00 mg/g sample for pinocembrin [24,33,34,39]. Furthermore, when comparing the quantitative composition of phenolic acids and flavonoids with other studies in the literature, our results for the caffeic acid and *p*-coumaric acid [8,22,29,30,38], ferulic acid [8,22,30,38], and kaempferol [3,29,38] contents were in agreement with the results of these works. Moreover, the ethanolic extracts of our propolis samples were mostly found to be richer in phenolic compounds (the highest values being 5.41, 5.33, and 5.88 mg/g sample for kaempferol, apigenin, and *p*-coumaric acid, respectively) than those of the samples. The caffeic acid content in propolis was reported by Pellati et al. (2011) in Italian poplar-type propolis samples ranging from 0.02 to 1.19 mg/g sample [39]. Ozdal et al. (2019) reported the caffeic acid content in a range between 0.04 and 0.61 mg/g sample in propolis collected from different parts of Türkiye. In our study, we determined the caffeic acid content, ranging from 0.88 to 7.38 mg/g sample. Caffeic acid and *p*-coumaric acid are the most abundant phenolic acids detected in propolis from *Populus* spp. in temperate zones [29]. Most individual phenolic acid and flavonoid contents in 24 Turkish propolis samples are in agreement with the results of other authors [3,8,24,33,34,41]. However, in our study, chlorogenic acid was detected only in the samples from Samsun, Kırklareli, and Hatay. The phenolic content in propolis varies depending on the geographical location; therefore, its biological activity is closely related to biogeographical factors such as local flora, climate, and seasonal influences [26,40]. The variability of the constituents of the propolis samples suggests that some of the propolis samples have different chemical compositions.

Our findings for the antimicrobial activity of Anatolian propolis samples showed that the three propolis extracts were significantly more effective against *S. aureus* than against *E. coli.* These results are in agreement with those reported in the literature [13,42,43], as there is a consistency between the results of all studies showing that propolis extracts are always more efficient against Gram+ than Gram− bacteria. The flavonoids, phenolic acids and esters, and vanillin found in the samples could be responsible for the antibacterial activity against Gram+ bacteria [42,43]. The zones of inhibition against *C. albicans* in the Petri dishes caused by ethanolic extracts (containing 0.1 g crude sample/mL) from Anatolian propolis are in agreement with the results of some authors: 8–14 mm found by Kartal et al. [13], 10.0–12.5 mm found by Silva et al. [44], 13–24 mm found by Aliyazicioglu et al. [45], and 14–18 mm found by Kujumginev et al. [9]. Aliyazicioglu et al. (2013) reported that an inhibitory zone ranging from 8 to 12 mm was obtained against *A. niger*, whereas, in our study, we obtained a zone of inhibition ranging between 8.75 and 12.50 mm [45].

This study showed that by applying chemometric methods, principal component analysis (PCA) on the biochemical composition and biological activity of propolis we were able to successfully group Anatolian propolis samples from different regions of Türkiye, except for the samples from the Black Sea region.

4. Materials and Methods

4.1. Materials

Folin–Ciocalteu reagent, neocuprine (Nc), 2,2-diphenyl-1-picrylhydrazyl (DPPH), Trolox (≥95%), gallic acid (≥98%), and quercetin (≥95%) were obtained from Sigma-Aldrich. Sodium hydroxide, sodium carbonate, sodium nitrite, potassium persulfate, ferric chloride hexahydrate, ethanol (≥99.8%), methanol (≥99.9%), dipotassium hydrogen phosphate, potassium dihydrogen phosphate, and Whatman® filter papers No. 1 were purchased from Merck (Darmstadt, Germany). Polytetrafluoroethylene (PTFE) filter 0.45 µm, copper (II) chloride, ammonium acetate, aluminum chloride, and potassium chloride were purchased from Fluka Chemie (Buchs, Switzerland). All chemicals used as standards in the HPLC analysis, including apigenin, chlorogenic acid, caffeic acid, ferulic acid, kaempferol, *p*-coumaric acid, pinocembrin, pinobanksin, trans-cinnamic acid, and vanillin, were obtained from Sigma-Aldrich Chemie GmbH (Steinheim, Germany).

4.1.1. Propolis Samples

The Anatolian propolis samples were collected from 4 regions (24 different cities) of Türkiye in June and April 2019. These regions were the Black Sea region (n = 9), the Central Anatolia region (n = 4), the Marmara region (n = 7), and the Mediterranean region (n = 4) of Türkiye. All propolis samples were collected from only one provider for each city. The hand-collected propolis samples were ground and pulverized and then stored in individual packages in freezing conditions (−18 °C) until processing for extract preparation. All analyses were performed in triplicate for the 24 propolis samples.

4.1.2. Culture Media and Test Microorganisms

Antimicrobial activities of samples were evaluated against *Staphylococcus aureus*—ATCC 25923—as Gram-positive bacteria, *Escherichia coli*—ATCC 25922—as Gram-negative bacteria, *Candida albicans—ATCC 10231*—as yeast, and *Aspergillus niger*—ATCC 16404—as a mold species. All microorganisms were purchased from the ATCC Culture collection and conserved and grown in the Food Engineering Department of Istanbul Technical University laboratories. The screening of antimicrobial activity was performed using Tryptic Soy Agar (TSA, Oxoid Ltd. Hampshire, UK) for bacteria and Sabouraud Dextrose Agar (SDA, Oxoid Ltd., Hampshire, UK) for yeast and fungi.

4.2. Methods

4.2.1. Extraction of Propolis

Extraction of propolis was carried out as described by other authors [3,8,46] with some minor modifications. All propolis samples were ground in liquid nitrogen on a lab-scale mill (IKA A11 basic analytical mill, Königwinster, Germany) before extraction. In total, 1 g of the finely ground crude propolis sample was mixed with 10 mL (70%) of aqueous ethanol under constant stirring to obtain a 0.1 g/mL solid extract at room temperature for 24 h. The suspension was then poured into a 50 mL falcon tube and centrifuged at rpm for 10 min to achieve complete separation. After centrifugation, the collected supernatant was filtered through the Whatman filter paper. The filtrate obtained was kept refrigerated and centrifuged again after 24 h at 11,000 rpm at −10 °C for 5 min to remove the wax present in the propolis extract. The final filtrates were kept in the freezer for further analysis. On the other hand, for the microbiological tests, the ethanolic extracts of the propolis samples were dried (first, the ethanol was removed with a rotary evaporator, and then, the aqueous solution was placed in a freeze dryer to remove the water and obtain a dry extract). Then, the dried propolis extracts were dissolved in 10% dimethyl sulfoxide (DMSO) to obtain a final concentration of 0.1 mg/mL dry extract for the microbiological tests.

4.2.2. Determination of Moisture Content

To determine the moisture content of the samples, a halogen infrared moisture analyzer device (HE53, Mettler&Toledo, Zaventem, Belgium) was used. On a tared measuring cup made of aluminum foil, the previously powdered propolis samples were weighed as 0.5 g and placed in the instrument, and the moisture content of the samples was determined by reading the value directly on the screen. All measurements were performed in triplicate.

4.2.3. Antimicrobial Activity Test

The antibacterial and antifungal activity of propolis samples was assayed by using the disc diffusion method on agar [47]. The antibacterial activity of propolis was tested using TSA for bacteria and using SDA (Oxoid) as inoculum for the antifungal activity of propolis samples. *S. aureus*, *E. coli*, *C. albicans*, and *A. niger* strains were used as test microorganisms. The antimicrobial activity was determined by calculating the diameter of the inhibitory zones in a Petri dish, which were discolored after 24 h of incubation at 37 °C. An inhibitory zone with a diameter of less than 5 mm was considered inactive (the diameter of the spot was 5 mm) [9]. The control experiments showed that the solvent used as a control had no activity. All microorganisms used as inoculum were cultured overnight at 37 °C in TSB. The turbidity of these suspensions was adjusted to approximately 1×10^6 CFU/mL by dilution with peptone water. Colonized agar Petri dishes were prepared (15–20 mL) and inoculated with 100µL of suspensions containing approximately 10 microorganisms per ml. Then, sterile paper discs (diameter = 6 mm) were positioned on the agar to load 100 µL of each propolis extract (a concentration of 0.1 mg propolis extract in 1 mL of 10% DMSO). All propolis extracts were dissolved in the 10% DMSO prepared previously. In addition, DMSO (10%) was used as a negative control, and commercial discs of Ampicillin (10 mg—Oxoid) were used as a positive control for antibacterial activity, while commercial discs of vorozanole (11 mg—Oxoid) were used as a positive control for antifungal activity. *S. aureus* and *E. coli* strains were incubated at 37 °C for 24 h, *C. albicans* at 37 °C for 48 h, and *A. niger* at 25 °C for 72 h. Subsequently, the zones of inhibition around the discs were determined in millimeters (mm). All test results were triple-checked.

4.2.4. Total Flavonoid and Phenolic Content and Antioxidant Activity Tests

Determination of Total Flavonoid Content

The total flavonoid content (TFC) of the propolis samples was determined by reading a prepared mixture in a spectrophotometer, as described by Kim, Jeong, and Lee (2003) [48] and modified by Uluata et al. [49] (2021) and Kızıltaş (2021) [50]. The absorbance of the mixture was measured at 510 nm. All test results were triplicated. A quercetin standard curve was prepared to determine the TFC of the extracts, and the results were expressed as mg quercetin equivalent (QE) per gram of crude sample.

Determination of Total Phenolic Content

The total phenolic content (TPC) of propolis was determined using the Folin–Ciocalteu method, as described by Chen et al. 2015, Hızır-Kadı et al. 2020, and Topal et al. 2021 [51–53]. After 45 min of storage at room temperature in a dark place, the absorbance of the mixtures was measured at 765 nm using a Biotek Synergy HTC multimode microplate reading spectrophotometer (Biotek Instruments Inc., USA). Phenolic content was calculated using a standard curve generated with gallic acid. All measurements were performed in triplicate.

TPC concentration was expressed as mg gallic acid equivalent (GAE) per gram of crude sample.

Determination of DPPH Activity

The DPPH assay was also used to determine the radical scavenging power of the propolis samples according to the method described by Apak et al. 2014 [54]. The violet free radical DPPH was measured at 517 nm to determine decolorization. All measurements were performed in triplicate. Trolox was used as a standard, and results were expressed as µg Trolox equivalent (TE) per gram of crude sample.

Determination of Cupric Ion-Reducing Antioxidant Capacity (CUPRAC)

The CUPRAC assay was performed on propolis samples according to Apak et al., 2004 [54], modified by Pasli et al., 2019 [55]. $CuCl_2$ solution, 10^{-2} mM, was prepared in distilled water. In total, 19.27 g of NH_4Ac was dissolved in distilled water and diluted to 250 mL to prepare an ammonium acetate buffer with a pH of 7.0 and a concentration of 1.0 M. A neocuproine (Nc) solution in ethanol (7.5×10^{-3} M) was freshly prepared. In total, 1 mL of the $CuCl_2$ solution, Nc solution, ammonium acetate buffer, and distilled water were added to 100 µL of the extract to provide a total of 4.1 mL of the mixture. After 30 min, absorbance was measured at 450 nm against a reagent blank. All measurements were performed in triplicate. Trolox was used as a standard, and results were expressed as mg TE per gram of crude sample.

4.2.5. Sample Preparation for HPLC Analysis

After completing the extraction, an aliquot extract (1 mL) was evaporated to dryness in a rotary evaporator (IKA RV10, Germany) at 40 °C. After evaporation, the solvent was dissolved in 10 mL of pure methanol. It was diluted 100-fold with MeOH and filtered through a 0.45 µm PTFE filter (Waters, Milford, CA, USA) immediately before injection into the HPLC system. All sample preparations were performed in triplicate; thus, quantification data are the average of the three results.

4.2.6. HPLC-PDA Analysis

For HPLC analyses, propolis samples at a concentration of 1 mg/mL were injected into a Shimadzu 20A Series ultrafast liquid chromatograph (UFLC, Shimadzu Corporation, Japan) equipped with a microvacuum degasser, autosampler, column oven, controller, and PDA detector. An ACE C18 column (250 mm × 4,6 mm, 3 µm) was used for chromatographic separations. LC Solution software (Shimadzu Corporation, Kyoto, Japan) was used for data acquisition and elaboration. The mobile phases were water with 0.75% formic acid (*v*/*v*) (solvent A) and HPLC-grade methanol with 0.75% formic acid (*v*/*v*) (solvent B). A gradient of mobile phase A and mobile phase B was used with a flowrate of 0.5 mL and 10 µL of injection volume for each standard mixture, and the column temperature was set at 40 °C [48]. The mobile phase was degassed in an ultrasonic bath and filtered through a 0.45 µm PTFE filter before use. A blank injection was also used to assess chromatographic interference at resolution. Stock solutions of the chemical standards were prepared in a final volume of 10 mL MeOH at a concentration of 1000 µg/mL. Each standard's stock solutions were prepared in methanol at 10 mg/mL and kept at −20 °C. The working standard solutions were prepared at 5 calibration levels with final concentrations of 6.25, 12.5, 25, 50, and 100 g/mL. The phenolic/aromatic acids, phenolic aldehyde, and flavonoids present in the samples were identified and quantified by comparing the retention time and the size of the peaks in the methanolic extracts with those of the standard components as follows: chlorogenic acid, caffeic acid, vanillin, apigenin, kaempferol, *p*-coumaric acid, ferulic acid, trans-cinnamic acid, pinobanksin, and pinocembrin.

4.2.7. Statistical Analysis

All statistical analyses of the data obtained were evaluated using R Statistical Program version 4.0.4 (R Core Team) and MINITAB Statistical Program version 19 (Minitab Inc.). One-way ANOVA and Tukey's post hoc test for multiple comparisons with statistical significance at a 95% confidence level ($p < 0.05$) were performed to identify group differences. Analytical data from Anatolian propolis samples were subjected to principal component analysis (PCA), multivariate statistical analysis [56]. Pearson's correlation test was used to determine the correlation between antioxidant activities (CUPRAC, DPPH) with TPC and TFC by using R 4.0.4.

Recent studies have revealed that variables such as antioxidants and antimicrobial activities, individual phenolic and flavonoid substances of bee products are significant, with the greatest discriminatory power in a PCA [4,57,58]. Therefore, these variables were selected to accurately group the Anatolian propolis samples according to their geographical origin.

5. Conclusions

This work clarified the phenolic composition and antioxidant and antimicrobial activities of 24 Anatolian propolis samples from Türkiye. The current study revealed the presence of phytochemicals, mainly caffeic acid, *p*-coumaric acid, ferulic acid, pinobanksin, and apigenin in Turkish propolis samples. We conclude that ethanolic extracts of propolis could be a useful adjunct to pharmaceutical products in improving human health by aiding the antioxidant defense system in combating free radical formation. Our results clearly demonstrate that Anatolian propolis samples have remarkable antioxidant and antimicrobial activities, which was expected since propolis is considered the bee's defense system against infections.

Our results and the data from the literature on propolis's chemical composition and biological action do not point to a single compound or class of compounds that could be responsible for this effect. The biochemical properties of Anatolian propolis appear to have broad therapeutic significance as a natural mixture rather than as a source of a novel antibacterial, antifungal, or antiviral chemical. The samples were well clustered using principal component analysis, with antioxidant and antimicrobial activity and phenolic compound values as parameters. The first two principal components were used to separate the Turkish propolis samples from each other and proved to be an efficient way to classify the propolis samples into groups according to the collection site.

Author Contributions: Conceptualization, Ü.A., İ.G. and B.Ö.; methodology, Ü.A. and B.Ö.; software, Ü.A., İ.G. and B.Ö.; investigation, Ü.A., İ.G. and B.Ö.; data curation, Ü.A., İ.G. and B.Ö.; writing—original draft preparation, Ü.A.; writing—review and editing, Ü.A., İ.G. and B.Ö.; visualization, Ü.A., İ.G. and B.Ö.; supervision, B.Ö. All authors have read and agreed to the published version of the manuscript.

Funding: This research received no external funding.

Institutional Review Board Statement: Not applicable.

Informed Consent Statement: Not applicable.

Data Availability Statement: Data are contained within the article.

Acknowledgments: The authors thank Assoc. Prof. Funda KARBANCIOĞLU GÜLER for supplying microorganisms to be used in antimicrobial tests and are grateful to Ayşe SAYGÜN, Dilara Nur DIKMETAŞ, and Dilara DEVECIOĞLU for their contributions to the evaluation of the antimicrobial test results.

Conflicts of Interest: The authors declare no conflict of interest.

References

1. Farooqui, T.; Farooqui, A.A. Beneficial effects of propolis on human health and neurological diseases. *Front. Biosci. Elite* **2012**, *4*, 779–793. [CrossRef]
2. Hernandez Zarate, M.S.; Abraham Juarez, M.R.; Ceron Garcia, A.; Ozuna Lopez, C.; Gutierrez Chavez, A.J.; Segoviano Garfias, N.; Avila Ramos, F. Flavonoids, phenolic content, and antioxidant activity of propolis from various areas of Guanajuato, Mexico. *Food Sci. Technol.* **2018**, *38*, 210–215. [CrossRef]
3. Jansen-Alves, C.; Maia, D.S.V.; Krumreich, F.D.; Crizel-Cardoso, M.M.; Fioravante, J.B.; da Silva, W.P.; Borges, C.D.; Zambiazi, R.C. Propolis microparticles produced with pea protein: Characterization and evaluation of antioxidant and antimicrobial activities. *Food Hydrocoll.* **2019**, *87*, 703–711. [CrossRef]
4. Lopez, B.G.C.; Schmidt, E.M.; Eberlin, M.N.; Sawaya, A.C.H.F. Phytochemical markers of different types of red propolis. *Food Chem.* **2014**, *146*, 174–180. [CrossRef] [PubMed]
5. Rivero-Cruz, J.F.; Granados-Pineda, J.; Pedraza-Chaverri, J.; Perez-Rojas, J.M.; Kumar-Passari, A.; Diaz-Ruiz, G.; Rivero-Cruz, B.E. Phytochemical Constituents, Antioxidant, Cytotoxic, and Antimicrobial Activities of the Ethanolic Extract of Mexican Brown Propolis. *Antioxidants* **2020**, *9*, 70. [CrossRef]
6. Seibert, J.B.; Bautista-Silva, J.P.; Amparo, T.R.; Petit, A.; Pervier, P.; dos Santos Almeida, J.C.; Azevedo, M.C.; Silveira, B.M.; Brandão, G.C.; de Souza, G.H.B.; et al. Development of propolis nanoemulsion with antioxidant and antimicrobial activity for use as a potential natural preservative. *Food Chem.* **2019**, *287*, 61–67. [CrossRef] [PubMed]
7. Yuan, Y.; Zheng, S.L.; Zeng, L.H.; Deng, Z.Y.; Zhang, B.; Li, H.Y. The Phenolic Compounds, Metabolites, and Antioxidant Activity of Propolis Extracted by Ultrasound-Assisted Method. *J. Food Sci.* **2019**, *84*, 3850–3865. [CrossRef]
8. Ozdal, T.; Ceylan, F.D.; Eroglu, N.; Kaplan, M.; Olgun, E.O.; Capanoglu, E. Investigation of antioxidant capacity, bioaccessibility and LC-MS/MS phenolic profile of Turkish propolis. *Food Res. Int.* **2019**, *122*, 528–536. [CrossRef] [PubMed]
9. Kujumgiev, A.; Tsvetkova, I.; Serkedjieva, Y.; Bankova, V.; Christov, R.; Popov, S. Antibacterial, antifungal and antiviral activity of propolis of different geographic origin. *J. Ethnopharmacol.* **1999**, *64*, 235–240. [CrossRef]
10. Asem, N.; Gapar, N.A.A.; Abd Hapit, N.H.; Omar, E.A. Correlation between total phenolic and flavonoid contents with antioxidant activity of Malaysian stingless bee propolis extract. *J. Apicultural Res.* **2020**, *59*, 437–442. [CrossRef]
11. Bhadauria, M.; Nirala, S.K.; Jaswal, A.; Raghuvansh, S.; Bhatt, R.; Shukla, S. Propolis: Therapeutic Perspectives against Silica Induced Toxic Manifestations. *Chapter* **2010**, 1–18.
12. Andrade, J.K.S.; Denadai, M.; de Oliveira, C.S.; Nunes, M.L.; Narain, N. Evaluation of bioactive compounds potential and antioxidant activity of brown, green and red propolis from Brazilian northeast region. *Food Res. Int.* **2017**, *101*, 129–138. [CrossRef] [PubMed]
13. Kartal, M.; Yildiz, S.; Kaya, S.; Kurucu, S.; Topcu, G. Antimicrobial activity of propolis samples from two different regions of Anatolia. *J. Ethnopharmacol.* **2003**, *86*, 69–73. [CrossRef] [PubMed]
14. Kartal, M.; Kaya, S.; Kurucu, S. GC-MS Analysis of propolis samples from two diferent regions of Turkey. *Z. Nat. C* **2002**, *57*, 905–909.
15. Popova, M.; Silici, S.; Kaftanoglu, O.; Bankova, V. Antibacterial activity of Turkish propolis and its qualitative and quantitative chemical composition. *Phytomedicine* **2005**, *12*, 221–228. [CrossRef]
16. Velikova, M.; Bankova, V.; Sorkun, K. Chemical composition and biological activity of propolis from Turkish and Bulgarian origin. *Mellifera* **2000**, *11*, 57–59.
17. Sorkun, K.; Süer, B.; Salih, B. Determination of chemical composition of Turkish propolis. *Z. Nat. C* **2001**, *56*, 666–668. [CrossRef]
18. Machado, B.A.S. Determination of Parameters for the Supercritical Extraction of Antioxidant Compounds from Green Propolis Using Carbon Dioxide and Ethanol as Co-Solvent. *PLoS ONE* **2015**, *10*, e0134489. [CrossRef]
19. Bonvehi, J.S.; Coll, F.V.; Jorda, R.E. The Composition, Active Components and Bacteriostatic Activity of Propolis in Dietetics. *J. Am. Oil Chem. Soc.* **1994**, *71*, 529–532. [CrossRef]
20. Machado, B.A.S.; Silva, R.P.D.; Barreto, G.; Costa, S.S.; da Silva, D.F.; Brandão, H.N.; Da Rocha, J.L.C.; Dellagostin, O.A.; Henriques, J.A.P.; Umsza-Guez, M.A.; et al. Chemical Composition and Biological Activity of Extracts Obtained by Supercritical Extraction and Ethanolic Extraction of Brown, Green and Red Propolis Derived from Different Geographic Regions in Brazil. *PLoS ONE* **2016**, *11*, e0145954. [CrossRef]
21. Bankova, V. Chemical diversity of propolis and the problem of standardization. *J. Ethnopharmacol.* **2005**, *100*, 114–117. [CrossRef] [PubMed]
22. Cai, Y.; Luo, Q.; Sun, M.; Corke, H. Antioxidant activity and phenolic compounds of 112 traditional Chinese medicinal plants associated with anticancer. *Life Sci.* **2004**, *74*, 2157–2184. [CrossRef] [PubMed]
23. Kumazawa, S.; Hamasaka, T.; Nakayama, T. Antioxidant activity of propolis of various geographic origins. *Food Chem.* **2004**, *84*, 329–339. [CrossRef]
24. Ahn, M.-R.; Kumazawa, S.; Usui, Y.; Nakamura, J.; Matsuka, M.; Zhu, F.; Nakayama, T. Antioxidant activity and constituents of propolis collected in various areas of China. *Food Chem.* **2007**, *101*, 1383–1392. [CrossRef]
25. Anjum, S.I.; Ullah, A.; Khan, K.A.; Attaullah, M.; Khan, H.; Ali, H.; Bashir, M.A.; Tahir, M.; Ansari, M.J.; Ghramh, H.A.; et al. Composition and functional properties of propolis (bee glue): A review. *Saudi J. Biol. Sci.* **2019**, *26*, 1695–1703. [CrossRef] [PubMed]

26. Christov, R.; Trusheva, B.; Popova, M.; Bankova, V.; Bertrand, M. Chemical composition of propolis from Canada, its antiradical activity and plant origin. *Nat. Prod. Res.* **2005**, *19*, 673–678. [CrossRef]
27. Yildirim, Z.; Hacievliyagil, S.; Kutlu, N.O.; Aydin, N.; Kurkcuoglu, M.; Iraz, M.; Durmaz, R. Effect of water extract of Turkish propolis on tuberculosis infection in guinea-pigs. *Pharmacol. Res.* **2004**, *49*, 287–292. [CrossRef]
28. Biscalia, D.; Ferreira, S.R.S. Propolis extracts obtained by low pressure methods and supercritical fluid extraction. *J. Supercrit. Fluids* **2009**, *51*, 17–23. [CrossRef]
29. Uzel, A.; Önçağ, Ö.; Çoğulu, D.; Gençay, Ö. Chemical compositions and antimicrobial activities of four different Anatolian propolis samples. *Microbiol. Res.* **2005**, *160*, 189–195. [CrossRef]
30. Marcucci, M.C. Propolis: Chemical composition, biological properties and therapeutic activity. *Apidologie* **1995**, *26*, 83–99. [CrossRef]
31. Lagouri, V.; Alexandri, G. Antioxidant Properties of Greek O. Dictamnus and R. Officinalis Methanol and Aqueous Extracts—HPLC Determination of Phenolic Acids. *Int. J. Food Prop.* **2013**, *16*, 549–562. [CrossRef]
32. Huang, S.; Zhang, C.-P.; Wang, K.; Li, G.Q.; Hu, F.-L. Recent Advances in the Chemical Composition of Propolis. *Molecules* **2014**, *19*, 19610–19632. [CrossRef] [PubMed]
33. Shi, H.; Yang, H.; Zhang, X.; Yu, L. Identification and Quantification of Phytochemical Composition and Anti-inflammatory and Radical Scavenging Properties of Methanolic Extracts of Chinese Propolis. *J. Agric. Food Chem.* **2012**, *60*, 12403–12410. [CrossRef] [PubMed]
34. Yesiltas, B.; Capanoglu, E.; Firatligil-Durmus, E.; Sunay, A.E.; Samanci, T.; Boyacioglu, D. Investigating the in-vitro bioaccessibility of propolis and pollen using a simulated gastrointestinal digestion System. *J. Apic. Res.* **2014**, *53*, 101–108. [CrossRef]
35. Wieczynska, A.; Wezgowiec, J.; Wieckiewicz, W.; Czarny, A.; Kulbacka, J.; Nowakowska, D.; Gancarz, R.; Wilk, K.A. Antimicrobial Activity, Cytotoxicity and Total Phenolic Content of Different Extracts of Propolis from the West Pomeranian Region in Poland. *Acta Pol. Pharm.* **2017**, *74*, 715–722.
36. Kubiliene, L.; Laugaliene, V.; Pavilonis, A.; Maruska, A.; Majiene, D.; Barcauskaite, K.; Kubilius, R.; Kasparaviciene, G.; Savickas, A. Alternative preparation of propolis extracts: Comparison of their composition and biological activities. *BMC Complement. Altern. Med.* **2015**, *15*, 156. [CrossRef]
37. Moreira, L.; Dias, L.G.; Pereira, J.A.; Estevinho, L. Antioxidant properties, total phenols and pollen analysis of propolis samples from Portugal. *Food Chem. Toxicol.* **2008**, *46*, 3482–3485. [CrossRef]
38. Ahn, M.-R.; Kumazawa, S.; Hamasaka, T.; Bang, K.-S.; Nakayama, T. Antioxidant Activity and Constituents of Propolis Collected in Various Areas of Korea. *J. Agric. Food Chem.* **2004**, *52*, 7286–7292. [CrossRef]
39. Pellati, F.; Orlandini, G.; Pinetti, D.; Benvenuti, S. HPLC-DAD and HPLC-ESI-MS/MS methods for metabolite profiling of propolis extracts. *J. Pharm. Biomed. Anal.* **2011**, *55*, 934–948. [CrossRef]
40. Erdogan, S.; Ates, B.; Durmaz, G.; Yilmaz, I.; Seckin, T. Pressurized liquid extraction of phenolic compounds from Anatolia propolis and their radical scavenging capacities. *Food Chem. Toxicol.* **2011**, *49*, 1592–1597. [CrossRef]
41. Sarikaya, A.O.; Ulusoy, E.; Öztürk, N.; Tuncel, M.; Kolayli, S. Antioxidant Activity and Phenolic Acid Constituents of Chestnut (Castania Sativa Mill.) Honey and Propolis. *J. Food Biochem.* **2009**, *33*, 470–481. [CrossRef]
42. Letullier, C.; Manduchet, A.; Dlalah, N.; Hugou, M.; Georgé, S.; Sforcin, J.M.; Cardinault, N. Comparison of the antibacterial efficiency of propolis samples from different botanical and geographic origins with and without standardization. *J. Apic. Res.* **2020**, *59*, 19–24. [CrossRef]
43. Seidel, V.; Peyfoon, E.; Watson, D.G.; Fearnley, J. Comparative study of the antibacterial activity of propolis from different geographical and climatic zones. *Phytother. Res.* **2008**, *22*, 1256–1263. [CrossRef]
44. Silva, F.R.G.; Matias, T.M.S.; Souza, L.I.O.; Matos-Rocha, T.J.; Fonseca, S.A.; Mousinho, K.C.; Santos, A.F. Phytochemical screening and in vitro antibacterial, antifungal, antioxidant and antitumor activities of the red propolis Alagoas. *Braz. J. Biol.* **2019**, *79*, 452–459. [CrossRef]
45. Aliyazicioglu, R.; Sahin, H.; Erturk, O.; Ulusoy, E.; Kolayli, S. Properties of Phenolic Composition and Biological Activity of Propolis from Türkiye. *Int. J. Food Prop.* **2013**, *16*, 277–287. [CrossRef]
46. Choi, Y.M.; Noh, D.O.; Cho, S.Y.; Suh, H.J.; Kim, K.M.; Kim, J.M. Antioxidant and antimicrobial activities of propolis from several regions of Korea. *LWT Food Sci. Technol.* **2006**, *39*, 756–761. [CrossRef]
47. Akkaya, N.E.; Ergun, C.; Saygun, A.; Yesilcubuk, N.; Akel-Sadoglu, N.; Kavakli, I.H.; Turkmen, H.S.; Catalgil-Giz, H. New biocompatible antibacterial wound dressing candidates; agar-locust bean gum and agar-salep films. *Int. J. Biol. Macromol.* **2020**, *155*, 430–438. [CrossRef]
48. Kim, D.-O.; Jeong, S.W.; Lee, C.Y. Antioxidant capacity of phenolic phytochemicals from various cultivars of plums. *Food Chem.* **2003**, *81*, 321–326. [CrossRef]
49. Uluata, S.; Altuntaş, U.; Özçelik, B. Characterization of Turkish Extra Virgin Olive Oils and Classification Based on Their Growth Regions Coupled with Multivariate Analysis. *Food Anal. Methods* **2021**, *14*, 1682–1694. [CrossRef]
50. Kızıltaş, H.; Bingol, Z.; Gören, A.C.; Kose, L.P.; Durmaz, L.; Topal, F.; Alwasel, S.H.; Gulcin, İ. LC-HRMS profiling and antidiabetic, anticholinergic, and antioxidant activities of aerial parts of kınkor (Ferulago stellata). *Molecules* **2021**, *26*, 2469. [CrossRef] [PubMed]
51. Chen, L.Y.; Cheng, C.W.; Liang, J.Y. Effect of esterification condensation on the Folin-Ciocalteu method for the quantitative measurement of total phenols. *Food Chem.* **2015**, *170*, 10–15. [CrossRef] [PubMed]

52. Hızır-Kadı, I.; Gültekin-Özgüven, M.; Altin, G.; Demircan, E.; Özçelik, B. Liposomal nanodelivery systems generated from proliposomes for pollen extract with improved solubility and in vitro bioaccessibility. *Heliyon* **2020**, *6*, e05030. [CrossRef] [PubMed]
53. Topal, M.; Ozturk Sarıkaya, S.B.; Topal, F. Determination of Angelica archangelica's Antioxidant Capacity and Mineral Content. *Chemistryselect* **2021**, *6*, 7976–7980. [CrossRef]
54. Apak, R.; Güçlü, K.; Özyürek, M.; Karademir, S.E. Novel total antioxidant capacity index for dietary polyphenols and vitamins C and E, using their cupric ion reducing capability in the presence of neocuproine: CUPRAC method. *J. Agric. Food Chem.* **2004**, *52*, 7970–7981. [CrossRef]
55. Pasli, A.; Yavuz-Düzgün, M.; Altuntas, U.; Altin, G.; Özçelik, B.; Firatligil, E. In vitro bioaccessibility of phenolics and flavonoids in various dried vegetables, and the determination of their antioxidant capacity via different spectrophotometric assays. *Int. Food Res. J.* **2019**, *26*, 793–800.
56. Menevseoglu, A. Evaluation of Portable Vibrational Spectroscopy Sensors as a Tool to Detect Black Cumin Oil Adulteration. *Processes* **2022**, *10*, 503. [CrossRef]
57. Bittencourt, M.L.; Ribeiro, P.R.; Franco, R.L.; Hilhorst, H.W.; de Castro, R.D.; Fernandez, L.G. Metabolite profiling, antioxidant and antibacterial activities of Brazilian propolis: Use of correlation and multivariate analyses to identify potential bioactive compounds. *Food Res. Int.* **2015**, *1*, 449–457. [CrossRef]
58. Kaygusuz, H.; Tezcan, F.; Erim, F.B.; Yildiz, O.; Sahin, H.; Can, Z.; Kolayli, S. Characterization of Anatolian honeys based on minerals, bioactive components and principal component analysis. *LWT Food Sci. Technol.* **2016**, *1*, 273–279. [CrossRef]

molecules

MDPI

Article

Assessment of the Effects of Roasting, Contact Grilling, Microwave Processing, and Steaming on the Functional Characteristics of Bell Pepper (*Capsicum annuum* L.)

Remigiusz Olędzki [1,2,*] and Joanna Harasym [1,2,*]

[1] Department of Biotechnology and Food Analysis, Wroclaw University of Economics and Business, Komandorska 118/120, 53-345 Wroclaw, Poland

[2] Adaptive Food Systems Accelerator-Science Centre, Wroclaw University of Economics and Business, Komandorska 118/120, 53-345 Wroclaw, Poland

* Correspondence: remigiusz.oledzki@ue.wroc.pl (R.O.); joanna.harasym@ue.wroc.pl (J.H.)

Abstract: Bell peppers (*Capsicum annuum* L.) in various stages of maturity are widely used in the diets of individuals and in the food industry; they are consumed both fresh and after thermal processing. However, every type of processing impacts the overall textural and bioactive characteristics of this plant-based food. In order to quantify the changes in the bioactive substances and color-structural characteristics that occur during selected heat treatments (contact grilling, roasting, roasting combined with microwaving, and steam cooking) of bell peppers at three maturity stages (green, yellow, and red), analyses of antioxidant activity, reducing sugar content, polyphenolic compound content, textural properties, and color coordinates in the L*a*b* system were carried out. Some of the processes used, such as contact grilling (15.43 mg GAE/g d.b.) and roasting combined with microwaving (15.24 mg GAE/g d.b.), proved to be beneficial as the total polyphenol content of green peppers (2.75 mg GAE/g d.b.) increased. The roasting (3.49 mg TE/g d.b.) and steaming (6.45 mg TE/g d.b.) methods decreased the antioxidant activity of yellow bell peppers (14.29 mg TE/g d.b.). Meanwhile, the roasting (0.88 mg Glc/g d.b.), contact-grilling (2.19 mg Glc/g d.b.), simultaneous microwaving and roasting (0.66 mg Glc/g d.b.), and steaming (1.30 mg Glc/g d.b.) methods significantly reduced the content of reducing sugars and reducing substances in red bell peppers (4.41 mg Glc/g d.b.). The studies proved that in order to preserve the antioxidant and bioactive properties of bell peppers, it is necessary to consider the use of appropriately selected heat treatments, depending on the different stages of maturity. The proper selection of adequate thermal treatment can not only increase digestibility, but also improve the bioavailability of bioactive substances from this raw material.

Keywords: bell pepper; steaming; roasting; microwave; *Capsicum annuum*; polyphenols; color; texture

Citation: Olędzki, R.; Harasym, J. Assessment of the Effects of Roasting, Contact Grilling, Microwave Processing, and Steaming on the Functional Characteristics of Bell Pepper (*Capsicum annuum* L.). *Molecules* **2024**, *29*, 77. https://doi.org/10.3390/molecules29010077

Academic Editors: José Pinela, Maria Inês Dias, Carla Pereira and José Ignacio Alonso-Esteban

Received: 28 November 2023
Revised: 18 December 2023
Accepted: 19 December 2023
Published: 22 December 2023

1. Introduction

The bell pepper (*Capsicum annuum* L.) is a vegetable that is cultivated almost worldwide and is consumed both fresh and after thermal processing; it is also used as a type of spice [1]. The bell pepper, which belongs to the Solanaceae family, includes many species and varieties, both sweet and spicy [2]. Regardless of the cultivar, annual bell pepper fruits are a rich source of polyphenolic compounds, which can be found in the range of 9.2–15.4 mg GAE/100 g of fresh weight (f.w.) [3]. Polyphenols have been confirmed to have an important function in protecting the body from reactive oxygen species (ROS), which are one of the main initiators of many diseases [4]. The consumption of flavan-3-ols, for example, has been proved to be associated with a reduced risk of myocardial infarction, stroke, and diabetes [5].

Polyphenols present in the diet have also been shown to help modify the blood lipid profile, normalize blood pressure, reduce insulin resistance, and reduce systemic inflammation [6,7], while the consumption of polyphenols such as quercetin and resveratrol has been linked to improved cardiovascular function [8]. Analyses using high-performance liquid chromatography coupled with tandem mass spectrometry and electrospray ionization (HPLC-ITMS) techniques have enabled the identification of a number of polyphenolic compounds in *Capsicum annuum* fruits, such as caffeic acid, coumaric acid, coumaroylquinic acid, 3-*O*-caffeoylquinic acid, ferulic acid, sinapic acid, apigenin-*O*-hexoside, and quercetin-*O*-rhamnosyl-*O*-hexoside [9]. The health-promoting and therapeutic potential of polyphenols present in the diet is also closely linked to the functioning of the gut microbiome, which converts polyphenolic compounds into a form that is highly bioactive and bioavailable [10,11]. According to a 2007 study by Anand et al. [12], the bioavailability of, for example, hexahydroxycurcumin (HHC), a metabolite formed by the biotransformation of curcumin, is significantly higher than the unmodified form of curcumin.

Some researchers reported that the consumption of bell peppers (especially in the presence of fats) is beneficial due to their high content of carotenoids, which are essential for the differentiation process of human epithelial cells [13]. It has been confirmed that bell pepper fruits contain high amounts of carotenoids reaching levels, depending on the cultivar, from 133.9 mg/100 g to 324.2 mg/100 g dry basis (d.b.) [13,14]. The content of such carotenoids as lutein, zeaxanthin, α-carotene, and β-carotene in annual bell pepper fruits ranges from 1.95 to 3.12, 0.08 to 4.6, 0.0 to 5.16, and 0.0 to 44.42 μg/g d.b., respectively [15]. The activity and impact of carotenoids on the human system are related to the neutralization of free oxygen radicals, mainly singlet oxygen, so these substances help prevent or alleviate many chronic conditions and diseases, such as glandular cancer, diabetes, diseases of the visual system, cardiovascular diseases (such as ischemic heart disease), and aging processes [16].

Bell peppers also have a high content of ascorbic acid (vitamin C). Results show that the ascorbic acid content depends on the variety and maturity stage of the bell pepper, and can range from 16.52 to 107.3 mg/100 g of f.w. for green bell peppers, from 129.6 to 159.61 mg/100 g of f.w. for yellow bell peppers, and from 81.19 to 154.3 mg/100 g of f.w. for red bell peppers, respectively [17,18]. Ascorbic acid is a potent antioxidant, a cofactor for many biosynthetic enzymes, and for enzymes regulating the expression of genetic material [19,20]. In addition, ascorbic acid, of which annual bell peppers are a rich source, also strengthens immune defenses by supporting the cellular functions of both the innate and acquired immune systems [21]. This is because, among other things, phagocytic cells (such as neutrophils) have the ability to accumulate ascorbic acid, which enhances chemotaxis, phagocytosis, and the production of the reactive oxygen species needed by these cells to kill microorganisms [22]. For this reason, the consumption of raw materials rich in ascorbic acid results in increased immunity and a lower susceptibility to infection. Studies indicate that the consumption of ascorbic acid-rich raw materials or vitamin C supplementation (at 100–200 mg/day) helps to prevent respiratory tract infections and shortens the duration of systemic infections [23].

Therefore, due to the valuable bioactive and health-promoting properties of bell peppers, this raw material is often consumed in many types of dishes. Thus, the raw material is subjected to a different thermal-processing techniques that are standard in the food industry, such as roasting, contact grilling, steam cooking, or microwave processing. However, the use of these techniques is associated with changes in the bioactive properties and antioxidant content of the processed samples, which alter its health-promoting characteristics [24]. There is still a lack of studies explaining in detail the possible impact of various conventional thermal-processing methods on the content of the bioactive substances and, consequently, on their functional and health-promoting quality.

The purpose of this article is to evaluate the effects of selected heat-treatment methods on the antioxidant properties and content of the bioactive substances of annual bell pepper fruits at different stages of maturity.

2. Results and Discussion

2.1. Total Polyphenols Content

The analysis of the raw materials showed that red bell peppers have the highest content of polyphenolic compounds, while green bell peppers have the lowest (Table 1). Similar results were obtained in Zhang et al.'s 2003 study [3], where it was confirmed that the total content of polyphenols in methanol extracts of green, yellow, and red bell peppers was 48.4, 54.8, and 64.5 mg GAE/100 g, respectively.

Table 1. The total polyphenol content, antioxidant activity, and reducing sugar content of bell peppers after different thermal-processing treatments.

R-Stage	Processing	SUGARS mg GlcE/g d.b.	TPC mg GAE/g d.b.	DPPH mg TE/g d.b.	ABTS mg TE/g d.b.	FRAP μM $FeSO_4$/g d.b.
G	RM	4.15 ± 0.17^{b}	2.75 ± 0.42^{a}	11.77 ± 0.37^{c}	0.68 ± 0.16^{a}	10.54 ± 0.37^{c}
	CG	1.99 ± 0.16^{a}	15.43 ± 1.15^{c}	7.27 ± 0.52^{b}	3.32 ± 0.09^{c}	1.37 ± 0.07^{b}
	R	2.27 ± 0.20^{a}	7.09 ± 0.57^{b}	1.57 ± 0.06^{a}	0.58 ± 0.13^{a}	1.43 ± 0.08^{b}
	MWR	1.88 ± 0.13^{a}	15.24 ± 0.01^{c}	1.94 ± 0.13^{a}	1.05 ± 0.10^{b}	0.76 ± 0.04^{a}
	S	2.07 ± 0.62^{a}	6.25 ± 0.47^{b}	7.18 ± 0.42^{b}	3.79 ± 0.18^{d}	1.44 ± 0.05^{b}
Y	RM	5.61 ± 0.08^{d}	17.00 ± 0.68^{e}	14.29 ± 0.51^{e}	5.03 ± 0.12^{d}	19.93 ± 1.64^{b}
	CG	1.45 ± 0.21^{c}	12.56 ± 0.68^{c}	8.73 ± 0.77^{d}	4.42 ± 0.11^{c}	1.79 ± 0.13^{a}
	R	1.12 ± 0.08^{b}	9.02 ± 0.28^{b}	3.49 ± 0.11^{b}	3.77 ± 0.06^{a}	0.82 ± 0.04^{a}
	MWR	1.16 ± 0.06^{b}	15.24 ± 0.35^{d}	1.94 ± 0.16^{a}	5.40 ± 0.04^{e}	1.08 ± 0.04^{a}
	S	0.80 ± 0.04^{a}	6.31 ± 0.11^{a}	6.45 ± 0.50^{c}	4.03 ± 0.04^{b}	1.17 ± 0.00^{a}
R	RM	4.41 ± 0.12^{d}	17.26 ± 0.64^{b}	20.00 ± 0.51^{c}	5.28 ± 0.10^{d}	39.79 ± 1.05^{b}
	CG	2.19 ± 0.35^{c}	17.20 ± 2.57^{b}	9.29 ± 0.81^{c}	4.34 ± 0.04^{c}	1.62 ± 0.08^{a}
	R	0.88 ± 0.04^{ab}	23.56 ± 0.35^{c}	1.40 ± 0.14^{a}	1.77 ± 0.06^{a}	1.13 ± 0.04^{a}
	MWR	0.66 ± 0.01^{a}	5.46 ± 0.36^{a}	1.27 ± 0.05^{a}	2.82 ± 0.14^{b}	1.29 ± 0.01^{a}
	S	1.30 ± 0.26^{b}	6.01 ± 0.69^{a}	7.91 ± 0.55^{b}	4.16 ± 0.18^{c}	1.46 ± 0.20^{a}
processing		***	***	***	***	***
R-stage		***	***	***	***	***
processing × R-stage		***	***	***	***	***

R-stage—ripening stage; G—green pepper; Y—yellow pepper; R—red pepper. For Processing: RM—raw material, CG—contact grilling, R—roasting, MWR—microwave plus roasting, S—steaming. TPC—total polyphenols content; GlcE—glucose equivalent; GAE—gallic acid equivalent; TE—Trolox equivalent. Lower-case letters in columns mean that a statistically significant difference was found between samples for same R-stage ($p = 0.05$), ***—$p < 0.001$ for interaction between factors.

The results confirm that red bell peppers may be a raw material with high health-promoting potential. It has been found that consuming red bell peppers is related to a reduction in the risk of developing metabolic syndrome and the risk of death from cardiovascular disease, among other risks [25]. It has been shown that all the thermal treatments (R, MWR, CG, S) caused a statistically significant increase in the total content of polyphenolic compounds in green bell peppers. The greatest increase in the content of polyphenolic compounds was observed for green bell peppers after contact grilling (increase from 2.75 to 15.43 mg GAE/g d.b.).

On the other hand, in green bell pepper samples processed with simultaneous microwaving and roasting, a significant, more than five-fold increase, in the content of polyphenolic compounds (from 2.75 to 15.24 mg GAE/g d.b.) was observed, while a significant, more than two-fold increase (from 2.75 to 6.25 mg GAE/g d.b.) was observed as a result of steam cooking. Similar results were reported in our previous study for green bell peppers of the 'Ożarowska' cultivar (from the 2022 crop season);using a microwave treatment, it was observed that the total content of polyphenolic compounds was increased by more than 2.7 times (7.40 GAE μM/g d.b). Further, by cooking the peppers in boiling water, the total content of polyphenolic compounds was increased by more than 3.5 times (9.76 GAE μM/g d.b.) (compared to the raw material content of 2.75 GAE μM/g d.b.) [26].

Different results were obtained in a study by Özgür et al. in 2011 [27], where the total content of polyphenolic compounds in green bell peppers was reduced from 96.04 to 55.47 mg GAE/g during conventional oven drying, and in a study by Hameed et al. in 2023 [28], where steamed fresh green broccoli (*Brassica oleracea*) showed a higher total content of polyphenolic compounds (191.75 mg/g d.b.) compared to samples after microwave treatments (180.03 mg/g d.b.) and a hot-air-drying process in an oven (176.85 mg/g d.b.) [28].

It was also found that all the heat-treatment methods used (R, MWR, CG, S) caused a reduction in the total content of polyphenolic compounds in yellow bell peppers. The greatest reduction was observed for yellow bell peppers that underwent a steam-cooking treatment (from17.00 to 6.31 mg GAE/g d.b). In contrast, oven treatment resulted only in a 47% reduction in total polyphenolic compounds (from 17.00 to 9.02 mg GAE/g d.b.).

Similar results were reported in an earlier study of yellow bell peppers of the 'Ożarowska' cultivar, where a significant reduction of more than 40% in the total content of polyphenolic compounds was observed as a result of microwave treatment (10.14 GAE μM/g d.b.) and cooking in boiling water (10.03 GAE μM/g d.b.) compared to the raw material (17.00 GAE μM/g d.b.) [26]. Also, a study by Reis et al. in 2013 [29] showed that the total content of polyphenolic compounds in yellow bell peppers of the 'Cumari-do-Para' bell pepper (Capsicum chinense) cultivar grown in Brazil was reduced due to drying (at 65 °C) from 9748.22 to 1415.44 mg GAE/kg.

For red bell peppers, a significant, more than three-fold reduction in the content of polyphenolic compounds was observed as a result of simultaneous microwaving and roasting (from17.26 to 5.46 mg GAE/g d.b.), as well as steam cooking (reduction from 17.26 to 6.01 mg GAE/g d.b.). Oppositely, grilling resulted in a significant 36% increase in the TPC of red bell peppers, from 17.26 to 23.56 mg GAE/g d.b. Similar results were recorded in an earlier study on red bell peppers, where a significant 11% reduction in the TPC was also observed as a result of microwave treatment (15.37 GAE μM/g d. b) as well as a significant more than nine-fold reduction in the TPC (1.86 GAE μM/g d.b.) as a result of boiling water cooking [26].

Different results were obtained in a study of red bell peppers of the 'Capia' bell pepper cultivar, which were subjected to conventional and microwave drying. Here, the TPC decreased from 1297.49 mg GAE/100 g to 392.95 mg GAE/100 g (by 69.71% for drying in a 720 W microwave oven for 16 min) and 348.64 mg GAE/100 g (by 73.13% for drying in a conventional oven at 120 °C for 100 min), respectively. On the other hand, a 69.05% reduction in the TPC of 'Capia' bell peppers to 401.54 mg GAE/100 g was observed when using a warm-air-drying process (drying process carried out in the open air for 1 day) [30]. Also, Özgür confirmed that the oven-drying process for cooking red bell peppers causes a reduction in total polyphenol content from 130.79 to 89.82 mg GAE/g [27].

Different results were obtained by Speranza et al. in 2019 [31] for red bell peppers ('Senise' cultivar) dried in an oven (with forced air circulation) for 48 h at 55 °C (and at an internal relative humidity of the air stream of about 15–17%). As a result of this experiment, it was shown that the drying process did not cause significant changes in the total content of polyphenolic compounds in the red bell peppers, which were grown in the Battipaglia region of southern Italy. In contrast, the same 'Senise' cultivar of bell peppers grown in the Montanaso region (northern Italy) had a 7.6% reduction in the total polyphenolic compounds after a heat-treatment process in a forced-air oven [31].

The microwave and roasting method (reduction from17.26 to 5.46 mg GAE/g d.b.), as well as the steaming method, resulted in a significant reduction in the total content of polyphenolic compounds in red bell peppers (reduction from 17.26 to 6.01 mg GAE/g d.b.). Our study showed that the contact-grilling method preserves more polyphenolic components than the microwave treatment combined with roasting method as well as the steam-cooking method. This can be attributed to more controlled heating conditions. In addition, steam cooking appears to cause the leaching of water-soluble antioxidants, which increases with the duration of the process. Polyphenols are substances that are mostly water-soluble. During the steaming process, polyphenol molecules may move into the particles of steam rising in the steamer. Since a significant decomposition of plant tissues occurs during heat treatment, there may be an intensive migration of cellular components (including polyphenols) from the bell pepper skin into the hot steam.

Arfaoui reported results that are highly consistent with our results about different heat-treatment methods, which proves that cooking causes the most drastic changes in the composition of polyphenolic compounds in processed vegetables [32]. For this reason, perhaps processes such as roasting, contact grilling, and microwaving combined with roasting resulted in a smaller reduction in the content of polyphenolic compounds than the steam-cooking process.

On the other hand, processes such as roasting and contact grilling may contribute to the degradation of polysaccharides such as hemicellulose and pectin, which fill the gaps between cellulose microfibrils in the cell wall of bell pepper cells [33]. This phenomenon leads to a shrinking of the cell wall, which may result in a retention of a large number of polyphenolic compounds inside the cells in the outer layers of the bell pepper skin.

At the same time, the observed changes may have been responsible for the decrease in the hardness parameter of all types of processed bell peppers (green, yellow, and red) as a result of the heat-treatment methods used.

A study by Zahoor et al. in 2023 [34] showed that the total polyphenol content of red bell peppers was significantly reduced from 86.39 to 75.10 mg GAE/100 g d.b. as a result of microwave treatment (with a microwave power of 320 W) at 50 °C. However, when the same process was carried out (at a microwave power of 320 W) at 70 °C, no significant reduction in the total content of polyphenolic compounds was observed, which was confirmed to be 85.23 mg GAE/100 g d.b. On the other hand, when the drying process was carried out at a microwave power of 480 W and at 70 °C, a significant increase in the total content of polyphenolic compounds was observed to 92.30 mg GAE/100 g d.b. [34].

In our study, heat treatment in the form of roasting had a beneficial effect on the total content of polyphenolic compounds in green and red bell peppers. In addition, in the case of yellow bell peppers, only a 10% reduction in the total content of polyphenolic compounds was observed as a result of the simultaneous roasting and microwaving process. The reason for the observed changes may be the release (under high temperature and microwave radiation) of polyphenolic compounds, which are strongly bound in the cell-wall matrix of unprocessed raw materials [35].

Presumably, the high temperature and microwave radiation(for the Y-MWR variant) caused the inactivation of polyphenol oxidase (PPO) and other oxidative enzymes, such as lipoxygenase, which were released from the damaged bell pepper cells [36]. This phenomenon may have significantly reduced the degradation of polyphenolic compounds in the yellow bell peppers due to the simultaneous roasting and microwaving process.

A partial confirmation of the results we obtained are found in the results of a study on the microwave treatment of black peppers (*Piper nigrum* L.), the treatment of which (lasting 15 min at a power of 300 W) caused a significant reduction (from 79.62 to 36.20 mg GAE/100 g d.b.) in the total polyphenol content [37]. Also, in a study by Kaur et al. in 2005 [38], it was shown that the total polyphenol content of yellow sweet 'Bachata' bell peppers (*Capsicum annuum* L.) that were subjected to hot-air convection drying (at three different temperatures of 40, 50, and 60 °C) was reduced from 784.01 mg GAE/100 g d.b. to 630.76 mg GAE/100 g d.b.

2.2. *Total Antioxidant Activity vs. DPPH Radical*

The analysis of the total antioxidant potential of raw bell peppers showed that red bell peppers had the highest antioxidant activity (20.00 mg TE/g d.b.), followed by yellow bell peppers (14.29 mg TE/g d.b.) and green bell peppers (11.77 mg TE/g d.b.).

Our results are confirmed by research conducted by Sun et al. in 2007 [39], where green bell peppers were shown to have an antioxidant activity of 2.1 μmol TE/g f.w., while red bell peppers had an antioxidant activity of 3.9 μmol TE/g f.w. In the current study, it was observed that all heat treatments reduced the total antioxidant activity (measured using the DPPH method) in green, yellow, and red bell peppers. The greatest reduction in antioxidant activity in green bell peppers occurred as a result of roasting treatments (Table 1).

The current study showed that the steam cooking of green, yellow, as well as red bell peppers preserves more antioxidant components than roasting and the combined microwave and roasting treatment. This can be attributed to more destructive heating conditions (than during steamer heating), which cause the degradation of thermolabile antioxidants like ascorbic acid [40].

Largely similar results were reported in earlier studies on green and yellow bell peppers of the 'Ożarowska' cultivar, where a significant 62.76% reduction in the total antioxidant activity due to microwave treatment was observed in green bell peppers (reduction from 74.91 to 47.02 TE µM/g d.b.) and a significant 43.83% reduction in the antioxidant activity in yellow bell peppers (reduction from 57.10 to 25.03 TE µM/g d.b.), relative to the raw material [26]. Also, in a study conducted by Hameed et al. in 2023 [28], it was shown that a steam treatment of fresh green broccoli (Brassica oleracea) preserves higher antioxidant activity (58.80 DPPH (%)) (expressed as a percentage of DPPH radical neutralization) than a microwave treatment (54.95 DPPH (%)). However, in the same experiment, a hot-air treatment (oven roasting) (expressed as a percentage of neutralized DPPH radicals) proved almost as beneficial as the steam treatment and gave the tested broccoli an antioxidant activity of 58.36 DPPH (%) [28].

The greatest reduction in antioxidant activity in yellow (reduction from 14.29 to 1.94 mg TE/g d.b.) and red (reduction from 20.00 to 1.27 mg TE/g d.b.) bell peppers was due to the combined microwaving and roasting treatment process. It was shown that the process of roasting red bell peppers reduced the antioxidant activity of the sample by as much as 97% relative to the raw material.

The intense red color of red bell pepper fruits is due to the presence of carotenoids, the biosynthesis of which occurs mainly during the ripening of the peppers [41]. Our results indicate that there is probably a greater loss of carotenoids due to roasting than during processing with other heat-treatment techniques. At the same time, contact grilling allows for the better retention of carotenoids (greater stability of bell peppers to heat treatment).

Similar results were obtained in a study of red bell peppers of the 'Capia' cultivar, where it was observed that the total antioxidant activity of bell peppers is reduced by drying in a conventional oven (at 120 °C), where there was a reduction in the antioxidant activity to 3.80 mmol TE/kg, from a level of 35.85 mmol TE/kg for the raw material. On the other hand, when a microwave oven-drying process was used (at 720 W for 16 min), a reduction in the antioxidant activity to 3.71 mmol TE/kg f.w. was observed. When drying in fresh and warm air (for 1 day) was applied, a reduction in the total antioxidant activity was observed in 'Capia' bell peppers to a level of 3.94 mmol TE/kg f.w. [30]. Substantially similar results were obtained in a study of red bell pepper fruits of the 'Senise' cultivar grown in the Battipaglia region (southern Italy) and Montanaso region (northern Italy). It was confirmed that the process of drying (roasting) in a forced-air oven caused a significant reduction in the total antioxidant activity measured using the DPPH method by 33 percent [30]. Another study showed that the antioxidant activity of fresh red bell peppers is reduced using conventional oven-drying (from 8720.70 to 694.55 µmol TE/g d.b.) [27].

Desai, in a study of black pepper (*Piper nigrum* L.) fruit, observed a significant reduction in the total antioxidant activity (expressed as a percentage of the DPPH radical inhibition from 85.12 to 53.97%) (measured using the DPPH method) when the raw material was microwaved (for 15 min at an appliance power of 300 W) [37].

Different results were obtained by Zahoor et al. in 2023 [34], where it was shown that the microwave-treatment process performed at a temperature of 60 °C resulted in a 60% increase in the antioxidant activity (expressed as the amount of extract required to neutralize 50% of the DPPH radicals) relative to the raw material.

2.3. Total Antioxidant Activity vs. ABTS Radical

Slightly different results were provided by the analysis of antioxidant activity using the ABTS method. The contact grilling (increase from 0.68 to 3.32 mg TE/g d.b.), simultaneous microwaving and grilling (increase from 0.68 to 1.05 mg TE/g d.b.), and steam cooking (increase from 0.68 to 3.79 mg TE/g d.b.) of green bell peppers were shown to significantly increase the antioxidant activity of the raw material.

A 2023 study by Hameed et al. [28] similarly showed that fresh green broccoli (Brassica oleracea), after steam treatment, exhibits a higher antioxidant activity (64.28 DPPH (%)) (expressed as percentage of the DPPH radical neutralization) than after microwave treatment (57.60 DPPH (%)). In the same experiment, hot-air treatment (baking) proved significantly less favorable than steam treatment and conferred antiradical activity on the tested broccoli (expressed as a percentage of the neutralized ABTS cation radicals at 63.69 ABTS (%)) [28].

On the other hand, in the case of yellow bell peppers, it was observed that contact grilling (reduction from 5.03 to 4.42 mg TE/g d.b.), roasting (reduction from 5.03 to 3.77 mg TE/g d.b.) and steaming (reduction from 5.03 to 4.03 mg TE/g d.b.) resulted in a significant reduction in the antioxidant activity as measured using the ABTS method. At the same time, it was shown that simultaneous microwaving and roasting processes (increase from 5.03 to 5.40 mg TE/g d.b.) result in a significant increase in antioxidant activity in yellow bell peppers. It was shown that all the heat-treatment processes used caused a significant reduction in the antioxidant activity of red bell peppers. The greatest reduction in the antioxidant potential of red bell peppers was observed due to roasting (reduction from 5.28 to 1.77 mg TE/g d.b.). In a study conducted by Desai, it was shown that there is a significant reduction (from 4.81 to 2.33 mmol TE/100 g d.b.) in the total antioxidant activity (measured using the ABTS method) in black pepper (*Piper nigrum* L.) fruits that were microwave-treated (lasting 15 min at a device power of 300 W) [37].

In contrast, the smallest reduction in antioxidant activity was caused by the contact grilling of red bell peppers (reduction from 5.28 to 4.34 mg TE/g d.b.). Perhaps this was due to the fact that contact grilling does not deprive the processed samples of antioxidants to the same extent as other processing methods [42].

2.4. Total Reducing Activity

It was shown that for all the samples, each of the thermal-processing methods used resulted in a significant reduction in the total oxidoreductive activity. In the case of green bell peppers, the greatest reduction in oxidoreductive activity was observed in the case of the combined microwaving and roasting process(from 10.54 to 0.76 μM $FeSO_4$/g d.b.). In the case of green bell peppers, the process of roasting and steam-cooking resulted in the smallest reduction in oxidoreductive activity (from 10.54 to 1.44 μM $FeSO_4$/g d.b.). Similarly, for yellow bell peppers, the roasting process caused the greatest reduction in oxidoreductive activity (from 19.93 to 0.82 μM $FeSO_4$/g d.b.). In the case of yellow bell peppers, the contact-grilling process caused the smallest reduction in oxidoreductive activity compared to other heat treatments. However, it should be noted that in the case of yellow and red bell peppers, all heat-treatment processes caused a reduction in the oxidoreductive activity, with a similar degree of attenuation of the value of this parameter. In the case of red bell peppers, all heat-treatment processes caused a reduction in the oxidoreductive activity. In the case of red peppers, it was the roasting process that caused the greatest reduction in oxidoreductive activity (from 39.79 to 1.13 μM $FeSO_4$/g d.b.). In contrast, for red bell peppers, the oxidoreductive activity was reduced to the least extent with contact grilling (from 39.79 to 1.62 μM $FeSO_4$/g d.b.), as shown in Table 1.

Partially concurrent results were obtained in a study by Fong-in, where it was found that fresh edible Astraeus odoratus mushrooms (665 mg ascorbic acid/100 g d.b.) lose oxidoreductive potential as a result of steam cooking (510 mg ascorbic acid/100 g d.b.) and microwave treatment (445 mg ascorbic acid/100 g d.b.). The same study showed that the baking process (795 mg ascorbic acid/100 g d.b.), on the other hand, increases the oxidoreductive potential relative to the raw material (665 mg ascorbic acid/100 g d.b.). In contrast, the traditional grilling process (675 mg ascorbic acid/100 g d.b.) caused no change in the oxidoreductive potential of the tested raw material [43].

All thermal-treatment processes, such as roasting, contact grilling, steaming, and simultaneous roasting and microwaving, resulted in a reduction in the reducing activity for all types of tested bell peppers. Additionally, our study did not show significant differences in the reducing activity for all tested samples between the roasting process and the steaming process.

Reductive activity is closely related to the ability of a substance to donate electrons, which is revealed by the reduction of Fe^{3+} iron ions to Fe^{2+} iron ions. Lower values of reducing activity in the case of the thermal-treatment techniques used may result from the penetration of polyphenolic compounds into the water fraction (water vapor), or their flow out of the raw material together with the cell juice during roasting or contact grilling. Additionally, during thermal treatments, polyphenolic compounds can form strong hydrogen bonds with the sulfur amino acids present in bell peppers (such as cysteine and methionine), thus losing the ability to donate electrons [44].

The same observed changes were probably accompanied by a loss of turgor in the bell pepper cells, which led to a decrease in the hardness and chewiness of all types of processed bell peppers (green, yellow, and red) as a result of the heat-treatment methods used. This, in turn, was observed as the processed samples became softer and less firm at the same time.

Slightly different results were obtained in research conducted by Nandasiri, where it was shown that baking using the air-fry technique (baking with hot air) allowed for the maintenance of a higher reducing activity of broccoli sprouts (measured using the FRAP method at the level of 0.37 mM TE/g d.b.), and Brussels sprouts (at the level of 0.26 mM TE/g d.b.), than during the steam-cooking process, where values of 0.13 mM TE/g d.b. were confirmed for these materials, 0.14 and 0.8 mM TE/g d.b., respectively.

Similarly, it was observed that the air-frying treatment allowed the maintenance of a higher reducing activity (measured using the FRAP method) of kale leaves (0.31 mM TE/g d.b.), red cabbage (0.265 mM TE/g d.b.), and green cabbage (0.16 mM TE/g d.b.). This was more beneficial than the steaming process for these raw materials, where a reduction activity was observed at levels of 0.15, 0.09, and 0.055 mM TE/g d.b., respectively [45].

2.5. Reducing Sugars Content

It was shown that all the heat-treatment methods used reduced the reducing sugars content in green, yellow, and red bell peppers. The greatest loss of reducing sugars in green bell peppers was observed from the simultaneous microwaving and grilling of this raw material (from 4.15 to 1.88 mg Glc/g d.b.). In the case of yellow bell peppers, the greatest reduction in reducing sugars content was observed for bell peppers that were steam-treated (from 5.61 to 0.80 mg Glc/g d.b.). On the other hand, the greatest decrease in reducing sugars content was observed for red bell peppers that were steam-treated (from 4.41 to 0.88 mg Glc/g d.b.).

Similar results were obtained in a study of red bell peppers of the 'Senise' cultivar grown in the Montanaso region (northern Italy), which were subjected to roasting in a forced-air oven, resulting in a significant reduction of 28% in total reducing sugars [31]. Also, in a study conducted by Bianchi and Lo Scalzo in 2018 [46], it was observed that the drying process of hot chili bell pepper powder significantly reduces the content of sugars such as glucose and fructose. In a study that used a forced hot-air stream to dry Hungarian red sweet bell peppers (*Capsicum annuum*), a significant reduction in the reducing sugars content such as glucose and fructose was observed [47]. Perhaps the pronounced decrease in the content of reducing sugars in red bell peppers that we observed as a result of roasting was due to the utilization of reducing monosaccharides (mainly fructose), which underwent the Maillard reaction during the initial stages of thermal processing [48].

2.6. Color Profile Change after Treatments

Significant changes in the color of the analyzed samples were observed between the raw material and heat-treated samples. The color analysis showed that the values of L*, a*, and b* were significantly higher for roasted green peppers (GR), green peppers grilled using contact grilling (GCG), green bell peppers that were microwaved and roasted (GMWR) (except for the parameter a* for the GMWR sample), and green peppers that were processed using water steaming (GS) compared to the raw material (Table 2).

Table 2. Color values of samples in the L*, a*, b* (CIELAB) color space.

R-Stage	Processing	L*	a*	b*	C	h
G	RM	54.88 ± 2.00 ^{a}	−13.80 ± 0.46 ^{b}	15.40 ± 0.63 ^{a}	20.68± 0.76 ^{b}	131.83 ± 0.50 ^{e}
	CG	59.57 ± 0.07 ^{b}	−3.23 ± 0.06 ^{d}	17.51 ± 0.02 ^{b}	17.81 ± 0.03 ^{a}	100.40 ± 0.17 ^{a}
	R	67.44 ± 0.40 ^{d}	−11.11 ± 0.39 ^{c}	28.61 ± 0.98 ^{d}	30.69 ± 1.05 ^{c}	111.13 ± 0.25 ^{b}
	MWR	61.07 ± 0.23 ^{b}	−16.14 ± 0.10 ^{a}	24.70 ± 0.21 ^{c}	29.50 ± 0.23 ^{c}	123.10 ± 0.10 ^{c}
	S	65.32 ± 1.58 c	−10.89 ± 0.66 ^{c}	15.49 ± 1.37 ^{a}	18.93 ± 1.50 ^{a}	125.10 ± 0.69 ^{d}
Y	RM	99.28 ± 0.49 ^{e}	3.36 ± 0.18 ^{c}	89.63 ± 0.84 ^{e}	89.69 ± 0.83 ^{e}	87.93 ± 0.15 ^{c}
	CG	89.80 ± 0.68 c	5.33 ± 0.16 ^{d}	72.89 ± 0.96 ^{c}	73.08 ± 0.97 ^{c}	85.87 ± 0.06 ^{b}
	R	51.79 ± 0.42 ^{a}	−13.34 ± 0.07 ^{a}	14.04 ± 0.22 ^{a}	19.36 ± 0.21 ^{a}	133.47 ± 0.31 ^{e}
	MWR	84.42 ± 1.15 ^{b}	−0.65 ± 0.07 ^{b}	65.94 ± 2.11 ^{b}	65.94 ± 2.11 ^{b}	90.53 ± 0.06 ^{d}
	S	97.78 ± 0.12 ^{d}	10.72 ± 0.10 ^{e}	85.23 ± 1.16 ^{d}	85.89 ± 1.17 ^{d}	82.90 ± 0.00 ^{a}
R	RM	64.25 ± 0.53 ^{c}	54.82 ± 0.85 ^{d}	36.65 ± 1.18 ^{d}	65.94 ± 1.35 ^{c}	33.70 ± 0.48 ^{c}
	CG	57.52 ± 1.02 ^{a}	47.68 ± 0.76 ^{a}	27.82 ± 0.71 ^{a}	55.20 ± 1.01 ^{a}	30.20 ± 0.27 ^{a}
	R	62.18 ± 0.76 ^{b}	51.49 ± 1.84 ^{c}	33.47 ± 1.64 ^{c}	61.41 ± 2.43 ^{b}	32.95 ± 0.37 ^{b}
	MWR	62.66 ± 0.09 ^{b}	45.20 ± 0.17 ^{b}	29.95 ± 0.16 ^{b}	54.22 ± 0.08 ^{a}	33.45 ± 0.25 ^{bc}
	S	95.11 ± 0.47 ^{d}	2.90 ± 1.08 ^{e}	80.12 ± 0.48 ^{e}	80.18 ± 0.50 ^{c}	88.00 ± 0.74 ^{d}

R-stage—ripening stage; G—green pepper; Y—yellow pepper; R—red pepper. For Processing: RM—raw material; CG—contact grilling; R—roasting; MWR—microwave plus roasting; S—steaming. Lower-case letters means that a statistically significant difference was found between samples for the same R-stage.

This indicates that processes such as roasting, simultaneous roasting and microwaving, and steam processing cause a loss of green color intensity (toward a brightening of the green color), while the contact-grilling process causes browning of the bell pepper surface.

Perhaps the particularly large increase in the b* value for green bell peppers due to the roasting process and the combined roasting and microwaving process was a result of the effects of reactive oxygen species, such as ozone, peroxides, and hydroxyl radicals, which oxidize the color substances contained in the skin of green bell peppers [49].

Reactive oxygen species that are generated during microwave processing cause the degradation of cell structures and the chlorophyll pigments they contain, such as chlorophyll a and b. Reactive oxygen species accelerate the degradation of chlorophyll during heat treatment to brown pheophytin, which is caused by the loss of the magnesium ion in the active center of the chlorophyll molecule [50].

On the other hand, in 2021 Darici et al. [51] reported that the microwave drying of green bell peppers (180 W) resulted in a decrease in L* and b* values, while a significant increase in a* values compared to fresh bell peppers was observed. This was indicated by the browning of the processed samples.

For green bell peppers, the smallest difference in color relative to the raw material was observed for the steam-treated bell peppers. After the combined microwave and roasting treatment, a delta E value of 11.4 was obtained for the green bell peppers (Table 3).

Similar results were obtained by Darıcı et al. in 2021 [51] (ΔE values of 13.97) for green bell peppers that were microwave-dried. On the other hand, in a study conducted by Eyarkai et al. in 2016 [52], it was shown that the blanching process for vegetables such as green peas, eggplant, and green bell peppers resulted in a reduction of all CIE color coordinates (L, a, b) in these raw materials, which was associated with a weakening of color as a result of the leaching and migration of color substances into hot water [53].

Table 3. Delta E values for the tested bell peppers.

R-Stage	RM-R	RM-MVR	RM-S	RM-CG
G	19.0 ± 2.2	11.4 ± 0.7	10.3 ± 3.2	11.7 ± 0.9
Y	90.2 ± 1.0	28.1 ± 3.0	8.5 ± 0.6	18.2 ± 2.8
R	5.1 ± 4.0	11.8 ± 1.5	74.4 ± 0.6	13.2 ± 1.2

For R-stage: G—green pepper; Y—yellow pepper; R—red pepper. For Processing: RM—raw material; CG—contact grilling; R—roasting; MWR—microwave plus roasting; S—steaming.

In the case of yellow bell peppers, on the other hand, it was observed that the heat-treatment methods used reduced the L* brightness value and the b* value. The a* value of yellow bell peppers was decreased in the case of the roasting process and the simultaneous roasting and microwaving process and increased due to contact grilling and steaming. The observed changes were related to the yellow bell peppers' loss of its bright yellow color toward a green color due to roasting, as well as the loss of its intense bright yellow color toward a dark yellow color due to contact grilling. There was a loss of an intense bright yellow color due to the simultaneous microwaving and roasting method, and its transformation to a green–yellow color was observed. The smallest difference in color was observed for yellow bell peppers after the steam treatment, as shown in Table 3.

Darici reported the same results from a study on yellow sweet bell peppers of the 'Charleston' cultivar (bell peppers with an elongated conical shape) and yellow sweet bell peppers. The microwave drying of these raw materials (180 W) resulted in a decrease in L* and b* values, while a significant increase in a* values compared to fresh bell peppers was observed, which was indicated by a strong browning of the raw material [51].

It is possible that the heat-treatment techniques used resulted in increased browning (non-enzymatic browning) with the simultaneous decomposition of pigments present in yellow and red bell peppers, such as β-carotene, zeaxanthin, quercetin, and luteolin. The decomposition of these pigments may have reduced the L* value in the yellow and red bell peppers [54]. In addition, it was indicated that the drying phenomenon that accompanies such thermal treatments as roasting and microwaving can cause the oxidation of ascorbic acid to dehydroascorbic acid, which in turn leads to a decrease in the brightness of the surface of the processed samples [55,56].

In the case of red bell peppers, there was a reduction in the L*, a* and b* values due to the roasting, contact-grilling, and combined microwaving and roasting processes. This resulted in a brightening (loss of brightness) of the red color of peppers when roasted, a darkening of the color when contact-grilled, and a dulling of the surface of peppers that were both roasted and microwaved. Also, in another study on red bell peppers of the 'Capia' cultivar, a reduction in the L* (brightness) value was noted after air-drying this raw material (from 35.56 to 28.30). When using conventional oven drying (at 120 °C) and drying using a microwave oven (at 720 W for 16 min), a reduction in L* values was observed to 31.86 and 31.04, respectively [27]. The same study confirmed that the a* and b* values of red raw bell peppers (22.67), as a result of the heat-treatment methods used (oven-drying, microwave-drying, and air-drying processes) are significantly reduced [27]. Also, in a study conducted by Deng et al. in 2018 [57], it was shown that the L* values of red peppers after the drying process decreased from a value of 41.60 to 26.64.

However, in the case of red bell peppers, the steam treatment resulted in an increase in L* and b* values and a simultaneous decrease in the a* value. In the case of red bell peppers, steam cooking caused a marked brightening of the surface of the red bell peppers toward yellow (Table 3). Different results were obtained in a study conducted by Eyarkai et al. in 2016 [52], where the blanching process of beets caused a reduction in all CIE color coordinates (L*, a* and b*), which was associated with a weakening of the color intensity of this raw material.

Statistically significant differences were found in the C values between raw green bell peppers and roasted green bell peppers, contact-grilled green bell peppers, and green bell peppers that were both microwaved and roasted. There were also statistically significant differences in C values between raw yellow bell peppers and yellow bell peppers that were roasted, yellow bell peppers that were contact-grilled, yellow bell peppers that were processed with both microwaving and roasting, and yellow bell peppers that were treated with steam cooking. There were also statistically significant differences in C values between raw red bell peppers and roasted red peppers, red peppers grilled by the contact-grilling method, red bell peppers that were processed with both microwaving and roasting, and red bell peppers that were treated with steam cooking. There were statistically significant differences in h values between the raw material and the samples subjected to all heat treatments for all types of bell peppers tested.

2.7. Textural Profile Changes after Treatments

Changes in textural parameters depending on the type of heat-treatment method used are shown in Table 4.

Table 4. Textural properties of the tested bell peppers.

R-Stage	Processing	Hardness [N]	Cohesiveness	Springiness	Chewiness [N]	Resilience
G	RM	76.7 ± 0.1 [e]	0.784 ± 0.023 [ab]	0.667 ± 0.000 [a]	40.01 ± 1.14 [e]	0.68 ± 0.09 [ab]
	CG	5.4 ± 0.1 [a]	0.726 ± 0.028 [a]	0.838 ± 0.028 [b]	3.22 ± 0.18 [a]	0.76 ± 0.11 [bc]
	R	8.7 ± 0.4 [b]	0.839 ± 0.065 [b]	0.926 ± 0.004 [b]	6.69 ± 0.19 [c]	0.57 ± 0.02 [a]
	MWR	10.4 ± 0.1 [c]	0.755 ± 0.005 [ab]	0.641 ± 0.037 [a]	5.02 ± 0.17 [b]	0.98 ± 0.06 [d]
	S	12.2 ± 0.9 [d]	0.841 ± 0.030 [b]	1.046 ± 0.064 [c]	10.63 ± 0.50 [d]	0.87 ± 0.02 [cd]
Y	RM	34.4 ± 2.9 [b]	0.818 ± 0.057 [a]	0.701 ± 0.018 [a]	19.73 ± 2.49 [b]	0.86 ± 0.18 [a]
	CG	10.4 ± 0.4 [a]	0.824 ± 0.036 [ab]	0.834 ± 0.235 [a]	7.11 ± 2.08 [a]	0.73 ± 0.21 [a]
	R	12.1 ± 1.0 [a]	0.868 ± 0.009 [ab]	0.727 ± 0.084 [a]	7.64 ± 1.45 [a]	0.68 ± 0.04 [a]
	MWR	10.6 ± 0.4 [a]	0.798 ± 0.018 [a]	0.691 ± 0.033 [a]	5.87 ± 0.63 [a]	0.66 ± 0.02 a
	S	12.1 ± 1.2 [a]	0.914 ± 0.036 [b]	0.731 ± 0.272 [a]	7.94 ± 2.50 [a]	0.77 ± 0.09 [a]
R	RM	48.6 ± 1.1[d]	0.720 ± 0.015 [a]	0.622 ± 0.030 [a]	21.73 ± 1.10 [d]	0.73 ± 0.12 [a]
	CG	3.3 ± 0.8 [a]	0.698 ± 0.125 [a]	1.084 ± 0.118 [b]	2.56 ± 1.30 [a]	0.67 ± 0.42 [a]
	R	11.5 ± 0.1 [b]	0.896 ± 0.008 [b]	0.763 ± 0.100 [ab]	7.86 ± 1.05 [b]	0.70 ± 0.10 [a]
	MWR	16.4 ± 2.9 [c]	0.786 ± 0.013 [ab]	0.818 ± 0.257 [ab]	10.24 ± 1.64 [bc]	0.74 ± 0.16 [a]
	S	19.6 ± 0.6 [c]	0.813 ± 0.035 [ab]	0.718 ± 0.072 [a]	11.42 ± 1.00 [c]	0.88 ± 0.14 [a]
processing		***	*	ns	***	ns
R-stage		***	***	*	***	ns
Processing × R-stage		***	ns	ns	***	ns

R-stage—ripening stage; G—green pepper; Y—yellow pepper; R—red pepper. For Processing: RM—raw material; CG—contact grilling; R—roasting; MWR—microwave plus roasting; S—steaming. Lower-case letters mean that a statistically significant difference was found between samples for the same R-stage. (p = 0.05), ns—statistically non-significant, *—$p < 0.05$, ***—$p < 0.001$.

Hardness is one of the most important textural parameters that change after heat treatment. Studies indicate that the hardness of the vegetable samples is strongly correlated with fiber content and is defined as the maximum value of the force measured at the first compression stage [58].

The results of the texture analysis showed that there was a significant reduction in the hardness of all types of processed peppers—green, yellow, and red—as a result of all four heat-treatment methods used (Table 4).

To the greatest extent, the analyzed raw materials were subjected to a reduction in hardness when they were subjected to contact grilling. Hardness reduction occurred by 93%, 69%, and 93%, respectively, for the raw green, yellow, and red bell peppers.

Similar results were obtained by Merve in a study of yellow sweet 'Charleston' peppers, green peppers, and yellow sweet bell peppers that were subjected to microwave drying, where a reduction in the hardness parameter was observed by 60%, 48%, and 14%, respectively, relative to the raw materials [51]. Studies indicate that heat-treatment processes such as microwave processing and roasting cause a strong reduction in the turgor potential (pressure) and solubilization of pectic substances, resulting in the collapse of cellular structures and the softening of highly hydrated tissues, such as those present in many varieties of annual peppers (*Capsicum annuum* L.) [59].

In our study, steam processing was shown to reduce the hardness of green, yellow, and red peppers by 84%, 64%, and 59%, respectively. Also, in a study conducted by Eyarkai et al. in 2016 [52], a significant reduction in the hardness of the vegetables studied was observed due to blanching. The greatest reduction in hardness (by 90%) was observed in green peppers, followed by eggplant (by 80%), green peas (by 60%), and a slight reduction (by 25%) in beets. Our study showed a significant reduction in the hardness of green, yellow, and red peppers due to roasting (roasting) by 88.65%, 64.82%, and 76.33%, respectively. Also, in a study on raw chickpea seeds (*Cicer arietinum* L.), it was shown that their hardness decreased significantly with an increase in roasting time [60].

No significant changes in cohesiveness were shown for green bell peppers as a result of all four heat treatments used. Similarly, no significant changes in cohesiveness were shown for yellow bell peppers as a result of the thermal-processing methods used, such as roasting, simultaneous microwaving and roasting, and contact grilling. Only as a result of steam processing was there a significant increase in consistency for yellow bell peppers relative to the raw material. There was also no significant change in consistency for red bell peppers as a result of the thermal treatments used, such as simultaneous microwaving and roasting, contact grilling, and steam processing. Only as a result of roasting was there a significant increase in consistency for red bell peppers relative to the raw material.

The significant increase in springiness was observed for green bell peppers as a result of roasting, contact-grilling, and steam-processing methods.

In contrast, no significant change in springiness was observed for yellow bell peppers as a result of all four heat treatments. For red bell peppers, a significant increase in chewiness (relative to the raw material) was observed only as a result of their processing with contact grilling. There was a significant reduction in the chewiness that was observed (relative to the raw material) in all processed types of samples—green, yellow, and red peppers—as a result of all four thermal-processing methods used. The reduction in chewiness occurred to the greatest extent for the contact grilling of green bell peppers, the simultaneous microwaving and roasting of yellow peppers, and the simultaneous microwaving and roasting of red bell peppers.

The results of the study showed that only the simultaneous microwaving and roasting method and steam-processing method significantly increased the resilience of green bell peppers. In contrast, no significant changes in resilience were observed for yellow and red peppers as a result of all four heat-treatment methods.

3. Materials and Methods

3.1. Analytical Reagents and Standards Used

The following analytical reagents were used for the analyses: glucose, Trolox (6-hydroxyl-2,5,7,8–tetramethylchromo-2-carboxylic acid, 2,2–diphenyl-1-picrylhydrazyl (DPPH), 2,2–azino–bis (3-ethyl benzothiazoline-6-sulphonic acid) (ABTS), TPTZ (2,3,5-triphenyltetrazolium chloride), gallic acid, and iron (III) sulphate hydrate (Pol-Aura, Zabrze, Poland). All solutions and buffers were prepared using distilled and deionized water.

3.2. Plant Research Material

The research material consisted of annual bell pepper fruits (*Capsicum annuum* L.), which were cultivated in the southern regions of Mazovia in Poland. The peppers were of the 'Ożarowska' cultivar. The bell pepper fruits were purchased in a fruit and vegetable store in Wrocław, Poland, in August 2022. The tested bell pepper fruits represented the three main stages of ripening (green, yellow, and red) of this cultivar. The fruits were stored for several dozen hours at a temperature of 4 °C until they were processed and analyzed.

3.3. *Heating Procedures*

The tested raw materials were pre-washed in running water and the bell pepper stalks were then removed by hand. Then, the raw materials were cut into fragments 7–8 cm long and about 4–5 cm wide, from which the seeds were removed. Then, the tested raw materials were subjected to four thermal treatments: roasting, roasting combined with microwaving, contact grilling, and steaming. The resulting samples are shown in Table 5.

Table 5. Bell pepper chunks before (raw material) and after each thermal treatment.

Processing	Green Bell Pepper	Yellow Bell Pepper	Red Bell Pepper
Raw Material—RM			
Steaming—S			
Roasting—R			
Roasting and Microwave—MWR			
Contact Grilling—CG			

3.3.1. Steam Cooking

The washed fragments (quarters) of the bell peppers were drained of moisture and were then placed on a sieve in the upper part of a steamer (Clatronic DG 3547, Clatronic, Kempen, Germany) so that the raw material did not come into contact with the boiling water. The steaming time recommended for annual bell peppers (10 min) was counted from the moment the water boiled in the lower part of the device and produced hot steam, which generated a temperature of 110–120 °C inside the steamer.

3.3.2. Contact Grilling

The washed fragments (quarters) of the bell peppers were drained of moisture and were then placed between the lower and upper grill plates (OptiGrill Elite XL GC760D electric grill with automatic programs, TEFAL, Rumilly, France). The recommended contact-roasting time for annual bell peppers (15 min) was counted from the moment the device heated up to a temperature of 200 °C.

3.3.3. Roasting

The washed fragments (quarters) of the bell peppers were drained of moisture and were then placed in the oven chamber on the rotating tray of a multifunction microwave device (GE83X, Samsung Electronics, Suwon, Republic of Korea). The recommended microwaving time for annual bell peppers (15 min) was counted from the moment the device heated up to 180 °C.

3.3.4. Simultaneous Roasting and Microwave

The washed fragments (quarters) of the bell peppers were drained of moisture and were then placed in the oven chamber on the rotating tray of a multifunction microwave device (GE83X, Samsung Electronics, Republic of Korea), which was thermally processed using a system of heaters and microwave radiation simultaneously. The recommended time for this type of processing (2 min and 30 s) was counted from the moment the device heated up to a temperature of 180 °C.

3.4. Extract Preparation

The raw materials and processed samples were homogenized using a manual homogenizer (CAT DI 18 Basic, Caterpillar, Irving, TX, USA). The resulting 5 g of homogenate was extracted with a 10 mL methanol and water (80/20 *v*/*v* %) mixture for one hour on a radial stirrer (MX-RD PRO, ChemLand, Stargard, Poland) at 60 rotations per minute, and then centrifuged (MPW-350, MPW, Warszawa, Poland) for 15 min. at 10,000 rpm and 5031 g. The supernatant obtained after centrifugation was analyzed for the total antioxidant potential, reduction potential, and content of bioactive substances.

3.5. Determination of Total Phenolic Compounds

The total content of polyphenolic compounds was determined spectrophotometrically (SEMCO, S91 E, Gdynia, Poland) using the Folin–Ciocalteu reagent according to the method of Yen et al. in 1995 [61] after introducing minor modifications. For this purpose, 0.1 mL of the Folin–Ciocalteu reagent and 1.58 mL of H_2O were added to the obtained extracts (0.02 mL). After 5 min of incubation, 0.3 mL of a saturated sodium carbonate solution (Na_2CO_3) was added. The total phenolic compounds were determined after 20 min of incubation at 38 °C in the dark. As it reacts with polyphenolic substances, the Folin–Ciocalteu reagent forms a blue complex that has a maximum absorbance at 765 nm. A standard curve was prepared for gallic acid [61]. All the samples were analyzed in duplicate. The content of polyphenolic compounds in the tested material was calculated from the calibration curve and presented in milligrams of gallic acid equivalent (GEA) per gram of dry basis (dry weight)—d.b.

3.6. Determination of Antioxidant and Oxidoreductive Activities

3.6.1. DPPH Test

The antioxidant capacity (in relation to stablethe 2,2-diphenyl-1-picrylhydrazyl radical ($DPPH^•$)) for the raw materials and the processed samples was measured spectrophotometrically (SEMCO, S91 E, Poland) using the method of Klymenko et al. in 2019 [62] with minor modifications. For this purpose, 0.035 mL of the test solution was measured and added to 1 mL of (0.1 mM) a methanolic DPPH solution. The mixture was shaken and left at room temperature for 20 min, after which the absorbance was measured at 517 nm. All the samples were analyzed in duplicate [62]. The total antioxidant activity in the tested material was calculated from the calibration curve and presented in milligrams of Trolox equivalent (TE) per gram of dry basis (dry weight)—d.b.

3.6.2. ABTS Test

The antiradical capacity of the raw materials and the processed samples against the cationic radical 2,2-azo-bis(3-ethylbenzothiazoline-6-sulfonic acid ($ABTS^{•+}$) was measured spectrophotometrically (SEMCO, S91 E, Poland)using the Sridhar method (2019) with minor modifications. The $ABTS^{•+}$ solution was prepared by mixing 7 mM of ABTS stock solution with 2.45 mM of potassium persulfate solution and incubating the mixture at room temperature (23 °C) in the dark for 16–24 h. The $ABTS^{•+}$ solution was diluted with phosphate buffer (0.1 M) to give an absorbance of 0.900 ± 0.05 at 734 nm. To 1.0 mL of the diluted $ABTS^{•+}$ solution, 0.02 mL of the extract to be tested was added. The absorbance at 734 nm was read exactly 10 s after mixing the test extract with the $ABTS^{•+}$ solution. All the samples were analyzed in duplicate [63]. The total antioxidant activity in the tested material was calculated from the calibration curve and presented in milligrams of Trolox equivalent (TE) per 1 g of dry basis (dry weight)—d.b.

3.6.3. FRAP Test

The reducing power (ability to reduce ferric ions Fe^{3+}) of the raw materials and the processed samples was measured spectrophotometrically (SEMCO, S91 E, Poland) according to the method of Re et al. in 1999 [64], with minor modifications. The test extract was added to 1 mL of FRAP solution (acetate buffer (300 μM, pH 3.6), a solution of 10 μM TPTZ in 40 μM HCl and 20 μM $FeCl_3$ in the ratio 10:1:1 (*v*/*v*)). The mixture was shaken and left at room temperature for 20 min, after which the absorbance was measured at 593 nm. All the samples were analyzed in duplicate [64]. The total reducing activity was calculated from the calibration curve and presented in milligrams of iron (II) sulfate equivalent $FeSO_4^{-} \cdot 7H_2O$ per 1 g dry basis (dry weight).

3.7. Measurement of Reducing Sugar Content

The sugar content of extracts for the raw materials and the processed samples was measured spectrophotometrically (SEMCO, S91 E, Poland) using a modified method according to Miller et al. in 1959 [65], taking advantage of the reducing properties of sugars towards 3,5-dinitrosalicylic acid (DNS). To 1 mL of the test sample, 1 mL of DNS reagent was added and mixed thoroughly. The resulting mixture was then heated in boiling water for 5 min. After the mixture cooled to room temperature, its absorbance at 535 nm was measured. All the samples were analyzed in duplicate [65]. The total content of reducing sugars in the tested material was presented in milligrams of glucose equivalent per gram of dry basis (dry weight).

3.8. Color Measurement

The color of the fresh and processed bell peppers' surface was assessed using a Konica Minolta CR-310 chroma meter (Konica Minolta, Ramsey, NJ, USA) in CIE color space (L*,a*, b* co-ordinates). Color co-ordinates (L*, a*, b*), and parameters (chroma, hue) were taken in triplicate and expressed as means ± standard deviation [26].

3.9. Texture Measurements

The texture of the bell peppers was examined using a TPA test (Texture Profile Analysis) with an AXIS texture analyzer FC200STAV500 (AXIS, Gdansk, Poland) with the software AXIS FM v.2.18, as previously reported by Olędzki & Harasym [26].

In order to determine the textural properties of bell peppers, an analysis of the texture profile was carried out on cylindrical samples with a diameter of 20 mm and a 5 mm height, taken with the skin from the middle fragments of the fruit. Measurements to the fresh bell peppers were performed on the skin (external) side.

Hardness was the force expressed in the unit of newton (N) at the maximum deformation, whereas cohesiveness, springiness, chewiness, and resilience were calculated from two peaks, corresponding, respectively, to the first and second sample compression cycle. The analysis was carried out in quadruplicate at 25 °C on bell pepper cylinders cut from different pieces of each bell pepper fruit. The results were reported as means (from four measurements from each sample) ± standard deviation [26].

3.10. Dry Basis Assessment

Four samples were taken from each raw pepper and processed pepper. Each sample was cut into small pieces and 2 g of the chopped sample was placed on an aluminum plate in a weighing–drying machine (MA-30, Sartorius GmbH, Göttingen, Germany). An automatic method was set in the machine's operating software, according to which the samples were dried at 105 °C. From the data obtained, the average of four measurements was calculated for each type of bell pepper. The obtained dry matter contents served as a reference point for presenting the results of the antioxidant activity, reducing activity, total content of polyphenolic compounds, and content of reducing sugars in the tested peppers.

3.11. Statistical Analysis

The results from all methods are given as mean values and their standard deviation. Data were subjected to one-way and two-way analysis of variance and mean values were compared using Tukey's test ($p < 0.05$). Values of $p < 0.05$ were considered statistically significant. The statistical analysis was performed using Statgraphics Centurion 19 (Statgraphics Technologies, Inc., The Plains, VA, USA) statistical software.

4. Conclusions

The study showed that the heat-treatment methods used have significant and differential effects on the antioxidant properties and bioactive content of fresh bell pepper fruits. For green bell peppers, the thermal processes that preserve more antioxidant components were contact grilling and steaming.

For yellow bell peppers, on the other hand, the thermal processes that proved to be beneficial in terms of polyphenol content were the simultaneous microwave and roasting method and contact-grilling method. However, due to the antioxidant activity, processes such as contact grilling and steaming appeared to be beneficial for yellow bell peppers. For red bell peppers, roasting proved to be an exceptionally favorable thermal-treatment process, which even increased the total content of phenolic compounds (relative to the raw material). On the other hand, in terms of antioxidant activity, contact-grilling and steaming processes proved to be favorable processes for red bell peppers.

The proper selection of the thermal-treatment process according to the ripeness and cultivar of bell peppers can increase not only their digestibility, but also improve the bioavailability of nutrients and bioactive substances contained in the raw material. Therefore, to ensure the optimal intake of antioxidants contained in processed bell pepper fruits, it is necessary to consider the use (during food preparation) of separate heat-treatment processes for different stages of bell pepper maturity (bell pepper color forms). The obtained results suggest that in the future it would be good to develop industry guidelines for the food and catering industry regarding thermal-processing methods for vegetable and fruit plant raw materials.

Author Contributions: Conceptualization: R.O. and J.H.; methodology: R.O.; investigation: R.O.; resources: R.O. and J.H.; data curation: R.O. and J.H.; writing—original draft preparation: R.O. and J.H.; writing—review and editing: R.O. and J.H.; supervision: R.O. and J.H.; funding acquisition: J.H. All authors have read and agreed to the published version of the manuscript.

Funding: This research received no external funding.

Institutional Review Board Statement: Not applicable.

Informed Consent Statement: Not applicable.

Data Availability Statement: The data presented in this study are available on request from the corresponding author.

Acknowledgments: The authors thanks to "Academic Mentoring" project proceeded under Strategy 2030 program of Wroclaw University of Economics and Business, Poland.

Conflicts of Interest: The authors declare no conflicts of interest.

References

1. van Zonneveld, M.; Ramirez, M.; Williams, D.E.; Petz, M.; Meckelmann, S.; Avila, T.; Bejarano, C.; Rios, L.; Pena, K.; Jager, M.; et al. Screening genetic resources of Capsicum peppers in their primary center of diversity in Bolivia and Peru. *PLoS ONE* **2015**, *10*, e0134663. [CrossRef] [PubMed]
2. Xavier, A.A.O.; Perez-Galvez, A. Peppers and chilies. In *Encyclopedia of Food and Health*; Caballero, B., Finglas, P.M., Toldra, F., Eds.; Elsevier: Kidlington, UK, 2016; p. 301.
3. Zhang, D.; Hamauzu, Y. Phenolic compounds, ascorbic acid, carotenoids and antioxidant properties of green, red and yellow bell peppers. *J. Food. Agric. Environ.* **2003**, *1*, 1–7.
4. Rana, A.; Samtiya, M.; Dhewa, T.; Mishra, V.; Aluko, R.E. Health benefits of polyphenols: A concise review. *J. Food. Biochem.* **2022**, *46*, e14264. [CrossRef] [PubMed]

5. Wang, S.; Du, Q.; Meng, X.; Zhang, Y. Natural polyphenols: A potential prevention and treatment strategy for metabolic syndrome. *Food Funct.* **2022**, *13*, 9734–9753. [CrossRef]
6. Agunloye, O.M.; Oboh, G.; Ademiluyi, A.O.; Ademosun, A.O.; Akindahunsi, A.A.; Oyagbemi, A.A.; Omobowale, T.O.; Ajibade, T.O.; Adedapo, A.A. Cardio-protective and antioxidant properties of caffeic acid and chlorogenic acid: Mechanistic role of angiotensin converting enzyme, cholinesterase and arginase activities in cyclosporine induced hypertensive rats. *Biomed. Pharmacother.* **2019**, *109*, 450–458. [CrossRef]
7. Bucciantini, M.; Leri, M.; Nardiello, P.; Casamenti, F.; Stefani, M. Olive polyphenols: Antioxidant and anti-inflammatory properties. *Antioxidants* **2021**, *10*, 1044. [CrossRef]
8. Agarwal, B.; Campen, M.J.; Channell, M.M.; Wherry, S.J.; Varamini, B.; Davis, J.G.; Baur, J.A.; Smoliga, J.M. Resveratrol for primary prevention of atherosclerosis: Clinical trial evidence for improved gene expression in vascular endothelium. *Int. J. Cardiol.* **2013**, *166*, 246–248. [CrossRef]
9. Giordano, D.; Facchiano, A.; Minasi, P.; D'Agostino, N.; Parisi, M.; Carbone, V. Phenolic Compounds and Capsaicinoids in Three *Capsicum annuum* Varieties: From Analytical Characterization to In Silico Hypotheses on Biological Activity. *Molecules* **2023**, *28*, 6772. [CrossRef]
10. Gade, A.; Kumar, M.S. Gut microbial metabolites of dietary polyphenols and their potential role in human health and diseases. *J. Physiol. Biochem.* **2023**, *79*, 695–718. [CrossRef]
11. Catalkaya, G.; Venema, K.; Lucini, L.; Rocchetti, G.; Delmas, D.; Daglia, M.; De Filippis, A.; Xiao, H.; Quiles, J.L.; Xiao, J.; et al. Interaction of dietary polyphenols and gut microbiota: Microbial metabolism of polyphenols, influence on the gut microbiota, and implications on host health. *Food Front.* **2020**, *1*, 109–133. [CrossRef]
12. Anand, P.; Kunnumakkara, A.B.; Newman, R.A.; Aggarwal, B.B. Bioavailability of curcumin: Problems and promises. *Mol. Pharm.* **2007**, *4*, 807–818. [CrossRef] [PubMed]
13. Campos, M.R.S.; Gómez, K.R.; Ordoñez, Y.M.; Ancona, D.B. Polyphenols, ascorbic acid and carotenoids contents and antioxidant properties of habanero pepper (*Capsicum chinense*) fruit. *Food Nutr. Sci.* **2013**, *4*, 47–54.
14. Tundis, R.; Loizzo, M.R.; Menichini, F.; Bonesi, M.; Conforti, F.; Statti, G.; De Luca, D.; De Cindio, B.; Menichini, F. Comparative study on the chemical composition antioxidant properties and hypoglycaemic activities of two *Capsicum annum* L. cultivars (Acuminatum small and Cerasiferum). *Plant Foods Hum. Nutr.* **2011**, *66*, 261–269. [CrossRef] [PubMed]
15. Carvalho, A.V.; de Andrade, M.R.; de Oliveira, R.A.; de Almeida, M.R.; Moresco, K.S.; de Souza Oliveira, T.C. Bioactive compounds and antioxidant activity of pepper (*Capsicum* sp.) genotypes. *J. Food Sci. Technol.* **2015**, *52*, 7457–7464. [CrossRef]
16. Rivera-Madrid, R.; Carballo-Uicab, V.M.; Cárdenas-Conejo, Y.; Aguilar-Espinosa, M.; Siva, R. Overview of carotenoids and beneficial effects on human health. In *Carotenoids: Properties, Processing and Applications*; Academic Press: Cambridge, MA, USA, 2020; pp. 1–40.
17. Martínez, S.; López, M.; González-Raurich, M.; Bernardo, A.A. The effects of ripening stage and processing systems on vitamin C content in sweet peppers (*Capsicum annuum* L.). *Int. J. Food. Sci. Nutr.* **2005**, *56*, 45–51. [CrossRef] [PubMed]
18. Marín, A.; Ferreres, F.; Tomás-Barberán, F.A.; Gil, M.I. Characterization and quantitation of antioxidant constituents of sweet pepper (*Capsicum annuum* L.). *J. Agric. Food. Chem.* **2004**, *52*, 3861–3869. [CrossRef] [PubMed]
19. Nishikimi, M.; Fukuyama, R.; Minoshima, S.; Shimizu, N.; Yagi, K. Cloning and chromosomal mapping of the human non-functional gene for l-gulono-gamma-lactone oxidase, the enzyme for l-ascorbic acid biosynthesis missing in man. *J. Biol. Chem.* **1994**, *269*, 13685–13688. [CrossRef]
20. Pullar, J.M.; Carr, A.C.; Vissers, M.C.M. The roles of vitamin C in skin health. *Nutrients* **2017**, *9*, 866. [CrossRef]
21. Maggini, S.; Wintergerst, E.S.; Beveridge, S.; Hornig, D.H. Selected vitamins and trace elements support immune function by strengthening epithelial barriers and cellular and humoral immune responses. *Br. J. Nutr.* **2007**, *98*, 29–35. [CrossRef]
22. Sharma, P.; Raghavan, S.A.; Saini, R.; Dikshit, M. Ascorbate-mediated enhancement of reactive oxygen species generation from polymorphonuclear leukocytes: Modulatory effect of nitric oxide. *J. Leukoc. Biol.* **2004**, *75*, 1070–1078. [CrossRef]
23. Carr, A.C.; Maggini, S. Vitamin C and Immune Function. *Nutrients* **2017**, *9*, 1211. [CrossRef] [PubMed]
24. Al-Juhaimi, F.; Ghafoor, K.; Özcan, M.M.; Jahurul, M.H.A.; Babiker, E.E.; Jinap, S.; Sahena, F.; Sharifudin, M.S.; Zaidul, I.S.M. Effect of various food processing and handling methods on preservation of natural antioxidants in fruits and vegetables. *J. Food Sci. Technol.* **2018**, *55*, 3872–3880. [CrossRef] [PubMed]
25. Sanati, S.; Razavi, B.M.; Hosseinzadeh, H. A review of the effects of *Capsicum annuum* L. and its constituent, capsaicin, in metabolic syndrome. *Iran. J. Basic Med. Sci.* **2018**, *21*, 439–448. [PubMed]
26. Olędzki, R.; Harasym, J. Boiling vs. microwave heating—The impact on physicochemical characteristics of bell pepper (*Capsicum annuum* L.) at Different Ripening Stages. *Appl. Sci.* **2023**, *13*, 8175. [CrossRef]
27. Özgür, M.; Özcan, T.; Akpinar-Bayizit, A.; Yilmaz-Ersan, L. Functional compounds and antioxidant properties of dried green and red peppers. *Afr. J. Agric. Res.* **2011**, *6*, 5638–5644. [CrossRef]
28. Hameed, A.; Fatima, N.; Iftikhar, H.; Mehmood, A.; Tariq, M.R.; Ali, S.W.; Ali, S.; Shafiq, M.; Ahmad, Z.; Ali, U.; et al. Effect of different drying and cooking treatments on phytochemicals and antioxidant activity in broccoli: An experimental in vitro study. *Food Sci. Technol.* **2023**, *43*, e101622. [CrossRef]
29. Reis, R.C.; Castro, V.C.; Devilla, I.A.; Oliveira, C.A.; Barbosa, L.S.; Rodovalho, R. Effect of drying temperature on the nutritional and antioxidant qualities of Cumari peppers from Para (*Capsicum chines* Jacqui). *Braz. J. Chem. Eng.* **2013**, *30*, 337–343. [CrossRef]

30. Özcan, M.M.; Uslu, N. Quantitative changes of bioactive properties and phenolic compounds in capia pepper (*Capsicum annuum* L.) fruits dried by the air, conventional heater, and microwave. *J. Food Process. Preserv.* **2022**, *46*, e16897. [CrossRef]
31. Speranza, G.; Lo Scalzo, R.; Morelli, C.F.; Rabuffetti, M.; Bianchi, G. Influence of drying techniques and growing location on the chemical composition of sweet pepper (*Capsicum annuum* L., var. Senise). *J. Food Biochem.* **2019**, *43*, e13031. [CrossRef]
32. Arfaoui, L. Dietary plant polyphenols: Effects of food processing on their content and bioavailability. *Molecules* **2021**, *26*, 2959. [CrossRef]
33. Lewicki, P.P. Effect of pre-drying treatment, drying and rehydration on plant tissue properties: A review. *Int. J. Food Prop.* **1998**, *1*, 1–22. [CrossRef]
34. Zahoor, I.; Ganaie, T.A.; Wani, S.A. Effect of microwave assisted convective drying on physical properties, bioactive compounds, antioxidant potential and storage stability of red bell pepper. *Food Chem. Adv.* **2023**, *3*, 100440. [CrossRef]
35. Ke, Y.; Deng, L.; Dai, T.; Xiao, M.; Chen, M.; Liang, R.; Liu, W.; Liu, C.; Chen, J. Effects of cell wall polysaccharides on the bioaccessibility of carotenoids, polyphenols, and minerals: An overview. *Crit. Rev. Food Sci. Nutr.* **2023**, *63*, 11385–11398. [CrossRef] [PubMed]
36. Schweiggert, U.; Schieber, A.; Carle, R. Inactivation of peroxidase, polyphenoloxidase, and lipoxygenase in paprika and chili powder after immediate thermal treatment of the plant material. *Innov. Food Sci. Emerg. Technol.* **2005**, *6*, 403–411. [CrossRef]
37. Desai, S.; Upadhyay, S.; Sharanagat, V.S.; Nema, P.K. Physico-functional and quality attributes of microwave-roasted black pepper (*Piper nigrum* L.). *Int. J. Food Eng.* **2023**, *19*, 561–572. [CrossRef]
38. Kaur, R.; Kaur, K.; Singh, J.S. Drying kinetics, chemical, and bioactive compounds of yellow sweet pepper as affected by processing conditions. *J. Food Process. Preserv.* **2022**, *46*, e16330. [CrossRef]
39. Sun, T.; Xu, Z.; Wu, T.C.; Janes, M.; Prinyawiwatkul, W.; No, H.K. Antioxidant activities of different colored sweet bell peppers (*Capsicum annuum* L.). *J. Food Sci.* **2007**, *72*, 98–102. [CrossRef] [PubMed]
40. Reda, S.Y. Evaluation of antioxidants stability by thermal analysis and its protective effect in heated edible vegetable oil. *Ciênc. Tecnol. Aliment. Camp.* **2011**, *31*, 475–480. [CrossRef]
41. Maiani, G.; Periago Castón, M.J.; Catasta, G.; Toti, E.; Cambrodón, I.G.; Bysted, A.; Schlemmer, U. Carotenoids: Actual knowledge on food sources, intakes, stability and bioavailability and their protective role in humans. *Mol. Nutr. Food Res.* **2009**, *53*, 194–218. [CrossRef]
42. Pérez-Burillo, S.; Rufián-Henares, J.Á.; Pastoriza, S. Effect of home cooking on the antioxidant capacity of vegetables: Relationship with Maillard reaction indicators. *Food Res. Int.* **2019**, *121*, 514–523. [CrossRef]
43. Fong-in, S.; Khwanchai, P.; Prommajak, T.; Boonsom, S. Physicochemical, nutritional, phytochemical properties and antioxidant activity of edible Astraeusodoratus mushrooms: Effects of different cooking methods. *Int. J. Gastron. Food Sci.* **2023**, *33*, 100743. [CrossRef]
44. Lemańska, K.; Szymusiak, H.; Tyrakowska, B.; Zieliński, R.; Soffers, A.E.; Rietjens, I.M. The Influence of PH on antioxidant properties and the mechanism of antioxidant action of hydroxyflavones. *Free Radic. Biol. Med.* **2001**, *31*, 869–881. [CrossRef]
45. Nandasiri, R.; Semenko, B.; Wijekoon, C.; Suh, M. Air-frying is a better thermal processing choice for improving antioxidant properties of brassica vegetables. *Antioxidants* **2023**, *12*, 490. [CrossRef] [PubMed]
46. Bianchi, G.; Lo Scalzo, R. Characterization of hot pepper spice phytochemicals, taste compounds content and volatile profiles in relation to the drying temperature. *J. Food Biochem.* **2018**, *42*, e12675. [CrossRef]
47. Vega-Galvez, A.; Di Scala, K.; Rodríguez, K.; Lemus-Mondaca, R.; Miranda, M.; López, J.; Perez-Won, M. Effect of air-drying temperature on physico-chemical properties, antioxidant capacity, colour and total phenolic content of red pepper (*Capsicum annuum* L. var. Hungarian). *Food Chem.* **2009**, *117*, 647–653. [CrossRef]
48. Dills, W.L. Protein fructosylation: Fructose and the Maillard reaction. *Am. J. Clin. Nutr.* **1993**, *58*, 779S–787S. [CrossRef]
49. Rana, J.N.; Mumtaz, S.; Choi, E.H.; Han, I. ROS production in response to high-power microwave pulses induces p53 activation and DNA damage in brain cells: Radiosensitivity and biological dosimetry evaluation. *Front. Cell Dev. Biol.* **2023**, *23*, 1067861. [CrossRef]
50. Li, Y.; Lu, F.; Wang, X.; Hu, X.; Liao, X.; Zhang, Y. Biological transformation of chlorophyll-rich spinach (*Spinacia oleracea* L.) extracts under in vitro gastrointestinal digestion and colonic fermentation. *Food Res. Int.* **2021**, *139*, 109941. [CrossRef]
51. Darıcı, M.; Süfer, Ö.; Simsek, M. Determination of microwave drying and rehydration kinetics of green peppers with the bioactive and textural properties. *J. Food Process. Eng.* **2021**, *44*, e13755. [CrossRef]
52. Eyarkai, N.V.; Gupta, R.K.; Kumar, S.; Sharma, P.C. Degradation kinetics of bioactive components, antioxidant activity, colour and textural properties of selected vegetables during blanching. *J. Food Sci. Technol.* **2016**, *53*, 3073–3082. [CrossRef]
53. Jaiswal, A.K.; Gupta, S.; Abu, G.N. Kinetic evaluation of colour, texture, polyphenols and antioxidant capacity of Irish York cabbage after blanching treatment. *Food Chem.* **2012**, *131*, 63–72. [CrossRef]
54. Bhat, R. Impact of ultraviolet radiation treatments on the quality of freshly prepared tomato (*Solanum lycopersicum*) juice. *Food Chem.* **2016**, *213*, 635–640. [CrossRef]
55. Ibarz, A.; Pagan, J.; Garza, S. Kinetic models for colour changes in pear puree during heating at relatively high temperatures. *ZJ. Food Eng.* **1999**, *39*, 415–422. [CrossRef]
56. Swain, S.; Samuel, D.V.K.; Bal, L.M.; Kar, A. Thermal kinetics of colour degradation of yellow sweet pepper (*Capsicum annuum* L.) undergoing microwave assisted convective drying. *Int. J. Food Prop.* **2014**, *17*, 1946–1964. [CrossRef]

57. Deng, L.Z.; Yang, X.H.; Mujumdar, A.S.; Zhao, J.H.; Wang, D.; Qian, Z.; Wang, J.; Gao, Z.J.; Xiao, H.W. Red pepper (*Capsicum annuum* L.) drying: Effects of different drying methods on drying kinetics, physicochemical properties, antioxidant capacity, and microstructure. *Dry. Technol.* **2018**, *36*, 893–907. [CrossRef]
58. Kaur, M.; Singh, N.; Sodhi, N.S. Physicochemical, cooking, textural and roasting characteristics of chickpea (*Cicer arietinum* L.) cultivars. *J. Food Eng.* **2005**, *69*, 511–517. [CrossRef]
59. Vega-Galvez, A.; Lemus-Mondaca, R.; Bilboa-Sainz, C.; Yagnam, F.; Rojas, A. Mass transfer kinetics during convective drying of red pepper var. Hungarian (*Capsicum annuum* L.): Mathematical modeling and evaluation of kinetic parameters. *J. Food Process Eng.* **2008**, *31*, 120–137. [CrossRef]
60. Barut, Y.T.; Koç, G.Ç.; Ergün, A.R. Effect of different roasting methods on the proximate by composition, flow properties, amino acid compositions, colour, texture, and sensory profile of the chickpeas. *Int. J. Food Sci. Technol.* **2023**, *58*, 482–492. [CrossRef]
61. Yen, G.C.; Chen, H.Y. Antioxidant activity of various tea extracts in relation to their antimutagenicity. *J. Agric. Food Chem.* **1995**, *43*, 27–32. [CrossRef]
62. Klymenko, S.; Kucharska, A.Z.; Sokół-Łętowska, A.; Piórecki, N. Antioxidant activities and phenolic compounds in fruits of cultivars of cornelian cherry (*Cornus mas* L.). *Agrobiodivers. Improv. Nutr. Health Life Qual.* **2019**, *3*, 484–499.
63. Sridhar, K.; Charles, A.L. In vitro antioxidant activity of Kyoho grape extracts in DPPH and ABTS assays: Estimation methods for EC50 using advanced statistical programs. *Food Chem.* **2019**, *275*, 41–49. [CrossRef] [PubMed]
64. Re, R.; Pellegrini, N.; Proteggente, A.; Pannala, A.; Yang, M.; Rice-Evans, C. Antioxidant activity applying an improved ABTS radical cation decolorization assay. *Free Radic. Biol. Med.* **1999**, *26*, 1231–1237. [CrossRef] [PubMed]
65. Miller, G.L. Use of dinitrosalicylic acid reagent for determination of reducing sugar. *Anal. Chem.* **1959**, *31*, 426–428. [CrossRef]

molecules

Article

Effect of Different Time/Temperature Binomials on the Chemical Features, Antioxidant Activity, and Natural Microbial Load of Olive Pomace Paste

Maria Manuela Sousa [1,†], Diana Melo Ferreira [1,†], Susana Machado [1], Joana C. Lobo [1], Anabela S. G. Costa [1], Josman D. Palmeira [2], Maria Antónia Nunes [1], Rita C. Alves [1], Helena Ferreira [2] and Maria Beatriz P. P. Oliveira [1,*]

1 REQUIMTE/LAQV, Department of Chemical Sciences, Faculty of Pharmacy, University of Porto, R. Jorge de Viterbo Ferreira 228, 4050-313 Porto, Portugal
2 REQUIMTE/UCIBIO-i4HB, Laboratory of Microbiology, Department of Biological Sciences, Faculty of Pharmacy, University of Porto, R. Jorge de Viterbo Ferreira 228, 4050-313 Porto, Portugal
* Correspondence: beatoliv@ff.up.pt
† These authors contributed equally to this work.

Abstract: Olive pomace is a by-product from olive oil production that can be further processed to obtain olive pomace paste. In this work, the influence of different time/temperature binomials (65 °C/30 min; 77 °C/1 min; 88 °C/15 s; and 120 °C/20 min) on the nutritional quality, chemical composition, and efficiency on control/elimination of natural microbial load of olive pomace paste was ascertained. The treatments significantly impacted the contents of ash, fat, vitamin E, phenolics (including hydroxytyrosol), flavonoids, and antioxidant activity, but not the fatty acids profile. The binomial 88 °C/15 s showed the greatest potential since it better preserved the phytochemical and antioxidant properties as well as the protein and fiber contents. This binomial is also faster and easy to be implemented at an industrial level, allowing the obtention of a safe functional ingredient to satisfy consumers' demands for novel sustainable products, simultaneously, responding to food safety and food security concerns.

Keywords: olive pomace; food security; heat treatment; vitamin E; fatty acids; hydroxytyrosol

Citation: Sousa, M.M.; Ferreira, D.M.; Machado, S.; Lobo, J.C.; Costa, A.S.G.; Palmeira, J.D.; Nunes, M.A.; Alves, R.C.; Ferreira, H.; Oliveira, M.B.P.P. Effect of Different Time/Temperature Binomials on the Chemical Features, Antioxidant Activity, and Natural Microbial Load of Olive Pomace Paste. *Molecules* **2023**, *28*, 2876. https://doi.org/10.3390/molecules28062876

Academic Editors: Mihai Brebu and Prokopios Magiatis

Received: 10 February 2023
Revised: 12 March 2023
Accepted: 20 March 2023
Published: 22 March 2023

1. Introduction

The world population is expected to reach 9.1 billion people by 2050, implying that 30% more food will be necessary [1]. Consequently, the food industry faces an emergent challenge: to ensure food security while avoiding environmental depletion. A new program called "Transforming our world: the 2030 Agenda for Sustainable Development" aimed at a crucial goal, sustainable food consumption and production [2]. This emphasizes the duty to adopt more sustainable measures, such as adding value to agri-food by-products through the production of healthy food for the growing human population while preventing environmental and natural resources exhaustion.

Portugal was the fourth major olive oil producer in the European Union in 2019 [3]. Therefore, considering that on average 35–40 kg of olive pomace (OP) are produced for each 100 kg of processed olives [4], it is evident that this by-product affects the Portuguese economy and the environment. Three common extraction techniques can be used in olive oil production: traditional pressing mills, three-phase systems, and two-phase systems [4]. Traditional pressing mills are mainly used in small olive mills. In a three-phase system, large amounts of water are added to the olive paste, which leads to the worldwide production of 30 million m^3 of olive mill wastewater every year. The two-phase system is an eco-friendly system where no water is added. Indeed, in this system, the olive paste is centrifuged, resulting in the production of olive oil and OP. Taking into account that this system reduces wastewater production and allows the obtention of OP, the two-phase system is becoming the most used one [4].

OP is a semi-solid biomass composed of olives' pulp, skin, and small stone fragments [5]. The interest in this by-product is due to its lipid fraction (rich in α-tocopherol and oleic acid) and its considerable content of bioactive compounds, especially hydroxytyrosol [1]. However, to use this by-product as a fresh ingredient in food products, the remaining stone pieces must be removed, a process that generates a homogenous biomass called olive pomace paste (OPP).

Taking all of this into consideration, in the present work, OP from a two-phase system from Trás-os-Montes, Portugal, was used. The sample was a mixture of the following olive varieties: Cobrançosa, Cordovil, Madural, and Verdeal Transmontana. This work aimed to manually remove the OP's stone to obtain OPP, characterize the chemical features of OPP, and evaluate the impact of different heat treatments (65 °C/30 min, 77 °C/1 min, 88 °C/15 s, and 120 °C/20 min) on the nutritional quality (proximate analysis, vitamin E and fatty acids (FA) profiles, phytochemicals contents, and antioxidant activity) as well as its efficiency on the control/elimination of the microbiological load of OPP to select the best process for obtaining a functional ingredient for incorporation into foodstuffs to satisfy both consumers' demands for novel sustainable products and to answer current food security concerns.

Some previous studies have also evaluated the impact of drying olive pomace through other methods but with slight differences in relation to the present work. Ahmad-Qasem and colleagues (2013) dried olive pomace in a forced air laboratory drier at different temperatures (50, 70, 90, 120, and 150 °C), using a sample from Spain which included the olive pits [6]. Uribe et al., in 2013 and 2014, used a convective dryer at different temperatures (40, 50, 60, 70, 80, and 90 °C), using olive pomace from Chile [7,8]. Uribe et al. (2013) used a mixture of the following olive varieties: Frantoio, Leccino, Racimo, Barnea, and Picual [7], while Uribe and colleagues (2014) used only the Picual olive variety [8]. Pasten and fellow researchers (2019) used a sample of exhausted olive pomace (the remaining olive oil was extracted using solvents) from Chile, which was also dried in a laboratory-scale convective hot-air dryer at 40, 50, 60, 70, 80, and 90 °C [9].

2. Results and Discussion

This section presents the results of the proximate analysis, vitamin E and FA profiles, phytochemicals contents, antioxidant activity, and microbiological load of OPP after treatment with different time/temperature binomials. The impact of the heat treatments on the composition of OPP is also discussed. The manual removal of the crushed olive stones from 5.3 kg of OP resulted in 2.3 kg of OPP, so this process had a yield of 43%. Thus, stone pieces make up 57% of OP, a part that contributes to the weight of the samples, but not for analysis as the aim of this study is to obtain a paste for food purposes. Even though this process had low profitability, its escalation into an industrial level is already implemented in some olive mills, e.g., with a stainless sieve and stainless rolls, not to obtain OPP but to recover the stones, which are then sold as biomass. In addition, OPP is a new and alternative approach to use OP, which has been studied by other authors to be handled by drying [6–9] as mentioned before. The results are expressed both in dry weight (dw) to assess the impact of the different applied treatments, and in fresh weight (fw) to evaluate the quality attributes of this new food ingredient.

2.1. Proximate Analysis

OP is a heterogeneous biomass with a moisture content of 50–60 g/100 g and large amounts of minerals, dietary fiber, and oligosaccharides [10]. The results from proximate analysis (Table 1) confirmed that fresh OP had a moisture content of 60.9 g/100 g, also presenting a residual fat content of 1.4 g/100 g (fw), an ash content of 1.1 g/100 g (fw), and a high-fiber content of 17.2 g/100 g (fw). Regarding fresh OPP, a significant 1.2-fold increase in moisture and ash as well as a 1.7-fold increase in total fat, were registered in comparison to the results obtained for fresh OP ($p < 0.05$). However, significant decreases of 56% in carbohydrates and 27% in total fiber were observed, relatively to the fw results. The differences between OP and OPP can be explained by stone removal.

Table 1. Proximate composition of olive pomace samples.

g/100 g	Sample	Moisture	Total Protein	Ash	Total Fat	Total Fiber	Remaining Carbohydrates
Dry weight	OP	-	6.3 ± 0.8 c	2.7 ± 0.0 d	3.6 ± 0.1 c	44.0 ± 0.9 b	43.4 ± 1.6 a
	OPP	-	9.6 ± 0.6 a	4.9 ± 0.0 c	9.3 ± 0.3 a	48.0 ± 0.9 a	28.2 ± 1.8 c
	OPPA	-	8.4 ± 0.2 ab	5.3 ± 0.0 bc	8.1 ± 0.4 ab	44.0 ± 0.4 b	34.1 ± 0.2 b
	OPPB	-	8.6 ± 0.5 ab	5.4 ± 0.0 b	8.2 ± 0.2 ab	44.3 ± 0.1 b	33.5 ± 0.8 bc
	OPPC	-	8.9 ± 0.1 ab	5.2 ± 0.0 bc	7.0 ± 0.0 b	44.6 ± 0.4 b	34.4 ± 0.4 b
	OPPD	-	7.0 ± 0.6 bc	5.9 ± 0.3 a	6.8 ± 0.9 b	43.5 ± 0.2 b	36.8 ± 2.1 b
Fresh weight	OP	60.9 ± 0.3 d	2.5 ± 0.3 a	1.1 ± 0.0 d	1.4 ± 0.0 c	17.2 ± 0.3 a	16.9 ± 0.3 a
	OPP	73.9 ± 0.1 ab	2.5 ± 0.2 a	1.3 ± 0.0 c	2.4 ± 0.1 a	12.5 ± 0.2 b	7.4 ± 0.4 d
	OPPA	73.1 ± 0.0 bc	2.2 ± 0.1 a	1.4 ± 0.0 b	2.2 ± 0.1 ab	11.8 ± 0.1 bc	9.2 ± 0.1 bc
	OPPB	74.7 ± 0.3 a	2.2 ± 0.1 a	1.4 ± 0.0 bc	2.1 ± 0.0 ab	11.2 ± 0.0 c	8.5 ± 0.5 cd
	OPPC	73.9 ± 0.6 ab	2.3 ± 0.0 a	1.3 ± 0.0 bc	1.8 ± 0.0 bc	12.3 ± 0.1 b	8.3 ± 0.5 cd
	OPPD	72.4 ± 0.0 c	1.9 ± 0.2 a	1.6 ± 0.1 a	1.9 ± 0.3 bc	11.4 ± 0.1 c	10.8 ± 0.6 b

OP, olive pomace; OPP, olive pomace paste; OPPA, olive pomace paste processed at 65 °C, 30 min; OPPB, olive pomace paste processed at 77 °C, 1 min; OPPC, olive pomace paste processed at 88 °C, 15 s; OPPD, olive pomace paste processed at 120 °C, 20 min. The results are presented in g/100 g of sample in fresh or dry weight, as mean ± standard deviation (n = 3). Within each column, different letters represent significant differences ($p < 0.05$) between samples, for results expressed in fresh or dry weight, separately.

Uribe et al. (2013) showed that, after drying, a loss of protein occurs due to solubility changes or denaturation [7]. Denatured protein is likely to be involved in Maillard reactions, which are reactions that occur between reactive carbonyl groups (reducing sugars) and nucleophilic amino groups (amino acids, peptides, or proteins), resulting in melanoidins formation [7,11]. These reactions occur spontaneously in food exposed to heat, especially if the temperature is above 100 °C [11]. In this study, the applied heat treatments decreased the total protein contents in comparison to OPP (13% in OPPA, 10% in OPPB, 7% in OPPC, and 27% in OPPD, relatively to the dw results). Only OPPD reached temperatures above 100 °C and this was the treatment where the greatest protein loss was detected ($p < 0.05$). On the contrary, the lower impact of OPPA, OPPB, and OPPC on protein content can be explained by the use of temperatures below 100 °C.

The applied heat treatments had a positive impact in the ash content, especially OPPD, where a 1.2-fold increment was observed ($p < 0.05$), regarding the dw results. Total ash increment may be explained by minerals release from the organic matter with the processing temperatures. Indeed, it was reported that processing can enhance the bioaccessibility of some minerals (e.g., Ca and Fe) in vegetable foods [12]. Therefore, it would be interesting to access the individual mineral composition of all samples since OP is rich in minerals, namely, K [7,9,13].

OPP residual fat content decreased with the different heat treatments, especially with OPPC and OPPD, which led to a significant loss of 25% and 27%, respectively, regarding the dw results, according to Pasten et al. (2019), temperature could promote fat degradation by hydrolysis or oxidation [9].

The results also showed high total fiber contents in all samples (43.5–44.6 g/100 g), but OPP presented the highest amount (48 g/100 g dw). This feature, typical of olive cakes, is related to the presence of olive skin and pulp [8] and confirms that OP is a rich source of dietary fiber [8,9]. The reported healthy properties of fiber in preventing hyperglycemia, decreasing cholesterol levels, reducing colon cancer risk, and heart diseases [7,14], make this fresh heat-treated OPP a product of great interest to the food sector. Regarding the impact of heat treatments in the fiber contents, small but statistically significant decreases were verified, with a maximum loss of 9% (in OPPD), regarding the dw results. Dhingra et al. (2012) reported that processing could change the physicochemical characteristics of dietary fiber [14]. As an example, different heat treatments applied to wheat bran formed heat-resistant fiber-protein complexes [14]. This means that heat treatment is a promising tool to be considered, as it can lead to various changes in food products.

All in all, the use of this heat-treated by-product as a food ingredient could allow the development of foodstuffs with "low-fat" and "high-fiber" claims, considering the proximate analysis results (Table 1) and the UE Regulation (EC) No. 1924/2006 of the European Parliament and Council of 20 December 2006, on nutrition and health claims made on foods. Therefore, this by-product is an undeniably interesting ingredient in answer to the rising interest of consumers in functional foods [15].

2.2. *Vitamin E Profile*

The term "vitamin E" refers to a set of fat-soluble compounds with unique antioxidant properties crucial for health: α-, β-, γ-, and δ-tocopherols, and α-, β-, γ-, and δ-tocotrienols [16]. Total vitamin E content (Table 2) ranged from 4.4 to 6.1 mg/100 g (dw). If we compare the total vitamin E contents in dw of OP and OPP a 1.4-fold increment occurred ($p < 0.05$), which can be explained by stone removal as previously mentioned, and the consequent increase in the total fat content. However, the location of vitamin E vitamers in plant cells should be considered, for example, α-tocopherol is inside chloroplasts; while β- and γ-tocopherols are outside organelles [9]. Therefore, the physical method used to obtain OPP can lead to the rupture of some intact cells and consequent release of a higher quantity of the vitamers. During processing, vitamins are often damaged and lost, due to their susceptibility to environmental factors (e.g., light, temperature, and oxygen), which can affect their stability [17]. This means that food processing can result in vitamin E loss due to exposure to the aforementioned degrading factors. In this study, all heat treatments had a slight negative impact on total vitamin E content ($p < 0.05$), with decreases ranging from 6% to 13% in relation to OPP (dw). Taking these results into consideration and the processing conditions (OPPC was exposed for 15 s to oxygen; while OPPD was processed in a closed glass container), exposure to oxygen may play an essential role in vitamin E degradation, as lower exposure to this factor has resulted in a lower reduction of its amount (only 6% in OPPD). Additionally, vitamin E is a powerful chain-breaking antioxidant [16], protecting long-chain FAs from oxidation [1]. Therefore, the observed reductions in this parameter could also be related to PUFA preservation, which will be discussed in Section 2.3. α-Tocopherol was the major vitamer present in all samples, which was also reported by Nunes et al. (2018) and Pasten et al. (2019) [1,9]. The amounts in fresh samples are similar to the ones reported for other seed oils [18]: guariroba, tamarind, and pinha (1.2, 1.2, and 1.4 mg/100 g, respectively), meaning that this by-product could be an interesting source of α-tocopherol, which is linked to the prevention of lipid peroxidation and scavenging of lipid peroxyl radicals [1]. Despite being heat-stable, α-tocopherol is sensitive to oxidation [7], which can explain the lowest decrease in its content especially in OPPD (7%, dw) in relation to OPP. The processing conditions (minimum oxygen exposure in OPPD) seem to explain these results since α-tocopherol has a considerable susceptibility to oxygen as previously mentioned. The obtained results confirm that OPP is an extremely valuable food ingredient, since their vitamin E contents could promote healthy aging and have potential in preventing cancer, arthritis, and cataracts [16].

Table 2. Vitamin E profile of olive pomace samples.

µg/100 g	Sample	α-Tocopherol	α-Tocotrienol	β-Tocopherol	γ-Tocopherol	δ-Tocopherol	Total Vitamin E
Dry weight	OP	4133 ± 138 ^d^	62 ± 1 ^c^	50 ± 1 ^c^	97 ± 2 ^c^	17 ± 0 ^c^	4360 ± 143 ^d^
	OPP	5715 ± 227 ^a^	109 ± 1 ^b^	78 ± 2 ^b^	162 ± 6 ^ab^	37 ± 0 ^b^	6101 ± 234 ^a^
	OPPA	4906 ± 65 ^c^	130 ± 1 ^a^	73 ±1 ^b^	146 ± 4 ^b^	36 ± 0 ^b^	5290 ± 65 ^c^
	OPPB	4983 ± 84 ^c^	127 ± 5 ^a^	73 ± 1 ^b^	146 ± 3 ^b^	37 ± 0 ^b^	5365 ± 87 ^c^
	OPPC	5066 ± 42 ^bc^	112 ± 3 ^b^	71 ± 0 ^b^	148 ± 5 ^b^	32 ± 0 ^b^	5430 ± 45 ^bc^
	OPPD	5335 ± 93 ^b^	nd	106 ± 6 ^a^	165 ± 11 ^a^	115 ± 4 ^a^	5722 ± 112 ^b^
Fresh weight	OP	1614 ± 54 ^a^	24 ± 0 ^d^	20 ± 1 ^b^	38 ± 1 ^bc^	7 ± 0 ^c^	1703 ± 56 ^a^
	OPP	1489 ± 59 ^b^	28 ± 0 ^c^	20 ± 1 ^b^	42 ± 2 ^ab^	10 ± 0 ^b^	1590 ± 61 ^b^
	OPPA	1318 ±17 ^c^	35 ± 0 ^a^	20 ± 0 ^b^	39 ± 1 ^bc^	10 ± 0 ^b^	1421 ± 18 ^c^
	OPPB	1262 ± 21 ^c^	32 ± 1 ^b^	18 ± 0 ^b^	37 ± 1 ^c^	9 ± 0 ^b^	1358 ± 22 ^c^
	OPPC	1324 ± 11 ^c^	29 ± 1 ^c^	18 ± 0 ^b^	39 ± 1 ^bc^	8 ± 0 ^b^	1419 ± 12 ^c^
	OPPD	1472 ± 26 ^b^	nd	29 ± 2 ^a^	46 ± 3 ^a^	32 ± 1 ^a^	1579 ± 31 ^b^

OP, olive pomace; OPP, olive pomace paste; OPPA, olive pomace paste processed at 65 °C, 30 min; OPPB, olive pomace paste processed at 77 °C, 1 min; OPPC, olive pomace paste processed at 88 °C, 15 s; OPPD, olive pomace paste processed at 120 °C, 20 min; nd, not detected. The results are presented in µg/100 g of sample in fresh or dry weight, as mean ± standard deviation (n = 3). Within each column, different letters represent significant differences ($p < 0.05$) between samples, for results expressed in fresh or dry weight, separately.

2.3. *Fatty Acids Profile*

All samples evidenced a rich composition in oleic acid (73–75%, Table 3), followed by palmitic (11%) and linoleic (9–10%) acids. These results are in agreement with Nunes et al. (2018) and Uribe et al. (2013) [1,7]. In general, OP and OPP had similar FA profiles but some statistical differences were found when comparing monounsaturated fatty acids (MUFA) and polyunsaturated fatty acids (PUFA) sums ($p < 0.05$). A minimal increase in MUFA was observed as well as a slight decrease in PUFA, particularly in linoleic acid. Since PUFA are more vulnerable to lipid peroxidation [19], oxygen exposure during OPP production probably caused their loss. Drying is also a cause of changes in OP's FA profile since high temperatures can induce lipid hydrolysis or oxidation [9] as stated before. However, in this study, neither the most abundant FAs (oleic, palmitic, and linoleic) nor the MUFA/PUFA ratio were significantly affected by processing. In another study, OP drying also resulted in only minor differences in the FA profile [7]. These results could be explained by the presence of vitamin E that protects PUFA against oxidative damage (Section 2.2). To conclude, the residual fat of this by-product can present noticeable health benefits similar to olive oil: high contents of oleic acid, a FA associated with the inhibition of cholesterol synthesis [9]; and a considerable amount of linoleic acid, a FA linked to blood pressure control [7].

Table 3. Fatty acids profile of olive pomace samples.

Fatty Acids (Relative %)	OP	OPP	OPPA	OPPB	OPPC	OPPD
Myristic (C14:0)	0.03 ± 0.00 ^a^	0.02 ± 0.00 ^b^	0.02 ± 0.00 ^b^	0.02 ± 0.00 ^b^	0.03 ± 0.00 ^a^	0.03 ± 0.00 ^a^
Palmitic (C16:0)	11.18 ± 0.08 ^a^	11.18 ± 0.04 ^a^	11.24 ± 0.01 ^a^	11.30 ± 0.15 ^a^	11.24 ± 0.02 ^a^	11.25 ± 0.02 ^a^
Palmitoleic (C16:1)	0.59 ± 0.03 ^a^	0.60 ± 0.03 ^a^	0.63 ± 0.00 ^a^	0.60 ± 0.03 ^a^	0.63 ± 0.01 ^a^	0.64 ± 0.00 ^a^
Heptanoic (C17:0)	0.10 ± 0.00 ^a^	0.10 ± 0.00 ^a^	0.10 ± 0.00 ^a^	0.10 ± 0.01 ^a^	0.10 ± 0.00 ^a^	0.10 ± 0.00 ^a^
Stearic (C18:0)	2.82 ± 0.15 ^a^	2.85 ± 0.20 ^a^	2.79 ± 0.01 ^a^	2.84 ± 0.21 ^a^	2.81 ± 0.01 ^a^	2.79 ± 0.04 ^a^
Oleic (C18:1n9c)	73.07 ± 0.40 ^b^	74.69 ± 0.26 ^a^	74.37 ± 0.12 ^a^	74.41 ± 0.11 ^a^	74.29 ± 0.03 ^a^	74.24 ± 0.24 ^a^
Linoleic (C18:2n6c)	9.97 ± 0.47 ^a^	8.52 ± 0.33 ^b^	8.79 ± 0.13 ^b^	8.68 ±0.28 ^b^	8.85 ± 0.05 ^b^	8.91 ± 0.07 ^b^
Arachidic (C20:0)	0.51 ± 0.05 ^a^	0.47 ± 0.03 ^a^	0.45 ± 0.00 ^a^	0.46 ± 0.03 ^a^	0.46 ± 0.01 ^a^	0.45 ± 0.04 ^a^
α-linolenic (C18:3n3)	0.92 ± 0.10 ^a^	0.90 ± 0.10 ^a^	0.96 ± 0.01 ^a^	0.89 ± 0.07 ^a^	0.93 ± 0.02 ^a^	0.93 ± 0.02 ^a^
cis-11-Eicosenoic (C20:1n9)	0.38 ± 0.01 ^a^	0.34 ± 0.02 ^a^	0.35 ± 0.00 ^a^	0.35 ± 0.01 ^a^	0.35 ± 0.01 ^a^	0.35 ± 0.02 ^a^
Behenic (C22:0)	0.28 ± 0.04 ^a^	0.21 ± 0.02 ^b^	0.21 ± 0.00 ^b^	0.22 ± 0.03 ^ab^	0.21 ± 0.00 ^b^	0.20 ± 0.02 ^b^
Lignoceric (C24:0)	0.16 ± 0.02 ^a^	0.12 ± 0.01 ^b^	0.11 ± 0.00 ^b^	0.12 ± 0.00 ^b^	0.11 ± 0.01 ^b^	0.12 ± 0.01^b^
∑ SFA	15.07 ± 0.15 ^a^	14.97 ± 0.22 ^a^	14.91 ± 0.01 ^a^	15.07 ± 0.35 ^a^	14.95 ± 0.03 ^a^	14.93 ± 0.12 ^a^
∑ PUFA	10.89 ± 0.46 ^a^	9.42 ± 0.34 ^b^	9.74 ± 0.10 ^b^	9.57 ± 0.28 ^b^	9.78 ± 0.05 ^b^	9.85 ± 0.07 ^b^
∑ MUFA	74.03 ± 0.31 ^b^	75.64 ± 0.20 ^a^	75.34 ± 0.10 ^a^	75.36 ± 0.10 ^a^	75.27 ± 0.02 ^a^	75.22 ± 0.18 ^a^
MUFA/PUFA	6.81 ± 0.33 ^b^	8.04 ± 0.31 ^a^	7.73 ± 0.09 ^a^	7.88 ± 0.23 ^a^	7.70 ± 0.04 ^a^	7.64 ± 0.07 ^a^

OP, olive pomace; OPP, olive pomace paste; OPPA, olive pomace paste processed at 65 °C, 30 min; OPPB, olive pomace paste processed at 77 °C, 1 min; OPPC, olive pomace paste processed at 88 °C, 15 s; OPPD, olive pomace paste processed at 120 °C, 20 min; SFA, saturated fatty acids; MUFA, monounsaturated fatty acids; PUFA, polyunsaturated fatty acids. The results are expressed in relative % as mean ± standard deviation (n = 3) in dry weight. Within each line, different letters represent significant differences ($p < 0.05$) between samples.

2.4. Phytochemicals Contents and Antioxidant Activity

OP contains many functional compounds [10]. Indeed, its high phenolic content makes it phytotoxic [1]. The following phytochemical contents and antioxidant activity assays were accessed to better understand OPP's phytochemical composition (Table 4).

Table 4. Phytochemicals contents and antioxidant activity of olive pomace samples.

	Sample	TPC g GAE/100 g	TFC g CE/100 g	HTC g/100 g	FRAP g FSE/100 g	DPPH•-SA g TE/100 g
Dry weight	OP	3.08 ± 0.13 d	2.69 ± 0.03 d	0.36 ± 0.00 cd	4.43 ± 0.57 d	1.53 ± 0.06 bc
	OPP	4.09 ± 0.11 a	3.44 ± 0.03 a	0.65 ± 0.04 a	6.10 ± 0.28 a	1.84 ± 0.10 a
	OPPA	3.46 ± 0.14 c	2.80 ± 0.19 cd	0.35 ± 0.02 d	4.98 ± 0.11 c	1.38 ± 0.05 c
	OPPB	3.50 ± 0.16 c	2.88 ± 0.09 c	0.40 ± 0.03 cd	5.17 ± 0.17 bc	1.46 ± 0.04 c
	OPPC	3.81 ± 0.15 b	3.10 ± 0.08 b	0.42 ± 0.02 c	5.47 ± 0.26 b	1.66 ± 0.12 b
	OPPD	3.81 ± 0.13 b	3.32 ± 0.20 a	0.54 ± 0.03 b	6.10 ± 0.43 a	1.92 ± 0.14 a
Fresh weight	OP	1.20 ± 0.05 a	1.05 ± 0.01 a	0.14 ± 0.00 b	1.73 ± 0.22 a	0.60 ± 0.02 a
	OPP	1.07 ± 0.03 b	0.90 ± 0.01 b	0.17 ± 0.01 a	1.59 ± 0.07 a	0.48 ± 0.03 c
	OPPA	0.93 ± 0.04 d	0.75 ± 0.05 d	0.09 ± 0.00 c	1.34 ± 0.03 b	0.37 ± 0.01 d
	OPPB	0.89 ± 0.04 d	0.73 ± 0.02 d	0.10 ± 0.01 c	1.31 ± 0.04 b	0.37 ± 0.01 d
	OPPC	1.00 ± 0.04 c	0.81 ± 0.02 c	0.11 ± 0.00 c	1.43 ± 0.07 b	0.43 ± 0.03 c
	OPPD	1.05 ± 0.04 bc	0.92 ± 0.05 b	0.15 ± 0.01 ab	1.68 ± 0.12 a	0.53 ± 0.04 b

OP, olive pomace; OPP, olive pomace paste; OPPA, olive pomace paste processed at 65 °C, 30 min; OPPB, olive pomace paste processed at 77 °C, 1 min; OPPC, olive pomace paste processed at 88 °C, 15 s; OPPD, olive pomace paste processed at 120 °C, 20 min; TPC, total phenolics content; TFC, total flavonoids content; HTC, hydroxytyrosol content; FRAP, ferric reducing antioxidant power; DPPH•-SA, 2,2-diphenyl-1-picrylhydrazyl radical scavenging ability; GAE, gallic acid equivalents; CE, catechin equivalents; FSE, ferrous sulphate equivalents; TE, trolox equivalents. The results are presented as mean ± standard deviation (n = 3). Within each column, different letters represent significant differences ($p < 0.05$) between samples, for results expressed in fresh or dry weight, separately.

2.4.1. Total Phenolics Content (TPC)

TPC in olive oil reaches a maximum of 53 mg gallic acid equivalents (GAE)/100 g [20], this is because only 2% of the phenolic compounds pass to olive oil during its production, so 98% remain in OP in two-phase extraction systems [1], which explains the considerable TPC found in OP (3.08 g GAE/100 g dw) in the present study. The production of OPP resulted in an increase of 1.3 times in TPC (dw) explained by stone removal. Pasten et al. (2019) showed that drying could have a negative impact and drying for long periods can lead to phenols aerial oxidation and enzymatic degradation [9]. In this study, all the applied treatments resulted in a slight but significant decrease in the TPC, ranging from 7% to 15% (dw). Additionally, TPC reduction seems to be correlated to oxygen exposure as OPPA and OPPB registered higher losses (15% and 14%, respectively) and were the treatments with prominent oxygen exposure. On the contrary, a minor reduction of 7% was observed in OPPC and OPPD submitted to the minimum oxygen exposure. Kim et al. (2021) reported that heat treatment of apple puree in the presence of oxygen led to a 33% reduction in TPC, while the heat treatment without oxygen preserved the TPC values, allowing to conclude that oxidation may be the cause of phenol degradation, but it can also occur by enzymatic action [21]. Nowadays, the consumption of polyphenol-rich foodstuffs is correlated to the prevention of some diseases, e.g., Parkinson's, Alzheimer's, or diabetes [15]. Taking into consideration the considerable TPC of heat-treated OPP and the rising interest for foodstuffs enriched in natural antioxidants, the incorporation of this ingredient in new products has great potential. Indeed, Difonzo et al. (2021) reported that the incorporation of 5% and 10% OP powder in pasta allowed for improved TPC and antioxidant activity results [10].

2.4.2. Total Flavonoids Content (TFC)

Flavonoids are one of the main phenol groups present in olive oil by-products [9], hence the considerable TFC registered in OP (2.69 g catechin equivalents—CE/100 g) (Table 4). Their consumption is linked to a reduction of neuroinflammation, cognition improvement, and weakening of the symptoms of Alzheimer's [15]. Therefore, the use of fresh heat-treated OPP as a natural food ingredient seems promising. All fresh heat-treated samples registered higher contents than the ones found in other fruits, namely, banana, mango, and mandarin (24.8, 22.7, and 27.4 mg CE/100 g, respectively) [22]. OPP production allowed a 1.3-fold increment in TFC ($p < 0.05$), which can be explained by stones removal and a possible release resulting from OP processing. Flavonoids are also heat-sensitive compounds [23]. In fact, a previous study by Pasten et al. (2019) showed that OP drying resulted in a significant reduction of TFC [9]. Here, all the applied treatments resulted in slight losses ($p < 0.05$), ranging from 3% (OPPD) to 19% (OPPA), regarding the dw results. Moreover, it is noticeable that OPPC and OPPD had the lower impact, being the treatments where oxygen exposure is minimum, whereas the higher oxygen exposure in OPPA and OPPB may explain the lowest contents. These results suggest that oxygen exposure while processing should be minimized, as oxygen seems to be the main degradative factor of the studied antioxidants (vitamin E, TPC, and TFC).

2.4.3. Hydroxytyrosol Content (HTC)

Hydroxytyrosol (HT), the major polyphenol present in OP draws the attention of the food industry due to its health-promoting traits and antioxidant potential [1]. HT is obtained, during olives ripening, from the hydrolysis of oleuropein [5]. Furthermore, during olive crushing and malaxation to produce olive oil, most of the HT glucoside is degraded to HT [24]. The solubility of HT in water [25] and OP's high moisture content (61 g/100 g) can explain the considerable HTC in all samples (0.35–0.65 g/100 g dw), when compared to the ones found in virgin olive oils: 0.3–29.3 mg/kg [20]. The significant 1.8-fold increase in HTC with OPP production could be explained by stone removal and possible release from cells. HT is one of the most powerful natural antioxidants, and its ability to scavenge reactive species and disrupt peroxidation chain reactions is due to its ortho-diphenolic group [25]. However, its numerous hydroxyl groups are extremely vulnerable to air and light exposure [25]. Therefore, it is possible to foresee that the applied heat-treatments could result in its degradation. In fact, similar to what happened for TPC, a decrease in HTC was registered in all treatments, especially in OPPA with a significant loss of 46% (dw). OPPD had the lowest impact, resulting in a 17% reduction ($p < 0.05$, dw). Once again, the processing conditions may explain the highest impact in OPPA and OPPB since they are more exposed to oxygen. On the other hand, OPPD occurred in a closed vase container, being the sample protected from oxidation, which avoided HT degradation. The same justification can be used for OPPC that was submitted to low-time processing (15 s). HT provides numerous health benefits: cardioprotective, anti-inflammatory, antitumoral, and neuroprotective activities [25]. For example, the enrichment of biscuits with HT allowed a reduction of low-density lipoprotein (LDL) blood levels [26]. Difonzo et al. (2021) reported that a traditional Italian snack showed a higher HTC when it was enriched with 20% fermented OP [10]. Hence, incorporating heat-treated OPP in foodstuffs may provide both healthy and tasty options to consumers.

2.4.4. Ferric Reducing Antioxidant Power (Frap) and 2,2-Diphenyl-1-picrylhydrazyl Radical Scavenging Ability (DPPH$^•$-SA)

The antioxidant capacity of all samples was determined by FRAP and DPPH$^•$-SA methods. In FRAP assay, 2,4,6-tripyridyl-s-triazine (TPTZ) is reduced to a colored product. In DPPH$^•$-SA assay, a radical is neutralized by reduction (electron transference) or quenching (hydrogen transference) [27]. These two methods are often used together due to their complementarity. The considerable FRAP and DPPH$^•$-SA values found in OP (4.43 g ferrous sulphate equivalents–FSE/100 g and 1.53 g Trolox equivalents—TE/100 g dw, respectively) are explained by the considerable amounts of phenolics, HT, and flavonoids, which were obtained in the present study and previously discussed. Difonzo et al. (2021) reported that OP incorporation in bread allowed for an increase in the antioxidant capacity [10]. This supports that the antioxidant attributes of this by-product can be exploited in the development of food products, especially those produced to be consumed raw, such as pâtés and toppings, since temperature due to cooking might result in antioxidants loss as seen in the previous results (vitamin E, TPC, TFC, and HTC), which consequently reduces the antioxidant capacity as will be discussed below. Table 4 shows that OPP and OPPD registered the highest values for FRAP and DPPH$^\bullet$-SA assays (around 6 g FSE/100 g and 2 g TE/100 g dw, respectively). A significant loss of antioxidant capacity with heat treatments in both assays was registered in this study. These results were expected since antioxidant contents were also reduced due to heat, oxygen, and light sensibility. Pasten et al. (2019) also showed that OP drying resulted in a significant loss of antioxidants [9]. Moreover, Kim et al. (2021) showed that heat-treated apple puree in the presence of oxygen had considerably lower antioxidant activities than fresh apple: FRAP and DPPH$^\bullet$-SA methods were reduced 4% and 6%, respectively [21]. Considering the processing conditions of this study, once again, oxygen exposure seems to have the bigger impact on antioxidants since OPPA and OPPB registered the lowest values in both assays. The minimal oxygen exposure can explain the antioxidant capacity registered in OPPD and OPPC in both assays. Nevertheless, there was a slight enhancement of the antioxidant properties in OPPD in FRAP and DPPH$^\bullet$-SA methods (fw), which could be explained by the formation of melanoidins, which are a result of Maillard reactions, as significant protein loss was also observed (Section 2.1).

2.5. Microbiological Analysis

Olive by-products exhibit antimicrobial activity against pathogenic bacteria and fungi strains due to phenolics' presence [28]. Therefore, considering the large TPC registered in all samples, OPP could exhibit antimicrobial activity even after a heat treatment. A microbiological analysis was still mandatory to ensure the effectiveness of the applied heat treatments to reduce the natural microbiota of this by-product and, therefore, guarantee consumers' safety. Indeed, heat treatment is an economic process that aims to reduce or destroy microorganisms, vegetative cells, yeasts, and molds. The total number of microorganisms was counted at 22 °C to understand the diversity of bacteria present at typical ambient temperatures and at 37 °C to encourage the growth of bacteria that can grow at body temperature [29]. According to the results (Table 5), all treatments effectively reduced the natural microbiological load of OPP, especially OPPA, OPPC, and OPPD. The results from OPPA, OPPB, and OPPC were expected to simulate pasteurization, a heat treatment which inactivates the non-spore-forming bacteria and the majority of vegetative spoilage microorganisms [30]. The treatment of OPPB was able to reduce the number of microorganisms in comparison to the control samples, but did not eliminate them entirely in the assay at 37 °C as happened in the other treatments, showing that it was not so effective. The effectiveness of OPPD was also expected since it replicates heat-sterilization, a successful process that kills all forms of microorganisms [31]. To conclude, the applied treatments are of great interest to ensure that OPP meets safety and quality standards and also allows shelf-life extension, being imperative to the sustainability of the food chain supply.

Table 5. Total count of microorganisms at 22 °C and 37 °C (48 h) of olive pomace samples.

Temperature	Sample	Dilution	Total Count of Microorganisms (CFU)
22 °C	OP	10^{-2}	3.6×10^3
	OPP	10^{-2}	4.4×10^3
	OPPA	10^{-1}	Ø
	OPPB	10^{-1}	Ø
	OPPC	10^{-1}	Ø
	OPPD	10^{-1}	Ø
37 °C	OP	10^{-1}	1.2×10^3
	OPP	10^{-1}	1.2×10^3
	OPPA	10^{-1}	Ø
	OPPB	10^{-1}	5.3×10^2
	OPPC	10^{-1}	Ø
	OPPD	10^{-1}	Ø

OP, olive pomace; OPP, olive pomace paste; OPPA, olive pomace paste processed at 65 °C, 30 min; OPPB, olive pomace paste processed at 77 °C, 1 min; OPPC, olive pomace paste processed at 88 °C, 15 s; OPPD, olive pomace paste processed at 120 °C, 20 min; CFU, colony forming unit; Ø, result below 30 CFU.

3. Materials and Methods

3.1. Chemicals

Ultra-pure water was obtained in a Milli-Q purification system (Millipore, Bedford, MA, USA). Chemicals and reagents were of analytical grade. Kjeldahl tablets, absolute ethanol, sodium carbonate (Na_2CO_3) decahydrate, sulfuric acid, sodium hydroxide (NaOH), *n*-hexane, and anhydrous sodium sulfate (Na_2SO_4) were obtained from Merck (Darmstadt, Germany). HPLC-grade solvents were acquired from Chem-Lab (Zedelgem, Belgium) and Merck (Darmstadt, Germany). The tocopherols and tocotrienols standards and tocol were obtained from Calbiochem (La Jolla, CA, USA) and Matreya Inc. (Pleasant Gap, PA, USA). Folin–Ciocalteus' reagent, gallic acid, catechin, heptahydrate ferrous sulphate, 2,2-diphenyl-1-picrylhydrazyl radical (DPPH$^{\bullet}$), Trolox, ferric chloride ($FeCl_3$), total dietary fiber assay kit, 2,4,6-tripyridyl-s-triazine (TPTZ), sodium nitrite ($NaNO_2$), aluminum chloride ($AlCl_3$), sodium acetate, boron trifluoride (BF_3) in methanol solution, butylated hydroxytoluene (BHT), Supelco 37 FAME Mix, and hydroxytyrosol (HT) standard were acquired from Sigma-Aldrich (St. Louis, MI, USA). Boric acid (4%) and potassium hydroxide (KOH) were purchased from Panreac (Barcelona, Spain). Methanol was acquired from Honeywell International, Inc. (Morris Plains, NJ, USA). Sodium chloride (NaCl), chloroform, and sand were acquired from VWR Chemicals (Alfragide, Portugal). Petroleum ether was purchased from Carlo Erba Reagents (Val de Reuil, France). Ethanol (96%) and dichloromethane were acquired from AGA (Prior Velho, Portugal) and Honeywell l Riedel-de Haën TM (Seelze, Germany), respectively. Total microorganism count was performed with Plate Count Agar, Liofilchem (Teramo, Italy).

3.2. Sample Preparation

The sample of OP (5.3 kg, Figure 1) was obtained from a two-phase extraction olive mill (Trás-os-Montes, Portugal). The sample was a mixture of the following olive varieties: Cobrançosa, Cordovil, Madural, and Verdeal Transmontana.

In the laboratory, after homogenization, the remaining crushed olive stones were manually removed using a stainless-steel sieve to obtain the OPP (Figure 2).

Then, OPP was homogenized and divided into 250 g samples, which were submitted to the different heat treatments (Table 6) to eliminate the natural microbiological load.

The designated treatments—OPPA, OPPB, and OPPC—were selected according to Fellows (2009) [32], and were performed in a Thermomix TM 31 (Vorwerk, Wuppertal, Germany, Figure 3a). The OPPD treatment took place in a SANYO Labo autoclave (Gemini BV, Apeldoorn, The Netherlands, Figure 3b). Control samples (OP and OPP) were not submitted to any heat treatment. All six samples (OP, OPP, OPPA, OPPB, OPPC, and OPPD) were stored at −80 °C separately, and lyophilized (Telstar Cryodos-80 Terrassa, Barcelona, Spain, Figure 3c).

Figure 1. Olive pomace sample.

Figure 2. Olive pomace paste.

Table 6. Time/temperature binomials applied to the olive pomace paste (OPP).

Samples	Temperature (°C)	Time
OPPA	65	30 min
OPPB	77	1 min
OPPC	88	15 s
OPPD	120	20 min

(a) (b) (c)

Figure 3. Equipment used for processing the olive pomace paste: (**a**) Thermomix TM 31; (**b**) SANYO Labo autoclave; (**c**) Telstar Cryodos-80 Terrassa.

3.3. Proximate Composition

Proximate analysis was carried out according to official methods [33]: total ash was analyzed in a muffle furnace at 500 °C (920.153 method); total fat was extracted with petroleum ether (991.36 method); total protein was determined by the Kjeldahl method (928.08 method), using 6.25 as the nitrogen conversion factor [34]; and total fiber was assessed by enzymatic-gravimetric procedures (985.29 method). Moisture content was determined in an infrared balance (DBS—KERN & SOHN GmbH, Balingen, Germany). The remaining carbohydrates were calculated by difference [34]. The results are presented in g/100 g of sample, both in fw and dw.

3.4. Vitamin E Profile

The lipid fraction of the samples was extracted according to Alves et al. (2009) [35], with minor modifications by Ferreira et al. (2023) [36]. An appropriate amount of sample was weighted to Falcon tubes to obtain 20 mg of fat, then 75 µL of 0.1% BHT solution (m/V), 50 µL of tocol (0.1 mg/mL, internal standard), and 1 mL of absolute ethanol were added. Tubes were stirred for 30 min (Heidolph Multi Reax Vibrating Shaker, VWR International, Radnor, PA, USA). Then, 2 mL of *n*-hexane was added. Tubes were stirred again (30 min) and 1 mL of 1% NaCl solution (*m*/*v*) was added, followed by agitation (2 min) and centrifugation (5000 rpm, 5 min). The supernatant was transferred to new tubes. A re-extraction was performed by adding 2 mL of *n*-hexane, followed by agitation (30 min) and centrifugation (5000 rpm, 5 min). Next, the supernatant was transferred to the second tube and anhydrous Na_2SO_4 was added. Tubes were centrifuged (5000 rpm, 5 min), the supernatant was transferred to new tubes, and concentrated under a nitrogen stream until 1 mL. Next, 500 µL of the supernatant was transferred to an amber injection vial for vitamin E profile analysis in a HPLC-DAD-FLD system (Jasco, Tokyo, Japan) equipped with an MD-2015 multiwavelength diode array detector coupled to an FP-2020 fluorescence detector (Jasco, Tokyo, Japan) programmed for excitation and emission at 290 and 330 nm, respectively. Compound separation was accomplished in a normal phase Supelcosil TM LC-SI column (75 mm × 3.0 mm, 3.0 µm, Supelco, Bellefonte, PA, USA). The eluent was 1.2% of 1,4-dioxane in *n*-hexane (*v*/*v*)—isocratic elution. The flow rate and injection volume were 0.7 mL/min and 20 µL, respectively. The UV spectra was used to identify the 8 vitamers previously mentioned. The comparison of their retention times with those of standards was used for quantification, using fluorescence signals and calibration curves obtained with standard stock solutions of each vitamer (α-tocopherol: $y = 0.159x - 0.007$, $R^2 = 0.9991$, 0.22–10.95 µg/mL; β-tocopherol: $y = 0.203x - 0.015$, $R^2 = 0.9995$, 0.21–10.38 µg/mL; γ-tocopherol: $y = 0.177x - 0.015$, $R^2 = 0.9994$, 0.21–13.48 µg/mL; δ-tocopherol: $y = 0.226x - 0.018$, $R^2 = 0.9995$, 0.23–11.71 µg/mL; α-tocotrienol: $y = 0.188x - 0.015$, $R^2 = 0.9990$, 0.19–9.65 µg/mL; β-tocotrienol: $y = 0.147x - 0.013$, $R^2 = 0.9994$, 0.25–12.36 µg/mL; γ-tocotrienol: $y = 0.167x - 0.021$, $R^2 = 0.9991$, 0.24–12.17 µg/mL, and δ-tocotrienol: $y = 0.177x + 0.016$, $R^2 = 0.9996$, 0.24–12.18 µg/mL). The results are presented in µg/100 g of sample in fw and dw.

3.5. Fatty Acids Profile

FAs were derivatized to fatty acid methyl esters (FAME), according to ISO 12966-2:2017 [37]. Briefly, the remaining 500 µL of the lipid fraction obtained in Section 3.4. were evaporated under a nitrogen stream and resuspended in 1 mL of dichloromethane. Afterwards, 1.5 mL of 0.5 M KOH in methanol was added, and tubes were stirred and placed in a heating block (100 °C, 10 min, Stuart SBH130D/3 block heater, Stafford, UK). After cooling (ice, 5 min), 2 mL of 14% BF_3 in methanol was added followed by another heating (100 °C, 30 min) and cooling (ice, 5 min) period. After adding 2 mL of deionized water and 4 mL of *n*-hexane, tubes were centrifuged (3000 rpm, 5 min). The supernatant was mixed with anhydrous Na_2SO_4, centrifuged (3000 rpm, 5 min), and 1 mL was transferred to injection vials for FA profile analysis in a gas chromatograph coupled to a flame ionization detector (GC-FID, Shimadzu GC-2010 Plus, Shimadzu, Tokyo, Japan) and a split/splitless AOC-20i auto-injector (Shimadzu, Tokyo, Japan). FAME separation was achieved in a

CP-Sil 88 silica capillary column (50 m × 0.25 mm i.d.; 0.20 μm film thickness, Varian, Middelburg, The Netherlands), using helium as carrier gas. The temperature program used was: 120 °C for 5 min; increase to 160 °C at 2 °C/min; hold for 15 min; and increase to 220 °C at 2 °C/min. Injector and detector were at 250 °C and 270 °C, respectively. A split ratio of 1:50 and an injection volume of 1.0 μL were used. FAME identification was carried out by comparing their retention times with those of standards (Supelco 37 Component FAME Mix, Supelco, Bellefonte, PA, USA). Data were analyzed based on relative peak areas, being the results expressed in relative % of each FA.

3.6. Phytochemicals Contents and Antioxidant Activity

For extracts preparation, 125 mg of each sample was mixed with 50 mL of methanol and deionized water (80/20; *v*/*v*) solution, at 40 ± 2 °C, under constant stirring (600 rpm, 60 min, MS-H-S10 magnetic stirrer, ChemLand, Stargard, Poland), in triplicate. Extracts were filtered (Whatman No. 4 filter paper) and stored at −20 °C until further analysis.

3.6.1. Total Phenolics Content (TPC)

TPC was determined according to Ferreira et al. (2023) [36]. In a microplate, 30 μL of each extract (Section 3.6) was mixed with 150 μL of Folin–Ciocalteu's reagent (1:10) and 120 μL of 7.5% (*m*/*V*) Na_2CO_3. The microplate was incubated at 45 °C for 15 min, followed by 30 min at room temperature (RT), light-protected. The sample's absorbance was measured in a microplate reader (Synergy HT GENS5, BioTek Instruments, Inc., Winooski, VT, USA) at 765 nm. A gallic acid calibration curve (y = 0.009x + 0.006, R^2 = 0.999, 5–100 mg/L) was used for quantification. Results are expressed in g of GAE/100 g of sample in fw and dw.

3.6.2. Total Flavonoids Content (TFC)

TFC was evaluated according to Ferreira et al. (2023) [36]. Thus, 1 mL of each extract (Section 3.6) was mixed with 4 mL of deionized water and 300 μL of 5% $NaNO_2$ (*m*/*V*). After 5 min at RT, 300 μL of 10% $AlCl_3$ (*m*/*v*) was added to the previous mixture. Later, after incubation at RT (1 min), 2 mL of 1 M NaOH and 2.5 mL of deionized water were added. The absorbance was measured at 510 nm in a microplate reader (Synergy HT GENS5, BioTek Instruments, Inc., Winooski, VT, USA). A catechin calibration curve (y = 0.002x + 0.001, R^2 = 0.998, 2.5–400 mg/L) was used for quantification. The sample's TFC is expressed in g of CE/100 g of sample in fw and dw.

3.6.3. Hydroxytyrosol Content (HTC)

HT analysis was carried out with 1 mL of each extract (Section 3.6) in an HPLC-DAD-FLD system (Jasco, Tokyo, Japan), consisting of a LC-NetII/ADC hardware interface, a pump (Jasco PU-2089), an automatic sampler (Jasco AS-2057 Plus), a multiwavelength diode array detector (Jasco MD-2018 Plus) coupled to a fluorescence detector (Jasco FP-2020 Plus) and a column thermostat (Jasco CO-2060 Plus). HT was evaluated by fluorescence and monitored at λ excitation and λ emission of 280 and 330 nm, respectively. A gradient elution program using as solvents acetic acid (A, 1%) and methanol (B, 100%) was employed: 0 min, 5% B; 30 min, 25% B; 50 min, 75% B; 55 min, 100% B; 60 min, 100% B; 63 min, 5% B. A Zorbax-SB-C18 (250 × 4.6 mm, 5 μm, Agilent Technologies, Amstelveen, The Netherlands) chromatographic column was used, at 20 °C, with a flow rate of 1 mL/min and an injection volume of 20 μL. A HT calibration curve was obtained (y = 10147x + 3486.5, R^2 = 0.9998, 0.25–200 μg/mL). Results are presented in g/100 g of sample in fw and dw.

3.6.4. Ferric Reducing Antioxidant Power (FRAP) and 2,2-Diphenyl-1-picrylhydrazyl Radical Scavenging Ability ($DPPH^\bullet$-SA)

FRAP assay was determined according to Ferreira et al. (2023) [36]. In a microplate, aliquots of 35 μL of each extract (Section 3.6) were mixed with 265 μL of the FRAP reagent: acetate buffer (0.3 M), TPTZ solution (10 mM), and $FeCl_3$ (20 mM). The microplate was incubated (37 °C, 30 min), light-protected, and the absorbance was measured in a microplate reader at 595 nm (Synergy HT GENS5, BioTek Instruments, Inc., Winooski, VT, USA). A calibration curve with ferrous sulphate (y = 0.002x + 0.080, R^2 = 0.999, 25–500 mg/L) was used for quantification. Results were expressed as g of FSE/100 g of sample in fw and dw.

DPPH•-SA assay was determined according to Ferreira et al. (2023) [36]. In a microplate, aliquots of 30 μL of each extract (Section 3.6) were mixed with 270 μL of fresh DPPH• solution in ethanol (6×10^{-2} mM). Absorbance (525 nm) was measured in a microplate reader (Synergy HT GENS5, BioTek Instruments, Inc., Winooski, VT, USA) every 2 min until reaction endpoint at 20 min to assess the kinetics of the reaction. A Trolox calibration curve was obtained (y = −0.007x + 0.540, R^2 = 0.999, 5.62–175.34 mg/L) and results presented in g of TE/100 g of sample in fw and dw.

3.7. Microbiological Analysis

The samples were serially diluted until 10^{-8} dilution, with ultrapure sterile water. Then, in all dilutions (from 10^{-1} to 10^{-8}), the total count of microorganisms was determined to evaluate the efficacy of the applied processes in the microbiological load reduction. The total count of microorganisms was achieved by the pour-plate method, meaning that 1 mL of each sample dilution was poured into different sterile Petri plate dishes, and then mixed with 20 mL of liquid plate count agar (PCA) cooled to about 50 °C. After solidification, plates were incubated at two different temperatures (22 °C and 37 °C) for 48 h (n = 3). In the analysis of the results, only the visual counting of the colony-forming units (CFU) that yield between 30 and 300 CFU was considered. The results obtained for the different heat-treatments (OPPA, OPPB, OPPC, and OPPD) were compared with OP and OPP (without treatment) to assess the efficiency of the applied treatments in the reduction of microbiological load. The results were expressed as the mean of CFU for each incubation temperature.

3.8. Statistical Analysis

Statistical analysis was performed using IBM SPSS v. 25 (IBM Corp., Armonk, NY, USA). Data are expressed as mean ± standard deviation. Significant differences between samples were assessed by one-way ANOVA, followed by Tukey's HSD to make pairwise comparisons between means. The level of significance for all hypothesis tests (p) was 0.05.

4. Conclusions

OPP (obtained from OP, a by-product of olive oil production) presented high moisture, residual fat, and considerable fiber contents. Moreover, its lipid fraction can be considered a quality attribute as a source of vitamin E (especially α-tocopherol), oleic acid (a MUFA related to healthy characteristics and food stability), and linoleic acid (a PUFA known for reducing total and LDL cholesterol blood levels). Additionally, it showed high contents of total phenolics, particularly hydroxytyrosol, and total flavonoids, allowing the development of food products with natural antioxidant properties. All of this is in agreement with new consumer trends: the search for functional foods rich in bioactive compounds.

In this study, OPP was submitted to four different heat treatments. All effectively reduced the microbial load, meaning that the use of heat-treated OPP as a functional ingredient will not compromise consumers' health. Overall, the selected treatments had a negative impact on the quality attributes of OPP, especially in vitamin E, namely, α-tocopherol, total phenolics, particularly hydroxytyrosol, total flavonoids, and antioxidant capacity. Nevertheless, none of them significantly affected OPP's FA profile, which remained a source of beneficial FA (oleic and linoleic acids).

An interesting conclusion of this study is that the heat treatment should be carefully planned and evaluated, especially in relation to oxygen exposure, while processing food products to maintain their nutritional quality. Indeed, undesirable changes in the antioxidant quality occurred with heat-treatment that can be related to the presence of oxygen. Therefore, oxygen exposure should be limited.

From the applied heat-treatments, OPPC (88 °C/15 s) and OPPD (120 °C/20 min) stood out due to the lower impact on the previously mentioned bioactive compounds and antioxidant properties. However, foreseeing the sustainable development and the cost of processing, OPPC was selected as the best treatment to apply for this product at an industrial level for further incorporation into foodstuffs as it is the fastest method and completely eliminated the microbiological load.

To conclude, even after heat treatment, OPP remains an interesting nutritional biomass due to its contents of total dietary fiber, vitamin E, and other bioactives. Therefore, incorporating heat-treated OPP at 88 °C for 15 s in new functional food products seems promising due to the health-related characteristics. Moreover, it contributes to the circular economy and sustainability of the olive oil sector, since it allows the use of OP, a by-product that otherwise represents an environmental burden.

Author Contributions: Conceptualization, M.M.S. and M.B.P.P.O.; methodology, D.M.F., S.M., A.S.G.C., R.C.A. and J.D.P.; validation, D.M.F., S.M., H.F. and M.A.N.; formal analysis, M.M.S., D.M.F. and J.D.P.; investigation, M.M.S., D.M.F., J.C.L., S.M. and J.D.P.; resources, A.S.G.C., H.F. and M.B.P.P.O.; data curation, M.M.S., D.M.F. and J.D.P.; writing—original draft preparation, M.M.S.; writing—review and editing, D.M.F. and R.C.A.; visualization, M.M.S. and D.M.F.; supervision, H.F. and M.B.P.P.O.; project administration, H.F. and M.B.P.P.O.; funding acquisition, H.F. and M.B.P.P.O. All authors have read and agreed to the published version of the manuscript.

Funding: This research was funded by FCT/MCTES (Fundação para a Ciência e Tecnologia/Ministério da Ciência, Tecnologia e Ensino Superior, Portugal) through the projects EXPL/SAU-NUT/0370/2021—Unveiling natural microbiota of fresh olive pomace for a new fermented food of potential synbiotic action, UIDB/50006/2020 and UIDP/50006/2020; and by the project AgriFood XXI I&D&I (NORTE-01-0145-FEDER-000041) cofinanced by the European Regional Development Fund, through the NORTE 2020 (Programa Operacional Regional do Norte 2014/2020).

Institutional Review Board Statement: Not applicable.

Informed Consent Statement: Not applicable.

Data Availability Statement: Not applicable.

Acknowledgments: D.M.F. is thankful for the PhD grant from FCT/MCTES (2022.13375.BD). S.M. is grateful to the project PTDC/SAU-NUT/2165/2021 for her research grant. J.C.L. acknowledges the grant provided by LAQV—Tecnologias e Processos Limpos—UIDB/50006/2020 (REQUIMTE 2018-11). J.D.P. thanks the project EXPL/SAU-NUT/0370/2021 for his researcher contract. R.C.A. thanks FCT for funding through the Scientific Employment Stimulus—Individual Call (CEECIND/01120/2017 contract).

Conflicts of Interest: The authors declare no conflict of interest.

Sample Availability: Samples of olive pomace are not available from the authors upon request, but they can be provided by the olive oil producer during the seasonal harvest period.

References

1. Nunes, M.A.; Costa, A.S.G.; Bessada, S.; Santos, J.; Puga, H.; Alves, R.C.; Freitas, V.; Oliveira, M.B.P.P. Olive pomace as a valuable source of bioactive compounds: A study regarding its lipid- and water-soluble components. *Sci. Total Environ.* **2018**, *644*, 229–236. [CrossRef]
2. Nations, U. Transforming Our World: The 2030 Agenda for Sustainable Development. Available online: https://sdgs.un.org/sites/default/files/publications/21252030%20Agenda%20for%20Sustainable%20Development%20web.pdf (accessed on 7 February 2023).
3. Commission, E. Olive Oil: Detailed Information on the Market Situation, Price Developments, Balance Sheets, Production and Trade. Available online: https://ec.europa.eu/info/food-farming-fisheries/plants-and-plant-products/plant-products/olive-oil_en (accessed on 7 February 2023).
4. Nunes, M.A.; Pimentel, F.B.; Costa, A.S.; Alves, R.C.; Oliveira, M.B.P. Olive by-products for functional and food applications: Challenging opportunities to face environmental constraints. *Innov. Food Sci. Emerg. Technol.* **2016**, *35*, 139–148. [CrossRef]
5. Nunes, M.A.; Páscoa, R.N.M.J.; Alves, R.C.; Costa, A.S.G.; Bessada, S.; Oliveira, M.B.P.P. Fourier transform near infrared spectroscopy as a tool to discriminate olive wastes: The case of monocultivar pomaces. *Waste Manag.* **2020**, *103*, 378–387. [CrossRef]

6. Ahmad-Qasem, M.H.; Barrajon-Catalan, E.; Micol, V.; Cárcel, J.A.; Garcia-Perez, J.V. Influence of air temperature on drying kinetics and antioxidant potential of olive pomace. *J. Food Eng.* **2013**, *119*, 516–524. [CrossRef]
7. Uribe, E.; Lemus-Mondaca, R.; Vega-Gálvez, A.; López, L.A.; Pereira, K.; López, J.; Ah-Hen, K.; Di Scala, K. Quality Characterization of Waste Olive Cake During Hot Air Drying: Nutritional Aspects and Antioxidant Activity. *Food Bioprocess Technol.* **2013**, *6*, 1207–1217. [CrossRef]
8. Uribe, E.; Lemus-Mondaca, R.; Vega-Gálvez, A.; Zamorano, M.; Quispe-Fuentes, I.; Pasten, A.; Di Scala, K. Influence of process temperature on drying kinetics, physicochemical properties and antioxidant capacity of the olive-waste cake. *Food Chem.* **2014**, *147*, 170–176. [CrossRef]
9. Pasten, A.; Uribe, E.; Stucken, K.; Rodríguez, A.; Vega-Gálvez, A. Influence of Drying on the Recoverable High-Value Products from Olive (cv. Arbequina) Waste Cake. *Waste Biomass Valorization* **2019**, *10*, 1627–1638. [CrossRef]
10. Difonzo, G.; Troilo, M.; Squeo, G.; Pasqualone, A.; Caponio, F. Functional compounds from olive pomace to obtain high-added value foods—A review. *J. Sci. Food Agric.* **2021**, *101*, 15–26. [CrossRef]
11. Ruan, D.; Wang, H.; Cheng, F.; Ruan, D.; Wang, H.; Cheng, F. *The Maillard Reaction*; Springer: Berlin/Heidelberg, Germany, 2018.
12. Cilla, A.; Barberá, R.; López-García, G.; Blanco-Morales, V.; Alegría, A.; Garcia-Llatas, G. Impact of processing on mineral bioaccessibility/bioavailability. In *Innovative Thermal and Non-Thermal Processing, Bioaccessibility and Bioavailability of Nutrients and Bioactive Compounds*; Elsevier: Amsterdam, The Netherlands, 2019; pp. 209–239.
13. Nunes, M.A.; Pawlowski, S.; Costa, A.S.G.; Alves, R.C.; Oliveira, M.B.P.P.; Velizarov, S. Valorization of olive pomace by a green integrated approach applying sustainable extraction and membrane-assisted concentration. *Sci. Total Environ.* **2019**, *652*, 40–47. [CrossRef]
14. Dhingra, D.; Michael, M.; Rajput, H.; Patil, R. Dietary fibre in foods: A review. *J. Food Sci. Technol.* **2012**, *49*, 255–266. [CrossRef]
15. de Araújo, F.F.; de Paulo Farias, D.; Neri-Numa, I.A.; Pastore, G.M. Polyphenols and their applications: An approach in food chemistry and innovation potential. *Food Chem.* **2020**, *338*, 127535. [CrossRef]
16. Rizvi, S.; Raza, S.T.; Ahmed, F.; Ahmad, A.; Abbas, S.; Mahdi, F. The role of vitamin E in human health and some diseases. *Sultan Qaboos Univ. Med. J.* **2014**, *14*, e157.
17. Dhakal, S.P.; He, J. Microencapsulation of vitamins in food applications to prevent losses in processing and storage: A review. *Food Res. Int.* **2020**, *137*, 109326. [CrossRef]
18. Shahidi, F.; Camargo, A.C.D. Tocopherols and Tocotrienols in Common and Emerging Dietary Sources: Occurrence, Applications, and Health Benefits. *Int. J. Mol. Sci.* **2016**, *17*, 1745. [CrossRef]
19. Ayala, A.; Muñoz, M.F.; Argüelles, S. Lipid Peroxidation: Production, Metabolism, and Signaling Mechanisms of Malondialdehyde and 4-Hydroxy-2-Nonenal. *Oxidative Med. Cell. Longev.* **2014**, *2014*, 360438. [CrossRef]
20. Bayram, B.; Esatbeyoglu, T.; Schulze, N.; Ozcelik, B.; Frank, J.; Rimbach, G. Comprehensive Analysis of Polyphenols in 55 Extra Virgin Olive Oils by HPLC-ECD and Their Correlation with Antioxidant Activities. *Plant Foods Hum. Nutr.* **2012**, *67*, 326–336. [CrossRef]
21. Kim, A.-N.; Lee, K.-Y.; Rahman, M.S.; Kim, H.-J.; Kerr, W.L.; Choi, S.-G. Thermal treatment of apple puree under oxygen-free condition: Effect on phenolic compounds, ascorbic acid, antioxidant activities, color, and enzyme activities. *Food Biosci.* **2021**, *39*, 100802. [CrossRef]
22. Silva, K.D.R.R.; Sirasa, M.S.F. Antioxidant properties of selected fruit cultivars grown in Sri Lanka. *Food Chem.* **2018**, *238*, 203–208. [CrossRef]
23. Gunathilake, K.D.P.P.; Ranaweera, K.K.D.S.; Rupasinghe, H.P.V. Effect of Different Cooking Methods on Polyphenols, Carotenoids and Antioxidant Activities of Selected Edible Leaves. *Antioxidants* **2018**, *7*, 117. [CrossRef]
24. Klen, T.J.; Vodopivec, B.M. The fate of olive fruit phenols during commercial olive oil processing: Traditional press versus continuous two- and three-phase centrifuge. *LWT Food Sci. Technol.* **2012**, *49*, 267–274. [CrossRef]
25. Leonardis, A.D.; Macciola, V.; Iacovino, S. Delivery Systems for Hydroxytyrosol Supplementation: State of the Art. *Colloids Interfaces* **2020**, *4*, 25. [CrossRef]
26. Mateos, R.; Martínez-López, S.; Baeza Arévalo, G.; Amigo-Benavent, M.; Sarriá, B.; Bravo-Clemente, L. Hydroxytyrosol in functional hydroxytyrosol-enriched biscuits is highly bioavailable and decreases oxidised low density lipoprotein levels in humans. *Food Chem.* **2016**, *205*, 248–256. [CrossRef]
27. Costa, A.S.G.; Alves, R.C.; Vinha, A.F.; Costa, E.; Costa, C.S.G.; Nunes, M.A.; Almeida, A.A.; Santos-Silva, A.; Oliveira, M.B.P.P. Nutritional, chemical and antioxidant/pro-oxidant profiles of silverskin, a coffee roasting by-product. *Food Chem.* **2018**, *267*, 28–35. [CrossRef]
28. Gullón, P.; Gullón, B.; Astray, G.; Carpena, M.; Fraga-Corral, M.; Prieto, M.A.; Simal-Gandara, J. Valorization of by-products from olive oil industry and added-value applications for innovative functional foods. *Food Res. Int.* **2020**, *137*, 109683. [CrossRef]
29. Ratnayaka, D.D.; Brandt, M.J.; Johnson, K.M. CHAPTER 6—Chemistry, Microbiology and Biology of Water. In *Water Supply*, 6th ed.; Ratnayaka, D.D., Brandt, M.J., Johnson, K.M., Eds.; Butterworth-Heinemann: Boston, MA, USA, 2009; pp. 195–266.
30. Deak, T. Chapter 17—Thermal Treatment. In *Food Safety Management*; Lelieveld, Y.M.H., Ed.; Academic Press: San Diego, CA, USA, 2014; pp. 423–442.
31. Berk, Z. *Food Process Engineering and Technology*, 3rd ed.; Academic Press: Cambridge, MA, USA, 2018.
32. Fellows, P.J. *Food Processing Technology: Principles and Practice*; Elsevier: Amsterdam, The Netherlands, 2009.
33. AOAC. *Official Methods of Analysis*, 21st ed.; Association of Official Analytical Chemists: Arlington, VA, USA, 2019.

34. Tontisirin, K. *Chapter 2: Methods of Food Analysis. Food Energy: Methods of Analysis and Conversion Factors: Report of a Technical Workshop*; Food and Agriculture Organization of the United Nations: Rome, Italy, 2003.
35. Alves, R.; Casal, S.; Oliveira, M.B.P. Determination of vitamin E in coffee beans by HPLC using a micro-extraction method. *Food Sci. Technol. Int.* **2009**, *15*, 57–63. [CrossRef]
36. Ferreira, D.M.; Nunes, M.A.; Santo, L.E.; Machado, S.; Costa, A.S.G.; Álvarez-Ortí, M.; Pardo, J.E.; Oliveira, M.B.P.P.; Alves, R.C. Characterization of Chia Seeds, Cold-Pressed Oil, and Defatted Cake: An Ancient Grain for Modern Food Production. *Molecules* **2023**, *28*, 723. [CrossRef]
37. *ISO 12966*; Animal and Vegetable Fats and Oils—Gas Chromatography of Fatty Acid Methyl Esters: Part 2: Preparation of Methyl Esters of Fatty Acids. ISO: London, UK, 2017.

molecules

Article

Effects of Drying Methods on the Antioxidant Properties of *Piper betle* Leaves

Kivaandra Dayaa Rao Ramarao [1,2], Zuliana Razali [1,2,*], Chandran Somasundram [1,2], Wijenthiran Kunasekaran [3] and Tan Li Jin [3]

1 Institute of Biological Sciences, Faculty of Science, University of Malaya, Kuala Lumpur 50603, Malaysia; kivaandra@gmail.com (K.D.R.R.); chandran@um.edu.my (C.S.)
2 The Center for Research in Biotechnology for Agriculture (CEBAR), University of Malaya, Kuala Lumpur 50603, Malaysia
3 Wari Technologies Sdn. Bhd., 2A-2, Galleria Cyberjaya, Jalan Teknokrat 6, Cyber 5, Cyberjaya 63000, Selangor, Malaysia; wijen@wari.tech (W.K.); lijin@wari.tech (T.L.J.)
* Correspondence: zuliana@um.edu.my

Abstract: *Piper betle* leaf powder is increasingly utilised as a health supplement. In this study, *P. betle* leaves were subjected to four different drying methods: convective air-drying, oven-drying, sun-drying, and no drying, with fresh leaves as control. Their antioxidant properties were then evaluated using colourimetric assays and GC-MS. Results showed that the sun-dried leaves had the highest ($p < 0.05$) total antioxidant capacity (66.23 ± 0.10 mg AAE/g), total polyphenol content (133.93 ± 3.76 mg GAE/g), total flavonoid content (81.25 ± 3.26 mg CE/g) and DPPH radical scavenging activity (56.48 ± 0.11%), and the lowest alkaloid content (45.684 ± 0.265 mg/gm). GC-MS analysis revealed that major constituents of aqueous extracts of fresh and sun-dried *P. betle* leaves were hydrazine 1,2-dimethyl-; ethyl aminomethylformimidate; glycerin; propanoic acid, 2-hydroxy-, methyl ester, (+/−)-; and 1,2-Cyclopentanedione. In conclusion, sun-dried leaves exhibited overall better antioxidant properties, and their aqueous extracts contained biologically active phytoconstituents that have uses in various fields.

Keywords: *Piper betle*; drying; total polyphenol content; total flavonoid content; DPPH radical scavenging activity; GCMS analysis

Citation: Ramarao, K.D.R.; Razali, Z.; Somasundram, C.; Kunasekaran, W.; Jin, T.L. Effects of Drying Methods on the Antioxidant Properties of *Piper betle* Leaves. *Molecules* **2024**, *29*, 1762. https://doi.org/10.3390/molecules29081762

Academic Editors: José Pinela, Carla Pereira, Maria Inês Dias and José Ignacio Alonso-Esteban

Received: 9 January 2024
Revised: 4 February 2024
Accepted: 7 February 2024
Published: 12 April 2024

1. Introduction

Piper betle leaves are from the genus Piper of the Piperaceae family (Table 1). There are over 100 varieties of which about 40 are found in India alone [1]. It is known as Betel leaves in English and commonly referred to as 'sirih' or 'sireh' by the locals in its place of origin, Malaysia. It is cultivated in India, Indonesia, the Philippines, and many other Southeast Asian and East African countries [2]. *P. betle* is an aromatic creeper with smooth, shiny dorsiventral heart-shaped leaves [3,4]. *P. betle* leaves are important for millions in India as their source of income depends on the supply chain of this crop, from production to processing to sales [5].

P. betle is a gem in traditional medicine where its uses range from the Ayurvedic field to TCM (traditional Chinese medicine) and even to Western medicine [6]. Its leaves are consumed after meals as a mouth freshener since it has a pleasant smell and as a digestive stimulant [7]. This edible leaf is known to possess medicinal properties such as anti-inflammatory and antioxidant activities [8]. It has seen uses in traditional medicine as a wound-healing agent and to improve digestion [9]. These biological activities are attributed to their high antioxidant activities and compounds [10]. The consumption of antioxidant-rich products like *P. betle* leaves can especially help to neutralise free radicals in our body, thus preventing or delaying the oxidative damage of lipids, proteins, and nucleic acids [11]. Various extractions, like aqueous extractions, were performed on the leaf, and gas chromatography-mass spectrometry (GC-MS) was commonly used to identify the phytochemical constituents [6]. These compounds include alcohols, esters, aldehydes, and alkanes [6,12].

Table 1. Taxonomic classification of *Piper betle*.

Level	Classification
Kingdom	Plantae
Division	Magnoliophyta
Class	Magnoliopsida
Order	Piperales
Family	Piperaceae
Genus	*Piper*
Species	*betle*

Drying is a process of moisture removal. It has been employed as a method for the drying of leaves for decades [10]. Drying methods involve heat application on a product to remove moisture, and this removal of water can help prevent the decomposition of phytochemicals and microbial contamination in the product [4]. The leaves' moisture content is usually reduced quickly to prevent enzymic reaction and oxidation, which often damages bioactive compounds in the leaf [10]. There is also a 35 to 75% loss of fresh leaf produce during storage and transport, which causes economic and environmental waste [13]. Therefore, drying is seen as a desirable method for the preservation of bioactive compounds like antioxidants in leaves. A recent study by Thi et al. [14] showed that dried *P. betle* leaves contained higher amounts of antioxidants than fresh leaves. This is because drying helps to concentrate the nutrients in the leaves [15]. Another study by Sahu et al. [16] showed that drying temperatures affect the antioxidant activity of *P. betle* leaves, specifically at high temperatures (80 °C) where polyphenol oxidases could have degraded, causing a reduction in the antioxidant activity. The methanolic extract of dried *P. betle* leaves was shown to contain potent antioxidant compounds like hydroxychavicol, which reduced inflammation by mediating the downregulation of the NF-κB and MAPK pathways [17]. It is also worth noting that these bioactive compounds in the leaves vary based on factors like geographical origin; therefore, it is important to not overgeneralise and continuously monitor their quality as health-promoting foods.

2. Results

The effects of various drying methods on the antioxidant properties of *P. betle* leaves were studied. The time taken for the betel leaves to dry was 4.5 h, 3 h, and 6 h in the oven, in the convective air-dryer, and under the sun respectively. The dried samples' percentage weight loss was 81.57% in the oven, 81.90% in the convective air-dryer, and 79.37% under the sun. Figure 1 presents the pictures of the leaves before drying (fresh leaves) and after they were subjected to each drying method, whereas Table 2 depicts the colour parameters as L*, a*, and b* values as measured using the chroma meter.

From Figure 1, it can be seen that the leaves darkened post-drying compared with the control (fresh leaves). This was supported by the data from the chroma meter's measurement where, as shown in Table 2, the L* values decreased ($p > 0.05$) in all three drying methods compared with fresh leaves, indicating that dried leaves became darker. In addition to that, the a* values were observed to have significantly increased ($p < 0.05$) while b* values significantly decreased ($p < 0.05$) for sun-dried leaves compared with fresh leaves.

Figure 1. Pictures of *Piper betle* leaves subject to various drying conditions: (**A**) fresh leaves; (**B**) oven-dried leaves; (**C**) convective air-dried leaves; (**D**) sun-dried leaves.

Table 2. Colour of *Piper betle* leaves subject to various drying methods.

Colour Parameter \ Drying Method	Fresh	Oven	Convective	Sun
L*	41.8 ± 4.53 ^{a}	33.17 ± 2.96 ^{a}	33.13 ± 2.91 ^{a}	33.47 ± 2.29 ^{a}
a*	−7.53 ± 0.54 ^{a}	−5.13 ± 0.86 ^{b}	−3.20 ± 0.67 ^{bc}	−2.70 ± 0.67 ^{c}
b*	28.07 ± 6.76 ^{a}	16.70 ± 2.07 ^{ab}	15.97 ± 2.38 ^{ab}	15.23 ± 1.53 ^{b}
a*/b*	−0.29 ± 0.10 ^{a}	−0.31 ± 0.03 ^{a}	−0.21 ± 0.07 ^{a}	−0.18 ± 0.05 ^{a}

Results were expressed as mean ± SD from three experiments (n = 3). 1 One-way ANOVA was carried out for each treatment ($p < 0.05$) and showed significance. Post hoc analysis, (Tukey HSD test) was used to identify which pair(s) in each column was/were statistically different. The same letter denotes mean values that are not significantly different ($p > 0.05$).

Based on Figure 2, TAC was the highest in sun-dried *P. betle* leaves (66.23 ± 0.10 mg AAE/g), while TAC in the fresh sample was significantly lower ($p < 0.05$) at 17.14 ± 1.44 mg AAE/g compared with oven-dried (63.30 ± 0.91 mg AAE/g) and convective air-dried (62.39 ± 0.11 mg AAE/g) samples. Overall, dried leaves exhibited a higher TAC when compared with fresh leaves. This observation was further noticed in TPC, DPPH radical scavenging activity, and TFC. Based on Figure 2, the total polyphenol content (TPC) in sun-dried samples (133.93 ± 3.76 mg GAE/g) was significantly higher ($p < 0.05$) compared with oven-dried, convective air-dried, and fresh leaf samples. This TPC value obtained was comparable to the TPC of *P. betle* leaves analysed by Dwijayanti et al. [8] at 128.92 ± 1.2 mg GAE/g and Vikrama et al. [18] at 130.00 ± 1.15 mg GAE/g. In addition to that, no significant difference ($p > 0.05$) was observed between oven-dried (122.69 ± 2.04 mg GAE/g) and convective air-dried (121.43 ± 2.64 mg GAE/g) samples. Figure 2 shows that the % inhibition in fresh *P. betel* leaves was the lowest, at 39.68 ± 0.10%, while sun-dried samples exhibited the highest % inhibition at 56.48 ± 0.11%. No significant difference ($p > 0.05$) was observed between oven-dried and convective air-dried samples with inhibition of 46.08 ± 0.48% and 42.46 ± 3.22%. Furthermore, the total flavonoid content (TFC) of *P. betle* ranged from 10.42 ± 0.25 to 81.25 ± 3.26 mg CE/g. Sun-dried samples had significantly higher ($p < 0.05$) TFC (81.25 ± 3.26 mg CE/g) compared with oven-dried and convective air-dried leaves at 68.19 ± 0.07 and 64.10 ± 0.16 mg CE/g respectively. As for alkaloid content, Figure 2 shows that the oven-dried samples possess the highest alkaloid content at 46.872 ± 0.153 mg/gm. No significant difference ($p < 0.05$) was observed between fresh and convective air-dried samples, while sun-dried had the lowest alkaloid content (45.684 ± 0.265 mg/gm) among the treatments. Alkaloids are widely found in plants and have various biological effects where they act primarily as defence components against pests or herbivores.

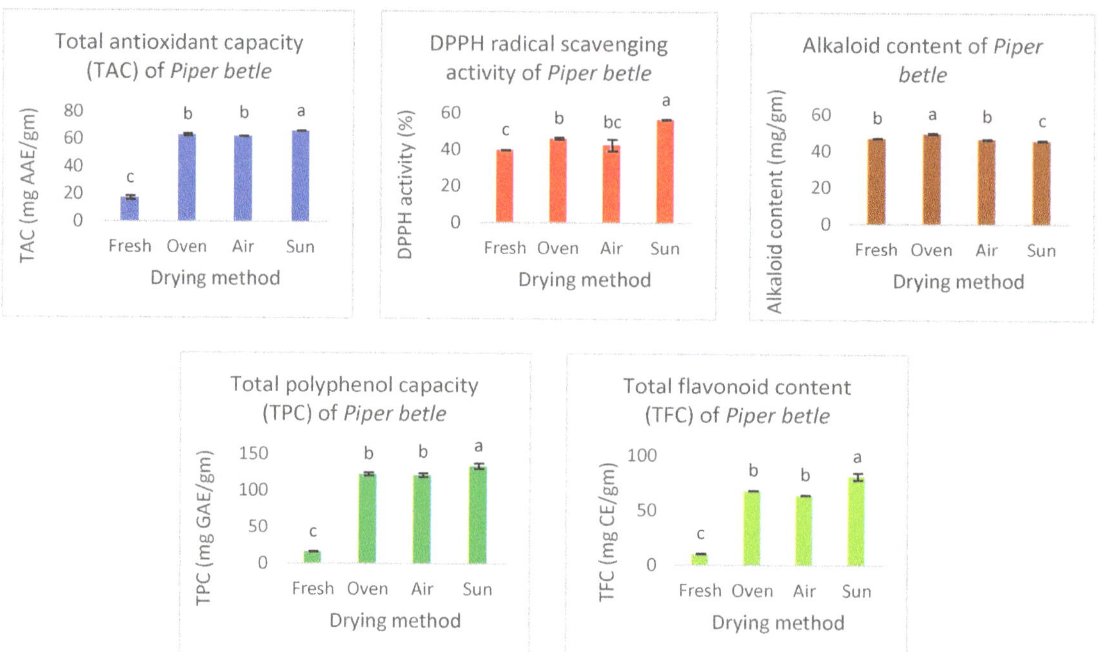

Figure 2. The TAC, DPPH radical scavenging activity, TPC, and TFC of *Piper betle* leaves subject to various drying treatments. Results were expressed as mean ± SD from three experiments ($n = 3$). One-way ANOVA was carried out for each treatment ($p < 0.05$) and showed significance. Post hoc analysis (Tukey HSD test) was used to identify which pair(s) in each column was/were statistically different. The same letter denotes mean values that are not significantly different ($p > 0.05$).

The phytocompounds and their compound nature in sun-dried and fresh samples extracted using water are presented in Table 3. A total of 27 types of compounds were identified using GC-MS and were classified into esters (6), ketones (6), heterocyclic compounds (4), alcohols (2), alkanes (2), ethers (2), amides (1), amines (1), carboxylic acids (1), phenolic compounds (1), and sulfur compounds (1). The percentages of each compound in the sun-dried and fresh leaves and the total from both are presented in a pie chart (Figure 3).

Table 3. List of phytocompounds identified by GC-MS in the aqueous extracts of *Piper betle* leaves using various drying and extraction methods.

Drying Method: Sun-Dried Extraction Method: First				
Peak	**R Time**	**Area %**	**Name**	**Nature of Compound**
1	3.144	43.56	Hydrazine, 1,2-dimethyl-	Hydronitrogen
2	3.262	4.04	Ethyl aminomethylformimidate	Ester
3	3.598	1.55	Acetic acid, hydroxy-, methyl ester	Ester
4	3.701	1.49	Acetoin	Ketone
5	3.974	10.35	Glycerin	Alcohol
6	4.119	2.30	Propanoic acid, 2-hydroxy-, methyl ester, (+/−)-	Ester
7	6.297	3.34	2-Cyclopenten-1-one	Ketone
8	8.110	0.87	1,3,5,7-Cyclooctatetraene	Heterocyclic compound
9	9.832	15.50	1,2-Cyclopentanedione	Ketone
10	12.656	2.21	Oxirane, [(2-propenyloxy)methyl]-	Ether
11	13.926	1.21	1,2-Cyclopentanedione, 3-methyl-	Ketone
12	14.086	2.20	1,2-Cyclopentanedione, 3-methyl-	Ketone
13	17.801	0.88	2-Cyclopenten-1-one, 3-ethyl-2-hydroxy-	Ketone
14	19.695	1.15	Silane, dimethyldi(but-3-enyloxy)-	Ether
15	21.990	0.86	Benzofuran, 2,3-dihydro-	Heterocyclic compound
16	23.041	1.88	Caprolactam	Amide
17	30.767	6.62	Benzoic acid, 2,5-dimethyl-	Carboxylic acid
Drying Method: Sun-Dried Extraction Method: Second				
Peak	**R time**	**Area %**	**Name**	**Nature of compound**
1	3.133	45.09	Hydrazine, 1,2-dimethyl-	Hydronitrogen
2	3.258	3.78	Ethyl aminomethylformimidate	Ester
3	3.599	1.46	Acetic acid, hydroxy-, methyl ester	Ester
4	3.699	1.56	Acetoin	Ketone
5	3.972	9.95	Glycerin	Alcohol
6	4.121	1.38	Propanoic acid, 2-hydroxy-, methyl ester, (+/−)-	Ester
7	6.297	2.77	2-Cyclopenten-1-one	Ketone
8	9.819	14.12	1,2-Cyclopentanedione	Ketone
9	12.651	1.88	Octane, 4-ethyl-	Alkane
10	13.921	1.19	1,2-Cyclopentanedione, 3-methyl-	Ketone
11	14.083	2.03	1,2-Cyclopentanedione, 3-methyl-	Ketone
12	17.799	1.09	2-Cyclopenten-1-one, 3-ethyl-2-hydroxy-	Ketone
13	19.693	1.04	Silane, dimethyldi(but-3-enyloxy)-	Ether
14	23.036	2.04	Caprolactam	Amide
15	25.080	0.54	2-Methoxy-4-vinylphenol	Phenolic compound
16	30.760	10.09	Benzoic acid, 2,5-dimethyl-	Carboxylic acid

Table 3. *Cont.*

Drying Method: None (Fresh Leaves) Extraction Method: First				
Peak	**R time**	**Area %**	**Name**	**Nature of compound**
1	3.031	16.59	Hydrazine, 1,2-dimethyl-	Hydronitrogen
2	3.253	9.74	Ethyl aminomethylformimidate	Ester
3	3.344	2.41	Trimethylsilyl ethaneperoxoate	Ester
4	3.985	29.41	Glycerin	Alcohol
5	4.133	3.87	Propanoic acid, 2-hydroxy-, methyl ester, (+/−)-	Ester
6	4.366	2.45	2-Propenoic acid, methyl ester	Ester
7	4.653	3.56	Alpha-monopropionin	Alcohol
8	6.341	4.63	2-Cyclopenten-1-one	Ketone
9	9.708	10.65	1,2-Cyclopentanedione	Ketone
10	9.804	5.77	1,2-Cyclopentanedione	Ketone
11	12.666	2.42	Decane	Alkane
12	14.070	6.26	2-Cyclopenten-1-one, 2-hydroxy-3-methyl-	Ketone
13	22.891	2.23	Caprolactam	Amide
Drying Method: None (Fresh Leaves) Extraction Method: Second				
Peak	**R time**	**Area %**	**Name**	**Nature of compound**
1	3.041	19.88	Hydrazine, 1,2-dimethyl-	Hydronitrogen
2	3.200	1.15	Ethyl aminomethylformimidate	Ester
3	3.261	9.83	Allyl mercaptan	Sulfur compound
4	3.342	2.20	Trimethylsilyl ethaneperoxoate	Ester
5	3.606	1.54	Acetic acid, hydroxy-, methyl ester	Ester
6	3.982	27.88	Glycerin	Alcohol
7	4.129	3.60	Propanoic acid, 2-hydroxy-, methyl ester, (+/−)-	Ester
8	4.427	2.08	Pyrrole	Heterocyclic compound
9	4.554	1.40	Glycerin	Alcohol
10	4.646	1.82	Propanoic acid, 1-methylpropyl ester	Ester
11	6.311	4.64	2-Cyclopenten-1-one	Ketone
12	8.115	0.97	Styrene	Heterocyclic compound
13	9.714	11.16	1,2-Cyclopentanedione	Ketone
14	9.792	4.65	1,2-Cyclopentanedione	Ketone
15	12.655	3.48	Decane	Alkane
16	13.922	1.70	1,2-Cyclopentanedione, 3-methyl-	Ketone
17	14.061	2.03	1,2-Cyclopentanedione, 3-methyl-	Ketone

R time = Retention time.

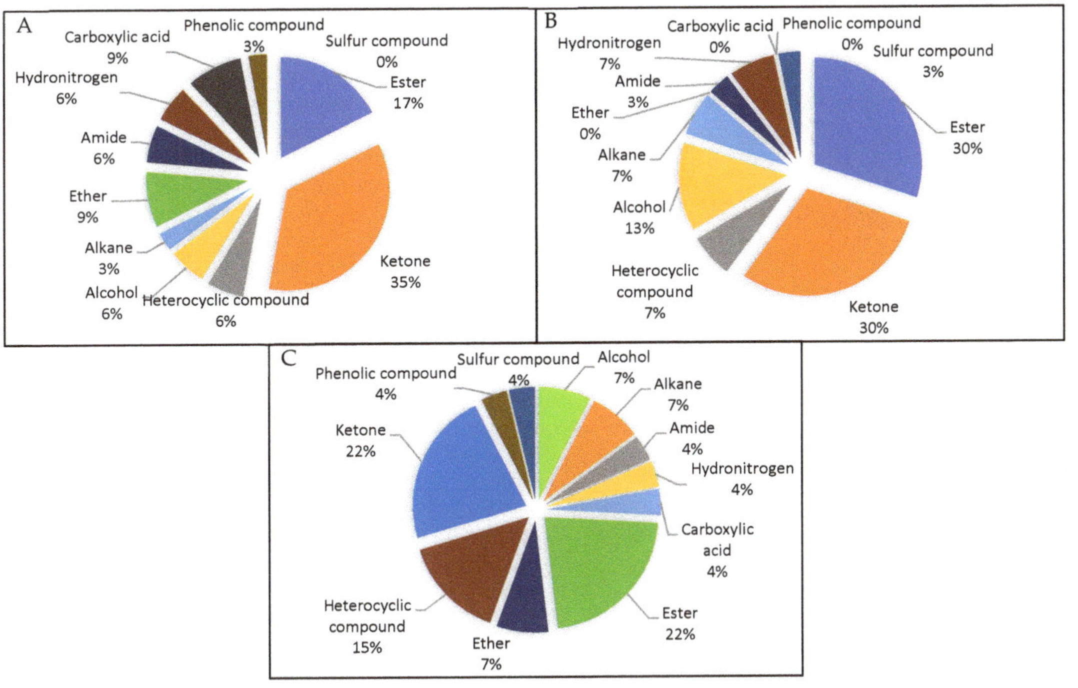

Figure 3. Pie diagram showing the percentage of phytochemical groups identified in *Piper betle* leaves for sun-dried samples (**A**), fresh samples (**B**), and the combined total (**C**).

3. Discussion

The decreased L* values in the dried leaves indicate that they became darker since L* values range from black at 0 to white at 100. However, the decrease was not significant ($p > 0.05$). On the other hand, the dried leaves also experienced an increase in a* and a decrease in b* values, indicating a loss of greenness as the a* measures red when positive and green when negative while b* measures yellow when positive and blue when negative. The natural greenness in leaves is associated with chlorophylls, but drying could lead to a loss of magnesium ions, which causes chlorophylls to be converted to pheophytins [19,20]. Higher L* values and lower a*/b* values are desirable in dried material [21]. Based on these criteria, both fresh and dried *P. betel* leaves presented desirable colour since there was no significant difference in the L* and a*/b* values between them.

From this experiment, dried *P. betle* leaves were found to contain higher amounts of antioxidants such as polyphenolics and higher antioxidant activity compared with fresh leaves, and this finding was consistent with a study by Thi et al. [14]. The dried samples had a higher TAC probably due to the drying treatment, which possibly induced structural changes in the leaf, thus increasing the extraction of antioxidant compounds by enhanced solvent and mass transfer [22]. Plant phenolics have garnered interest due to their effectiveness as free radical scavengers and antioxidants defending against ultraviolet radiation or pathogens [1,23,24]. The low TPC in fresh leaf samples was probably caused by the presence of an active enzyme called polyphenol oxidase that degrades the phenolic compounds, and in the dried samples, low water activity might have led to the inactivation of these enzymes resulting in higher levels of phenolic compounds in the samples [25]. Reports in the literature about the effects of drying on the TPC of plant samples vary. Some suggest that drying aids in breaking down the cell wall of plant materials, which helps release phenolics into the extracting solvent; conversely, drying can change the chemical structures of the phenolic compounds and cause them to adhere to other cellular components, which makes their extraction difficult [26]. Flavonoids and their derivatives are excellent free radical scavengers. Sun-dried samples had the highest TFC in the *P. betle* leaves. This could be due to the temperature in sun-dried samples being lower (38–45 °C) since the loss of flavonoids generally occurs at higher temperatures due to thermal degradation [27]. This might have led to a significant ($p < 0.05$) decrease in TFC value in oven-dried and convective air-dried samples. The drying time of sun-dried samples was also the longest, indicating a slower water loss rate, which might cause phenolics like flavonoids to increase as suggested by Zhang et al. [28].

DPPH is a stable free radical when interaction with an antioxidant receives electron or hydrogen atoms to neutralise its free radical character [29]. The DPPH free radical scavenging assay allows a preliminary assessment of a compound or sample of interest [9]. Based on our results, the sun-dried samples that had a significantly higher ($p < 0.05$) TPC and TFC also had a significantly higher ($p < 0.05$) DPPH radical scavenging activity. This trend was also followed at the opposite end of the spectrum, i.e., the fresh leaves had a significantly lower ($p < 0.05$) TPC and TFC, which corresponded to a significantly lower ($p < 0.05$) DPPH radical scavenging activity. This indicates that the drying influence of TPC and TFC of the leaf samples affected their antioxidant activity. Lou et al. [30] suggested that high TPC might contribute to the high antioxidant activity, which is probably due to the high hydrogen-donating ability of phenolic compounds [31]. The antioxidant activity of a leaf sample is also strongly related to its flavonoid content where the higher the content, the higher the radical scavenging activity [1,32,33]. Future studies can explore this further by testing multiple concentrations of the sun-dried leaf extract to optimise its antioxidant content and DPPH radical scavenging activity.

Based on the results, the major constituents found in the *P. betle* leaf extracts were esters (2%), ketones (22%), heterocyclic compounds (15%), alcohols (7%), alkanes (7%), and ethers (7%). The were differences in the type of compounds observed in the dried and fresh leaves upon extraction. For instance, an absence of phenolic type and carboxylic acid compounds was noticed in the aqueous extracts of the fresh leaves compared with the dried leaves. This could be due to their low amount in the fresh samples, coupled with the fact that carboxylic acids are usually extracted from aqueous solutions using organic solvents by the principle of reactive extraction [34]. In the dried samples, the drying process could have helped concentrate these compounds; therefore, their presence was more prominent. Conversely, there was almost double the percentage of alcohol- and ester-type compounds in the fresh leaf extracts compared with dried extracts. This could be due to the low boiling point of alcohols and esters, which have high heating sensitivity [35]. Furthermore, this was also consistent with a report by Zhang et al. [36] who observed that drying reduces ester content as they are hydrolysed during heat treatments.

Hydrazine, 1,2-dimethyl-; Ethyl aminomethylformimidate; glycerin; propanoic acid, 2-hydroxy-, methyl ester, (+/−)-; and 1,2-Cyclopentanedione were noticed in all of the sample extracts. Hydrazine, 1,2-dimethyl- is the only hydronitrogen compound found in our sample extracts. It has a molecular formula of $C_2H_8N_2$ and a molecular weight of 60.1 g/mol. Hydrazine and its derivatives have been utilised to prevent the corrosion of boiler plants [37] and are widely utilised in the production of polymers, pharmaceuticals, and agricultural pesticides [38,39]. It is worth noting that daminozide, a type of plant growth regulator, can degrade to form dimethylhydrazine [40]. Its presence in our results could shed light on potential contaminants and safety risks posed by small-time vendors' products since the leaf samples were obtained from a market. Pesticide residue has been reported in *P. betle* leaves recently [41]. Further studies are, therefore, recommended to explore this possibility in the samples obtained. Glycerin is also known as glycerol and has multiple uses across industries like pharmaceuticals and food [42]. To the best of our knowledge, this paper is the first to report on the presence of glycerin in the aqueous extract of the *P. betle* leaves. It is an alcoholic compound with a molecular weight of 92.09 g/mol and a molecular formula $C_3H_8O_3$. It has been used as an alternative energy source for animal feeding. It possesses antioxidant properties as shown in a study by Araújo et al. [43] who demonstrated that supplementation with glycerine in diets of broilers increased its expression of uncoupling protein (UCP), which is related to mitochondrial function and glutathione peroxidase (GPx), which combats reactive oxygen species. Glycerin also has antimicrobial and anti-inflammatory properties [44]. Glycerin and Alpha-monopropionin are alcohol-related compounds found in our sample extracts. Ethyl aminomethylformimidate is an ester compound with 102.14 g/mol molecular weight and a molecular formula of $C_4H_{10}N_2O$. Other ester compounds detected in the sample extracts were 2-Propenoic acid, methyl ester; acetic acid, hydroxy-, methyl ester; ethyl aminomethylformimidate; propanoic acid, 1-methylpropyl ester; propanoic acid, 2-hydroxy-, methyl ester, (+/−)-; and trimethylsilyl ethaneperoxoate. Oxirane, [(2-propenyloxy)methyl]- and silane, dimethyldi(but-3-enyloxy)- were ether-related products found in the extracts. 1,2-Cyclopentanedione is a ketone-based compound with a molecular weight of 98.1 g/mol and a molecular formula of $C_5H_6O_2$. Other ketone compounds found in the extracts were acetoin; 1,2-Cyclopentanedione; 1,2-Cyclopentanedione, 3-methyl-; 2-Cyclopenten-1-one; 2-Cyclopenten-1-one, 2-hydroxy-3-methyl-; and 2-Cyclopenten-1-one, 3-ethyl-2-hydroxy-. 1,2-Cyclopentanedione, 3-methyl- has been shown to have anti-inflammatory properties via the suppression of pro-inflammatory gene expression through NF-κB signalling pathway modulation [45]. It has also been shown to possess radical scavenging properties by decreasing ONOO−, which is a by-product of reactive oxygen and nitrogen species that causes tissue damage via the formation of nitrotyrosine adducts glutathione (GSH) reductase [46]. We have also identified alkanes such as decane and octane, 4-ethyl-. Caprolactam and benzoic acid, 2,5-dimethyl- were amide and carboxylic acid compounds, respectively. Caprolactam is, more specifically, a cyclic amide with a molecular weight of 113.16 g/mol and a molecular formula of $C_6H_{11}NO$. Its derivatives are utilised in biomedical fields for drug delivery systems [47]. 2-Methoxy-4-vinylphenol is a phenolic compound while allyl mercaptan is a sulfur-based compound found in the extracts. 2-Methoxy-4-vinylphenol has a molecular weight of 150.17 g/mol and a molecular formula of $C_9H_{10}O_2$. It is also a well-known styrene metabolite [48,49]. A study by Jeong et al. [50] showed that it possesses potent anti-inflammatory properties by inhibiting nitric oxide (NO), prostaglandins (PGE2), inducible NO synthase (iNOS), and cyclooxygenase-2 (COX-2) in cells. Several heterocyclic compounds were also identified in the sample extracts, such as 1,3,5,7-Cyclooctatetraene; benzofuran, 2,3-dihydro-; pyrrole; and styrene. While styrene is primarily a synthetic compound that can raise concerns about potential contaminants/ micropollutants, it has been identified in various natural plants, for example, in cinnamon by Fragnière et al. [51]. The authors acknowledged that incidences of styrene in food may not be related to exogenous contamination since it occurs naturally in foods. Overall, the bioactive compounds identified in the *P. betle* aqueous extracts possess medical properties such as antioxidant and anti-inflammatory activities that could have contributed to its role as a health-promoting agent and its role in the treatment of various ailments as claimed by traditional health practitioners.

4. Materials and Methods

4.1. Sample Preparation

Piper betle leaves were purchased from a local market in Petaling Jaya, Malaysia located approximately 6 km away from the Postharvest Laboratory, University of Malaya, Malaysia. They were immediately transported early in the morning (7 am MYT) in plastic bags and processed in the lab.

4.2. Drying Methods

Leaves were weighed, and 50 g was used for each drying experiment: oven-dried (55 °C), convective air-dried (60 °C), sun-dried (38–45 °C), and fresh leaves. The temperatures of the drying systems were measured using a thermometer. For sun-drying, the experiment was conducted from 10 am to 4 pm (sunlight directly reached the leaf samples in a tray without obstruction). Convective air-drying was conducted using a prototype self-built convective air-dryer [52]. The experiment was conducted in triplicates. The samples were spread evenly on the drying trays and left to dry until the sample weight was consistent for three readings. The dried leaves were pounded into powder under liquid nitrogen using a pestle and mortar.

4.3. Colour Measurement

Leaf colour was measured before and after drying with fresh leaves serving as a control. A chroma meter (Minolta CR-20, Tokyo, Japan) was used to measure the L*, a*, and b* values of fresh and dried *P. betle* leaves. The chroma meter was standardised with a white standard plate before three random measurements were taken for each sample. The values were recorded as L*, a* and b* and a*/b* based on Ali et al. [53] where L* measures the whiteness, with ranges from (black at 0 to white at 100), a* measures red when positive and green when negative, while b* measures yellow when positive and blue when negative.

4.4. Antioxidant Analysis

4.4.1. Preparation of Extract

Leaves were prepared based on a method by Uribe et al. [54]. Briefly, a solid/liquid mixture with the ratio of 1:4 comprising a powdered sample and 80% methanol was prepared. The mixture with 80% methanol was chosen based on Jaiswal et al. [55]. The mixture was placed on an orbital shaker (Shellab Orbital Shaking Incubator S14, Cornelius, OR, USA) for 30 min at 200 rpm (room temperature). Next, the mixture was centrifuged at 6500 rpm for 10 min at 4 °C, and the resulting supernatant was used for subsequent analysis.

4.4.2. Total Polyphenol Content

The samples' total polyphenol content (TPC) was determined using Folin–Ciocalteu assay modified to a microscale [56]. A total of 0.79 mL of distilled water was added to 0.01 mL of the sample (gallic acid solution of known concentrations replaced the sample for the standard curve), followed by 0.05 mL of Folin–Ciocalteu reagent. After 1 min, 0.15 mL of sodium carbonate was added, and the mixture was left to stand at 25 °C for 2 h. The absorbance was measured at a wavelength of 750 nm using a spectrophotometer (Shimadzu UV-200-RS, MRC, Petah-Tikva, Israel). The equation from the gallic acid standard curve was $y = 0.0056x$, R2 = 0.9955, and results were expressed as mg of gallic acid equivalent (GAE) per gram of sample.

4.4.3. Total Antioxidant Capacity

Total antioxidant capacity (TAC) was determined by the phosphomolybdenum method [57]. First, 1 mL of solution (0.6 M sulphuric acid, 4 mM ammonium molybdate, 28 mM sodium phosphate) of equal volumes was prepared. Subsequently, 0.01 mL of sample was added to the reagent mixture (80% methanol was used to replace the sample for blank), and the tubes were incubated for 90 min at 95 °C. The absorbance was read at 695 nm against blank once the sample cooled to room temperature. The equation of the ascorbic acid standard curve was $y = 0.0018x$, R2 = 0.9981. The results were expressed as milligrams of ascorbic acid equivalent (AAE) per gram of plant material.

4.4.4. DPPH Radical Scavenging Assay

The DPPH assay was carried out via a method described by Bae and Suh [54]. 0.1 mM DPPH solution was prepared with 80% methanol. A total of 1 mL of the solution was added to 500 µL of samples. The radical scavenging activity was calculated using the formula:

$$\%\ \text{DPPH inhibition} = (\text{Acontrol} - \text{Asample}/\text{Acontrol}) \times 100$$

where Acontrol: absorbance of control; Asample: absorbance of the sample.

The results were reported as % inhibition.

4.4.5. Total Flavonoid Content

Total flavonoid content (TFC) was determined using the colourimetric method based on Sakanaka et al. [58]. Briefly, 1.25 mL of distilled water was added to the 0.25 mL sample. The sample was replaced with (+)-standard catechin solution for standard curve construction. After that, 75 μL of a 5% sodium nitrite solution was added and left at room temperature for 6 min. Then, 150 μL of a 10% aluminium chloride solution was added, and the mixture was incubated for 5 min; then, 0.5 mL of 1 M sodium hydroxide was added. Distilled water was used to bring the mixture up to 2.5 mL. The absorbance was measured at 510 nm. The catechin standard curve had an equation of y = 0.0135x, R2 = 0.9943, and the results obtained were reported as milligrams of catechin equivalent (CE) per gram of sample.

4.5. Gas Chromatography-Mass Spectroscopy (GCMS) Analysis

Various households tend to prepare *P. betle* leaf extracts using water for consumption. Therefore, based on the antioxidant results, we extended our study by comparing phytocompounds found in aqueous extracts of sun-dried and fresh leaves using GCMS.

4.5.1. Sample Preparation

P. betle leaves were purchased from a local market in Petaling Jaya, Malaysia, located approximately 6 km away from the Postharvest Laboratory, University of Malaya, Malaysia. Leaves (10 g) were subjected to (a) the drying treatment under the sun and (b) no treatment (fresh) for GC-MS analysis. The leaves were ground into powder under liquid nitrogen using a pestle and mortar. Then, the powdered samples were extracted in aqueous solutions via 2 different methods. The first method was based on Madi et al. [59] with slight modifications where powdered leaves were soaked in hot distilled water (100 °C) in a 1:10 solid/liquid ratio for 30 min. Once the suspension settled at room temperature, it was filtered using Whatman No. 1 filter paper. The second method involved soaking the leaves in distilled water (1:10 solid/liquid ratio) and leaving them overnight. The following day, the extract was filtered using Whatman No. 1 filter paper. Filtered extracts from both the first and second extraction methods were then concentrated in a rotary evaporator and resuspended at 1 mg/mL.

4.5.2. Screening of Compounds

The characterisation of the phytochemicals in the leaves was performed using GC-MS QP2010 Plus (Shimadzu, Tokyo, Japan). GC was conducted in the temperature programming mode with an Rtx-5MS column (0.25 mm, 30 m). The initial column temperature was 40 °C for 1 min. The injection temperature was 300 °C (splitless mode); the oven temperature was programmed from 40 °C and held for 5 min to 160 °C at a rate of 4 °C/min, then to 280 °C at a rate of 5 °C/min and held for 15 min. Identifications were based on mass spectral matching with standard compounds in the NIST library with a similarity index of at least 80% [33]. The relative amounts of individual components were expressed as percent peak areas relative to the total peak area.

4.6. Statistical Analysis

Results were expressed as mean ± SD from three experiments ($n = 3$). One-way ANOVA was performed using the Statistical Package for the Social Sciences (SPSS) v23.0 (IBM, Armonk, NY, USA) for each treatment. If significance was shown ($p < 0.05$), a post hoc analysis, (Tukey HSD test) was performed to identify which pair(s) in each column was/were statistically different. The same letter denotes mean values that are not significantly different ($p > 0.05$).

5. Conclusions

Sun-drying has often been employed by the masses to dry products because it is easily accessible, requires low skill, and is low cost. This is crucial to those who rely on small-scale plantations of this crop like home-grown or small farms as a reliable source of income. However, it would be disadvantageous if the quality of the leaves is not proportionate to the work that was invested, especially for a health-conscious consumer market. To this end, many studies have shown varying reports on sun-drying's efficacy for leaf products, owing to the possible thermal and UV-induced degradation of bioactive compounds in them. Therefore, our study provided important evidence in support of the use of sun-drying as a preparation method for *P. betle* leaves. This is because the sun-dried samples showed a significantly higher ($p < 0.05$) TAC, TFC, TPC, and alkaloid content, and DPPH radical scavenging activity compared with all other drying treatments and fresh samples. This is important for locals who wish to market sun-dried *P. betle* leaves as a high-antioxidant natural product. Furthermore, the aqueous extracts revealed important bioactive compounds like 1,2-Cyclopentanedione, 3-methyl-, glycerin, and 2-Methoxy-4-vinylphenol that have strong antioxidant and anti-inflammatory properties, while the others have important applications in various fields ranging from pharmaceuticals to the food industry. Future studies are recommended to assess the individual composition of each leaf sample and evaluate the microbial and pesticide residue to provide a more comprehensive understanding of the safety of these leaves for consumption.

Author Contributions: Conceptualization, Z.R. and C.S.; methodology, K.D.R.R.; software, K.D.R.R. and T.L.J.; validation, Z.R., C.S. and W.K.; formal analysis, K.D.R.R.; investigation, K.D.R.R.; resources, W.K. and T.L.J.; data curation, K.D.R.R.; writing—original draft preparation, K.D.R.R.; writing—review and editing, K.D.R.R., Z.R. and C.S.; visualization, T.L.J.; supervision, Z.R. and C.S.; project administration, K.D.R.R. and T.L.J.; funding acquisition, C.S., W.K. and T.L.J. All authors have read and agreed to the published version of the manuscript.

Funding: This research was funded by the University of Malaya Science Faculty (Research Grant: GPF001B-2018).

Institutional Review Board Statement: Not applicable.

Informed Consent Statement: Not applicable.

Data Availability Statement: The data presented in this study are available in article.

Acknowledgments: The authors wish to thank Wari Technologies Sdn. Bhd. for their support and assistance throughout this work.

Conflicts of Interest: Authors Wijenthiran Kunasekaran and Tan Li Jin were employed by the company Wari Technologies Sdn. Bhd. The remaining authors declare that the research was conducted in the absence of any commercial or financial relationships that could be construed as a potential conflict of interest.

References

1. Sarma, C.; Rasane, P.; Kaur, S.; Singh, J.; Singh, J.; Gat, Y.; Dhawan, K. Antioxidant and antimicrobial potential of selected varieties of *Piper betle* L. (Betel leaf). *An. Acad. Bras. Cienc.* **2018**, *90*, 3871–3878. [CrossRef]
2. Umar, R.A.; Zahary, M.N.; Rohin, M.A.K.; Ismail, S. Chemical Composition and The Potential Biological Activities of *Piper Betel*–A Review. *Malays. J. Appl. Sci.* **2018**, *3*, 1–8.
3. Madhumita, M.; Guha, P.; Nag, A. Extraction of betel leaves (*Piper betle* L.) essential oil and its bio-actives identification: Process optimization, GC-MS analysis and antimicrobial activity. *Ind. Crops Prod.* **2019**, *138*, 111578. [CrossRef]
4. Pin, K.Y.; Chuah, T.G.; Rashih, A.A.; Law, C.L.; Rasadah, M.A.; Choong, T.S.Y. Drying of Betel Leaves (*Piper betle* L.): Quality and Drying Kinetics. *Drying Technol.* **2009**, *27*, 149–155. [CrossRef]
5. Das, S.; Parida, R.; Sandeep, I.S.; Nayak, S.; Mohanty, S. Biotechnological intervention in betelvine (*Piper betle* L.): A review on recent advances and future prospects. *Asian Pacific J. Trop. Med.* **2016**, *9*, 938–946. [CrossRef]
6. Biswas, P.; Anand, U.; Saha, S.C.; Kant, N.; Mishra, T.; Masih, H.; Dey, A. Betelvine (*Piper betle* L.): A comprehensive insight into its ethnopharmacology, phytochemistry, and pharmacological, biomedical and therapeutic attributes. *J. Cell Mol. Med.* **2022**, *26*, 3083–3119. [CrossRef]
7. Rathee, J.S.; Patro, B.S.; Mula, S.; Gamre, S.; Chattopadhyay, S. Antioxidant activity of *Piper betel* leaf extract and its constituents. *J. Agric. Food Chem.* **2006**, *54*, 9046–9054. [CrossRef] [PubMed]
8. Dwijayanti, D.R.; Puspitarini, S.; Widodo, N. *Piper betle* L. Leaves Extract Potentially Reduce the Nitric Oxide Production on LPS-Induced RAW 264.7 Cell Lines. *J. Exp. Life Sci.* **2023**, *13*, 78–83. [CrossRef]

9. Aulia, H.R.; Wienaldi, W.; Fioni, F. Effectiveness of green betel leaf extract cream in healing cut wounds. *J. Prima Medika Sains* **2023**, *5*, 187–195. [CrossRef]
10. Abrahim, N.N.; Kanthimathi, M.S.; Abdul-Aziz, A. *Piper betle* shows antioxidant activities, inhibits MCF-7 cell proliferation and increases activities of catalase and superoxide dismutase. *BMC Complement. Altern. Med.* **2012**, *12*, 220. [CrossRef] [PubMed]
11. Nur Sazwi, N.; Nalina, T.; Rahim, Z.H.A. Antioxidant and cytoprotective activities of *Piper betle*, *Areca catechu*, *Uncaria gambir* and betel quid with and without calcium hydroxide. *BMC Complement. Altern. Med.* **2013**, *13*, 351. [CrossRef]
12. Sahoo, B.C.; Singh, S.; Sahoo, S.; Kar, B. Comparative Metabolomics of High and Low Essential Oil Yielding Landraces of Betelvine (*Piper betel* L.) by Two-Dimensional Gas Chromatography Time-of-Flight Mass Spectrometry (GC× GC TOFMS). *Anal. Lett.* **2023**, *57*, 1–17.
13. Babu, A.K.; Kumaresan, G.; Raj, V.A.A.; Velraj, R. Review of leaf drying: Mechanism and influencing parameters, drying methods, nutrient preservation, and mathematical models. Renew. *Sustain. Energy Rev.* **2018**, *90*, 536–556. [CrossRef]
14. Thi, P.V.; Thi, T.N.; Nguyen, M.T.; Thi, D.N.P.; Trong, N.K.T.; Thi, C.H.C. The Comparison of Antioxidative and Cytotoxic Activities of Fresh and Dried *Piper betle* L. leave Extracts on MCF-7, HELA and SK-LU-1. *Asian J. Res. Bot.* **2023**, *6*, 339–351.
15. Vernekar, A.A.; Vijayalaxmi, K.G. Nutritional composition of fresh and dehydrated betel leaves. *J. Pharm. Innov.* **2019**, *8*, 602–605.
16. Sahu, C.K.; Balan, A.; Bayineni, V.K.; Banerjee, S. Study of physicochemical and antioxidant synergy efficacy of betel leaf dried paste powder. *Coatian J. Food Sci. Technol.* **2021**, *13*, 155–159.
17. Seo, J.; Lee, U.; Seo, S.; Wibowo, A.E.; Pongtuluran, O.B.; Lee, K.; Cho, S. Anti-inflammatory and antioxidant activities of methanol extract of *Piper betle* Linn. (*Piper betle* L.) leaves and stems by inhibiting NF-κB/MAPK/Nrf2 signaling pathways in RAW 264.7 macrophages. *Biomed. Pharmacother.* **2022**, *155*, 113734. [CrossRef] [PubMed]
18. Vikrama Chakravarthi, P.; Murugesan, S.; Arivuchelvan, A.; Sukumar, K.; Arulmozhi, A.; Jagadeeswaran, A. GC-MS profiling of methanolic extract of *Piper betle* (Karpoori Variety) leaf. *J. Pharmacogn. Phytochem.* **2018**, *7*, 2449–2452.
19. Therdthai, N.; Zhou, W. Characterisation of microwave vacuum drying and hot air drying of mint leaves (*Mentha cordifolia* Opiz ex Fresen). *J. Food Eng.* **2009**, *91*, 482–489. [CrossRef]
20. Rudra, S.G.; Singh, H.; Basu, S.; Shivhare, U. Enthalpy entropy compensation during thermal degradation of chlorophyll in mint and coriander puree. *J. Food Eng.* **2008**, *86*, 379–387. [CrossRef]
21. Doymaz, İ.; Tugrul, N.; Pala, M. Drying characteristics of dill and parsley leaves. *J. Food Eng.* **2006**, *77*, 559–565. [CrossRef]
22. Ali, A.; Chua, B.L.; Ashok, G.A. Effective Extraction of Natural Antioxidants from *Piper betle* with The Aid of Ultrasound: Drying And Extraction Kinetics. *J. Eng. Sci. Technol.* **2018**, *13*, 1–16.
23. Alara, O.R.; Ukaegbu, C.I.; Abdurahman, N.H.; Alara, J.A.; Ali, H.A. Plant-sourced Antioxidants in Human Health: A State-of-Art Review. *Curr. Nutr. Food Sci.* **2023**, *19*, 817–830. [CrossRef]
24. Stiller, A.; Garrison, K.; Gurdyumov, K.; Kenner, J.; Yasmin, F.; Yates, P.; Song, B.H. From fighting critters to saving lives: Polyphenols in plant defense and human health. *Int. J. Mol. Sci.* **2021**, *22*, 8995. [CrossRef] [PubMed]
25. Roshanak, S.; Rahimmalek, M.; Goli, S.A.H. Evaluation of seven different drying treatments in respect to total flavonoid, phenolic, vitamin C content, chlorophyll, antioxidant activity and color of green tea (*Camellia sinensis* or *C. assamica*) leaves. *J. Food Sci. Technol.* **2016**, *53*, 721–729. [CrossRef]
26. Salamatullah, A.M.; Ahmed, M.A.; Hayat, K.; Husain, F.M.; Arzoo, S.; Alzahrani, A.; Ahmad, S.R. Different drying techniques effect on the bioactive properties of rose petals. *J. King Saud. Univ. Sci.* **2024**, *36*, 103025. [CrossRef]
27. Martínez-Las Heras, R.; Heredia, A.; Castelló, M.; Andrés, A. Influence of drying method and extraction variables on the antioxidant properties of persimmon leaves. *Food Biosci.* **2014**, *6*, 1–8. [CrossRef]
28. Zhang, X.; Wang, X.; Wang, M.; Cao, J.; Xiao, J.; Wang, Q. Effects of different pretreatments on flavonoids and antioxidant activity of *Dryopteris erythrosora* leaves. *PLoS ONE* **2019**, *14*, e0200174. [CrossRef]
29. Wanigasekera, W.; Joganathan, A.; Pethiyagoda, R.; Yatiwella, L.; Attanayake, H. Comparison of antioxidant activity, Phenolic and Flavonoid contents of selected medicinal plants in Sri Lanka. *Ceylon J. Sci.* **2019**, *48*, 155–162. [CrossRef]
30. Lou, S.-N.; Lai, Y.-C.; Huang, J.-D.; Ho, C.-T.; Ferng, L.-H.A.; Chang, Y.-C. Drying effect on flavonoid composition and antioxidant activity of immature kumquat. *Food Chem.* **2015**, *171*, 356–363. [CrossRef]
31. Bahloul, N.; Boudhrioua, N.; Kouhila, M.; Kechaou, N. Effect of convective solar drying on colour, total phenols and radical scavenging activity of olive leaves (*Olea europaea* L.). *Int. J. Food Sci. Technol.* **2009**, *44*, 2561–2567. [CrossRef]
32. Romero-Márquez, J.M.; Navarro-Hortal, M.D.; Forbes-Hernández, T.Y.; Varela-López, A.; Puentes, J.G.; Pino-García, R.D.; Quiles, J.L. Exploring the Antioxidant, Neuroprotective, and Anti-Inflammatory Potential of Olive Leaf Extracts from Spain, Portugal, Greece, and Italy. *Antioxidants* **2023**, *12*, 1538. [CrossRef]
33. Ramarao, K.D.R.; Somasundram, C.; Razali, Z.; Kunasekaran, W.; Jin, T.L. The antioxidant properties and microbial load of *Moringa oleifera* leaves dried using a prototype convective air-dryer. *Saudi J. Biol. Sci.* **2022**, *29*, 103290. [CrossRef] [PubMed]
34. Djas, M.; Henczka, M. Reactive extraction of carboxylic acids using organic solvents and supercritical fluids: A review. *Sep. Purif. Technol.* **2018**, *201*, 106–119. [CrossRef]
35. ElGamal, R.; Song, C.; Rayan, A.M.; Liu, C.; Al-Rejaie, S.; ElMasry, G. Thermal Degradation of Bioactive Compounds during Drying Process of Horticultural and Agronomic Products: A Comprehensive Overview. *Agronomy* **2023**, *13*, 1580. [CrossRef]
36. Zhang, Y.R.; Lü, C.W.; Liu, T.; Zhen, J.M. Effect of different drying methods on volatile flavor components in *Agaricus blazei*. *Food Sci.* **2016**, *37*, 116–121.

37. George, M.; Nagaraja, K.S.; Balasubramanian, N. Spectrophotometric determination of hydrazine. *Anal. Lett.* **2007**, *40*, 2597–2605. [CrossRef]
38. Haque, A.M.J.; Kumar, S.; del Río, J.S.; Cho, Y.K. Highly sensitive detection of hydrazine by a disposable, Poly (Tannic Acid)-Coated carbon electrode. *Biosens. Bioelectron.* **2020**, *150*, 111927. [CrossRef]
39. Matsumoto, M.; Kano, H.; Suzuki, M.; Katagiri, T.; Umeda, Y.; Fukushima, S. Carcinogenicity and chronic toxicity of hydrazine monohydrate in rats and mice by two-year drinking water treatment. *Regul. Toxicol. Pharmacol.* **2016**, *76*, 63–73. [CrossRef] [PubMed]
40. Newsome, W.H. Determination of daminozide residues on foods and its degradation to 1, 1-dimethylhydrazine by cooking. *J. Agric. Food. Chem.* **1980**, *28*, 319–321. [CrossRef]
41. Prodhan, M.D.H.; Afroze, M.; Begum, A.; Ahmed, M.S.; Sarker, D. Optimization of a QuEChERS based analytical method for the determination of organophosphorus and synthetic pyrethroid pesticide residues in betel Leaf. *Int. J. Environ. Anal. Chem.* **2023**, *103*, 1292–1303. [CrossRef]
42. Konwar, L.J.; Mikkola, J.-P.; Bordoloi, N.; Saikia, R.; Chutia, R.S.; Kataki, R. Chapter 3—Sidestreams from Bioenergy and Biorefinery Complexes as a Resource for Circular Bioeconomy. In *Waste Biorefinery*; Bhaskar, T., Pandey, A., Mohan, S.V., Lee, D.J., Khanal, S.K., Eds.; Elsevier: Amsterdam, The Netherlands, 2018; pp. 85–125.
43. Araújo, R.S.; Sousa, K.R.S.; Sousa, F.C.B.; Oliveira, A.C.; Dourado, L.R.B.; Guimarães, S.E.F.; Silva, W.; Biagiotti, D.; Bayão, G.F.V.; Araujo, A.C. Effect of glycerin supplementation on the expression of antioxidant and mitochondrial genes in broilers. *Anim. Prod. Sci.* **2018**, *59*, 408–415. [CrossRef]
44. Jananie, R.K.; Priya, V.; Vijayalakshmi, K. Determination of bioactive components of Cynodon dactylon by GC-MS analysis. *N. Y. Sci. J.* **2011**, *4*, 1–5.
45. Chung, J.H.; Choi, S.Y.; Kim, J.Y.; Kim, D.H.; Lee, J.W.; Choi, J.S.; Chung, H.Y. 3-Methyl-1, 2-cyclopentanedione down-regulates age-related NF-κB signaling cascade. *J. Agric. Food Chem.* **2007**, *55*, 6787–6792. [CrossRef]
46. Kim, A.R.; Zou, Y.; Kim, H.S.; Choi, J.S.; Chang, G.Y.; Kim, Y.J.; Chung, H.Y. Selective peroxynitrite scavenging activity of 3-methyl-1, 2-cyclopentanedione from coffee extract. *J. Pharm. Pharmacol.* **2002**, *54*, 1385–1392. [CrossRef]
47. Rao, K.M.; Rao, K.S.V.K.; Ha, C.S. Stimuli responsive poly (vinyl caprolactam) gels for biomedical applications. *Gels* **2016**, *2*, 6. [CrossRef]
48. Alexakis, A.E.; Ayyachi, T.; Mousa, M.; Olsén, P.; Malmström, E. 2-Methoxy-4-Vinylphenol as a Biobased Monomer Precursor for Thermoplastics and Thermoset Polymers. *Polymers* **2023**, *15*, 2168. [CrossRef]
49. Yue, G.G.; Lee, J.K.; Kwok, H.F.; Cheng, L.; Wong, E.C.; Jiang, L.; Yu, H.; Leung, H.W.; Wong, Y.L.; Leung, P.C.; et al. Novel PI3K/AKT targeting anti-angiogenic activities of 4-vinylphenol, a new therapeutic potential of a well-known styrene metabolite. *Sci. Rep.* **2015**, *8*, 11149. [CrossRef]
50. Jeong, J.B.; Hong, S.C.; Jeong, H.J.; Koo, J.S. Anti-inflammatory effect of 2-methoxy-4-vinylphenol via the suppression of NF-κB and MAPK activation, and acetylation of histone H3. *Arch. Pharm. Res.* **2011**, *34*, 2109–2116. [CrossRef]
51. Fragnière, C.; Aebischer, J.N.; Dudler, V.; Sager, F. A short study on the formation of styrene in cinnamon. *Mitteilungen Leb. Hyg.* **2003**, *94*, 609–620.
52. Ramarao, K.D.R.; Razali, Z.; Somasundram, C. Mathematical models to describe the drying process of *Moringa oleifera* leaves in a convective-air dryer. *Czech J. Food Sci.* **2021**, *39*, 444–451. [CrossRef]
53. Ali, M.; Yusof, Y.; Chin, N.; Ibrahim, M.; Basra, S. Drying kinetics and colour analysis of *Moringa oleifera* leaves. *Agric. Agric. Sci. Procedia* **2014**, *2*, 394–400. [CrossRef]
54. Uribe, E.; Delgadillo, A.; Giovagnoli-Vicuña, C.; Quispe-Fuentes, I.; Zura-Bravo, L. Extraction Techniques for Bioactive Compounds and Antioxidant Capacity Determination of Chilean Papaya (*Vasconcellea pubescens*) Fruit. *J. Chem.* **2015**, *8*, 347532. [CrossRef]
55. Jaiswal, S.G.; Patel, M.; Saxena, D.K.; Naik, S.N. Antioxidant properties of *Piper betel* (L) leaf extracts from six different geographical domain of India. *J. Bioresour. Eng. Technol.* **2014**, *1*, 18–26.
56. Bae, S.H.; Suh, H.J. Antioxidant activities of five different mulberry cultivars in Korea. *LWT-Food Sci. Technol.* **2007**, *40*, 955–962. [CrossRef]
57. Prieto, P.; Pineda, M.; Aguilar, M. Spectrophotometric quantitation of antioxidant capacity through the formation of a phosphomolybdenum complex: Specific application to the determination of vitamin E. *Anal. Biochem.* **1999**, *269*, 337–341. [CrossRef] [PubMed]
58. Sakanaka, S.; Tachibana, Y.; Okada, Y. Preparation and antioxidant properties of extracts of Japanese persimmon leaf tea (kakinoha-cha). *Food Chem.* **2005**, *89*, 569–575. [CrossRef]
59. Madi, N.; Dany, M.; Abdoun, S.; Usta, J. *Moringa oleifera*'s nutritious aqueous leaf extract has anticancerous effects by compromising mitochondrial viability in an ROS-dependent manner. *J. Am. Coll. Nutr.* **2016**, *35*, 604–613. [CrossRef]

molecules

MDPI

Article

Chemical Composition, Functional and Antioxidant Properties of Dietary Fibre Extracted from Lemon Peel after Enzymatic Treatment

Vanesa Núñez-Gómez *, Marta San Mateo, Rocío González-Barrio and Mª Jesús Periago

Department of Food Technology, Food Science and Nutrition, Faculty of Veterinary Sciences, Regional Campus of International Excellence "Campus Mare Nostrum", Biomedical Research Institute of Murcia (IMIB-Arrixaca-UMU), University of Murcia, 30100 Murcia, Spain; marta.sanm@um.es (M.S.M.); rgbarrio@um.es (R.G.-B.); mjperi@um.es (M.J.P.)

* Correspondence: vanesa.nunez@um.es

Abstract: Lemon peel represents an interesting by-product owing to its content of dietary fibre (DF) and (poly)phenols, which is of great importance for its valorisation. Hence, the objective of this study was to characterise the DF, total phenolic content (TPC), and antioxidant capacity of two lemon-peel-derived ingredients using two different methods (drying with warm air and enzymatic hydrolysis with pectinesterase). The analysis included a DF assessment, followed by neutral sugars characterisation through GC-FID and uronic acids determination via colorimetry. Subsequently, TPC and antioxidant capacity using the FRAP method were quantified through spectrophotometry. The swelling capacity (SWC), water retention capacity (WRC), and fat absorption capacity (FAC) were also determined as functional properties. It was observed that pectinesterase treatment led to a reduction in soluble DF and an increase in insoluble DF. This treatment also affected the pectin structure, thereby diminishing its ability to absorb water and fat within its matrix. The TPC was also reduced, resulting in a decrease in antioxidant capacity. Conversely, employing warm air exhibited a noteworthy increase in antioxidant capacity. This underscores its crucial contribution to the valorisation of lemon peel, not only by diminishing the environmental impact but also by enabling the acquisition of fibre ingredients with a noteworthy antioxidant capacity.

Keywords: soluble dietary fibre; (poly)phenols; by-product; swelling capacity; water retention capacity; functional ingredient; valorisation

Citation: Núñez-Gómez, V.; San Mateo, M.; González-Barrio, R.; Periago, M.J. Chemical Composition, Functional and Antioxidant Properties of Dietary Fibre Extracted from Lemon Peel after Enzymatic Treatment. *Molecules* **2024**, *29*, 269. https://doi.org/10.3390/molecules29010269

Academic Editors: Francesco Cacciola, Carla Pereira, José Ignacio Alonso-Esteban, Maria Inês Dias and José Pinela

Received: 15 November 2023
Revised: 29 December 2023
Accepted: 2 January 2024
Published: 4 January 2024

1. Introduction

The growing generation of by-products in the agro-industry has heightened the necessity for developing techniques that enable their reintegration into the food chain [1]. This is crucial for implementing the increasingly prevalent circular economy models [2]. Notably, lemons, being a globally consumed fruit, contain components with health-enhancing properties, including dietary fibre, vitamins, minerals, and bioactive compounds, such as (poly)phenols and carotenoids [3].

Concerning by-products from lemons, they typically originate from the peel or pulp [1]. This is largely due to the predominant use of the fruit for juice production [4]. Citrus peel, in general, is rich in dietary fibre, flavonoids, carotenoids, and essential oils [5]. Specifically, the albedo, constituting the inner part and being the primary component of the peel, serves as the main source of fibre in the fruit. It is considered high-quality fibre owing to its association with bioactive compounds integrated into its matrix [6]. When treating these by-products, the extraction conditions are crucial. This includes solvents, enzymes, or physical conditions, all of which significantly influence the characteristics and the composition of the ingredients obtained from these by-products [7–9]. Furthermore, depending on the method selected, there are several advantages and disadvantages. In this sense, enzymatic methods have a higher yield and allow for obtaining fibre fractions with higher purity. On the other hand, these methods require more laborious protocols, which need a greater number of reagents and are therefore more expensive. Conventional drying processes, in turn, are cheaper and do not require reagents, thus reducing the use of water and energy, but, in contrast, the fibres obtained are of lower purity [10].

Dietary fibre has several health effects, such as intestinal motility, an effect on post-prandial glucose and insulin response, a cholesterol-lowering effect, and a prebiotic effect, among others [11]. In addition, dietary fibre may have an antioxidant effect, as it may have (poly)phenols attached through hydrogen bonds to the polysaccharides, which are known as non-extractable (poly)phenols (NEPPs) [12,13]. Extractable (poly)phenols (EPP), on the other hand, are those that are not bound to the fibre matrix and can therefore be extracted with organic solvents, while NEPPs must be extracted by acid or alkaline hydrolysis. As for EPPs, they can be absorbed in the first parts of the gastrointestinal tract because of their free form. NEPPs, on the other hand, must be released from the fibre, and this release occurs to a greater extent in the colon after fermentation of the dietary fibre [14,15].

Therefore, dietary fibre bound to (poly)phenols might have an antioxidant effect associated with the presence of these compounds. Upon ingestion, these (poly)phenols facilitate the neutralisation of free radicals by donating additional electrons. In this process, they prevent cell damage and serve as a protective barrier against chronic diseases associated with oxidative stress, including heart disease, diabetes, and specific types of cancer [16].

The objective of this study was to analyse the composition of dietary fibre and the (poly)phenol content and antioxidant capacity of two different samples, which were obtained from lemon peel by-products dried using hot air or treated through enzymatic hydrolysis with pectinesterase, in order to evaluate their nutritional value and their functional properties as fibre-rich food ingredients.

2. Results and Discussion

2.1. Dietary Fibre Content

Figure 1 illustrates the percentages of total dietary fibre (TDF), soluble dietary fibre (SDF), and insoluble dietary fibre (IDF) expressed in g/100 g, revealing statistically significant differences between the two samples of lemon peel (LP) and lemon peel treated with pectinesterase (LPp). The TDF content was 45% for the LP sample and 61% for the LPp, which was notably higher, indicating a significant reduction in fibre content in the sample that did not undergo pectinesterase hydrolysis or that the treatment with pectinesterases in the LPp sample results in the removal of other components that comprise the peel when the supernatant is removed after the enzymatic treatment, leading to fibre concentration. The TDF content observed for both samples closely aligns with that documented by the USDA for lemon peel, which is 58% in dry weight (calculated based on the water content) [17].

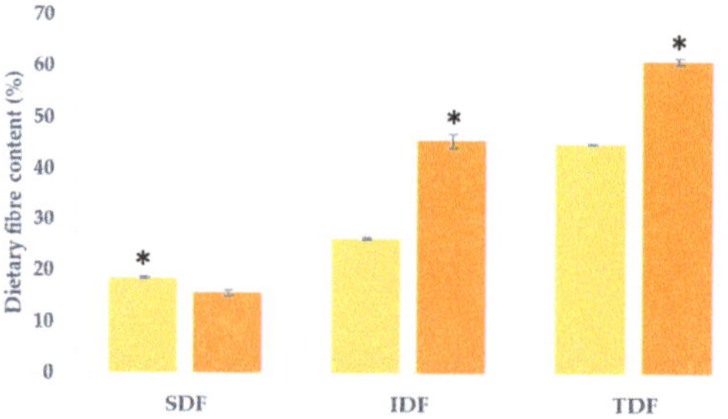

Figure 1. Dietary fibre composition expressed as percentage g/100 g of dry weight (d.w.) for lemon peel (LP) and lemon peel hydrolysed by pectinesterase (LPp). SDF (soluble dietary fibre); IDF (insoluble dietary fibre); TDF (total dietary fibre). Values are expressed as mean ± SD (n = 3). * Indicates significant differences ($p < 0.05$) among the samples for each fibre fraction (SDF, IDF, and TDF).

In terms of the SDF and IDF content, significant disparities were noted between the two samples. In both instances, the IDF fraction prevailed over SDF. When assessed as a percentage of the total fibre, the LP sample exhibited a higher proportion of SDF (41.5%) compared to the LPp sample (25.5%). Conversely, LPp displayed a greater percentage of IDF at 74.5% as opposed to LP, which registered 58.5%. The variance in SDF proportion can be attributed to the enzymatic treatment with pectinesterases to which the LPp sample has been subjected. These enzymes hydrolysed the polysaccharide primarily contributing to this fibre fraction, pectin, thus yielding smaller polymers exhibiting characteristics reminiscent of SDF.

A literature review estimated the amount of fibre in dried citrus peel to be 57%, with the results obtained in our study falling near this value [18]. Other authors estimated slightly higher average values. Czech et al. [19] described a TDF amount of 64%, while Rafiq et al. [6] showed values in a range between 60% and 68%. Another study reported similarity with the LPp sample in terms of the total percentage and different fractions; these authors described a TDF content for peels from different citrus fruits between 62% and 64%, of which the SDF was between 13% and 14% and the IDF was between 49% and 50% [20]. Moreover, when comparing these results with those obtained by other authors for soy fibre extracted with an enzymatic cellulase treatment, the results were also similar, reporting a mean content of 63% of IDF and 6.5% of SDF [21]. This outcome suggests that citrus-fruit-peel-derived products could be of interest to the food industry given their potential uses as functional components in confectionery, bakery items, or the production of high-fibre foods.

2.2. Dietary Fibre Characterisation by GC-FID

Table 1 presents the neutral sugars and uronic acid profiles of both lemon peel samples expressed as percentages. Moreover, Figure S1 shows the chromatograms obtained for the neutral sugars in both samples and in the standard mix. Notable differences in the content of all analysed sugars and uronic acids were observed between samples. In the LP sample, the primary sugars were arabinose, followed by galactose, while glucose, xylose, and mannose exhibited percentages ranging from 7% to 9%. Rhamnose and fucose were present in lower concentrations. In contrast, the LPp sample featured galactose, rhamnose, and glucose as the major sugars, with intermediate percentages of arabinose and mannose falling within the 4% to 9% range, and lower proportions of fucose and xylose. These findings in the LP sample align with those of previous researchers who reported that dried citrus peel samples predominantly contain arabinose, galactose, and glucose, underscoring the direct impact of the treatment on the proportion of neutral sugars [20]. The uronic acid content in LP was higher than the LPp due to the fact that LP was not treated with pectinesterase. However, the values for both samples were between those previously reported by other authors (12–24%) [9,20,22]. When comparing the results for both fractions with those obtained by other authors for a broccoli by-product fibre extracted with an anzymatic method and another as a control, the results agree with those obtained in our study, showing that the arabinose, xylose, and mannose content were higher in the control samples, as is observed for these neutral sugars in the LP sample [23].

The proportions of the primary dietary fibre polysaccharides, namely cellulose, hemicellulose, and pectin, were calculated from the neutral sugars, as shown in Table 1. In both samples, pectin emerged as the dominant polysaccharide, constituting approximately 73% of the composition, with no significant differences observed between the two samples. This outcome may be attributed to the pectin percentage calculation, which considers monosaccharides. In the case of the hydrolysed peel, these monosaccharides exist in a free state rather than forming more complex structures, such as pectin. However, notable variations were detected in the percentages of cellulose and hemicellulose. In the LP sample, hemicellulose prevailed at 18.7%, overcoming cellulose at 8.2%. Conversely, in the LPp sample, cellulose was the dominant component, accounting for 18.7%, while hemicellulose constituted 7.5% of the total polysaccharide content. These results were in agreement with those showed for the dietary fibre composition (Table 1), where it can be seen that LPp sample had more insoluble fibre, mainly due to the cellulose content. In accordance with findings from other authors, citrus residues generally display a composition expressed as a percentage, with approximately 43% pectin, 11% hemicellulose, and 9% cellulose, values that closely resemble those obtained in our samples [18].

Table 1. Percentages (%) of neutral sugars, uronic acids, cellulose, hemicellulose, and pectin in lemon peel (LP) and lemon peel hydrolysed by pectinesterase (LPp) §.

Composition (%)	LP	LPp
Rhamnose	3.3 ± 0.2	24.1 ± 1.2 *
Fucose	1.1 ± 0.3 *	0.2 ± 0.2
Arabinose	31.6 ± 1.4 *	9.3 ± 0.9
Xylose	9.4 ± 1.3 *	0.8 ± 0.2
Mannose	7.3 ± 0.3 *	4.4 ± 0.3
Galactose	21.4 ± 0.6	25.5 ± 0.8 *
Glucose	9.1 ± 0.2	20.8 ± 1.2 *
Uronic acids	16.9 ± 0.1 *	14.9 ± 0.1
Cellulose [1]	8.2 ± 0.2	18.7 ± 1.0 *
Hemicellulose [2]	18.7 ± 1.2	7.5 ± 0.4 *
Pectin [3]	73.1 ± 1.1	73.8 ± 0.7

§ Values are expressed as mean ± SD (n = 3). * Indicates significant differences ($p < 0.05$) among the samples. [1] Cellulose: glucose × 0.9; [2] hemicellulose: (fucose + xylose + mannose + (glucose × 0.1)); [3] pectin: (rhamnose + arabinose + galactose + uronic acids).

When calculating other parameters related to the fibre composition (Table 2), significant differences are observed in the mannose-to-xylose ratio, which reflects the proportion of mannose relative to xylose in hemicellulose. Previous studies, such as the investigation by Peng et al., have reported that water-soluble hemicelluloses from maize stems exhibit a higher mannose content and a lower xylose content [24]. This could elucidate why the hemicellulose in LPp, with a greater mannose contribution, was primarily composed of water-soluble hemicelluloses. Conversely, in the case of LP, which exhibited the lowest mannose contribution and thus the highest xylose content, non-water-soluble hemicelluloses were the predominant constituents.

Table 2. Sugar ratios for characterisation of pectin and hemicellulose from lemon peel (LP) and lemon peel hydrolysed by pectinesterase (LPp) §.

Parameter	LP	LPp
Mannans to hemicelluloses contr [1]	0.8 ± 0.1	5.7 ± 1.6 *
Linearity of pectin [2]	0.3 ± 0.0	0.3 ± 0.0
Rhamnose and uronic acid contr [3]	0.2 ± 0.0	1.6 ± 0.1 *
RG-I Branching [4]	16.1 ± 1.3 *	1.5 ± 0.1

§ Values are expressed as mean ± SD (n = 3). * Indicate significant differences ($p < 0.05$) among the samples. Lemon peel (LP); lemon peel hydrolysed by pectinesterase (LPp). [1] Contribution of mannans to hemicelluloses: mannose/xylose; [2] linearity of pectin: uronic cids/(fucose + rhamnose + arabinose + galactose + xylose); [3] contribution of rhamnose and uronic acids to pectins: rhamnose/uronic acids; [4] branching of RG-I: (arabinose + galactose)/rhamnose.

Moreover, additional indices, such as pectin linearity, the contribution of rhamnose and uronic acids to pectins, and RG-I branching, were computed, and significant differences were also observed for them. The results indicated that pectin linearity was similar, with no significant differences observed. This parameter holds significance as it determines the ability to create emulsions, which is favoured when linearity is higher due to the enhanced hydration properties of pectin [25]. Nevertheless, it is expected that the pectin chains may differ between the samples due to the enzymatic action.

Additionally, the ratio of rhamnose to uronic acid was significantly higher in LPp compared to LP, indicating the presence of longer RG-I domains. The RG-I domains represent around 7–14% of the whole pectin and are composed of rhamnose and galacturonic acids. These are the ones that give the branched structure, and they are crucial as they confer elasticity and viscosity among other gelling properties. Concerning RG-I branching, which reflects the extent of branching in the pectin molecules, LP exhibited the highest values. These findings align with the research of other authors, who have demonstrated that variations in the extraction process can directly influence the structure of the obtained pectin [26].

The composition of LPp suggests higher solubility, as indicated by the mannose-to-xylose ratio. Conversely, in terms of the contribution of rhamnose and uronic acids to pectin, LPp exhibited higher values, which means longer branches. This signifies that pectinesterase treatment led to the hydrolysis of pectin, thereby releasing uronic acids, maintaining rhamnose, and increasing this ratio. The RG-I branching index, which signifies the presence of galactose and arabinose side chains attached to the rhamnose of the pectin, shows more chain domains. To sum up, LP had more branches, but they are shorter, and LPp had less branches but longer ones. As such, LPp molecules were more flexible, which allowed them to interact less with each other, resulting in less rigid structures that allow, for example, a better stabilisation of emulsions [25,27]. Another parameter to characterise the pectin structure in the isolated samples was the length of the side chains, which are mainly formed by arabinose and galactose units. The length of the side chains was highest in LP, indicating that pectin from this sample may have stronger molecular interactions, thus leading to more consistent structures and making this characteristic interesting at an industrial level [27]. This outcome is due to LP preserving the integrity of lemon pectin, as it has not undergone pectinesterase treatment, thereby retaining the branching structure.

2.3. Functional Properties

Swelling capacity, water retention capacity, and fat absorption capacity were measured to evaluate the functional properties of LP and LPp (Table 3). There were significant differences in all of the functional properties determined, with the LP sample showing higher mean values compared to the LPp sample. The LP sample exhibits better hydration properties and contains a higher amount of SDF, allowing it to retain more water and fat within its matrix. These properties suggest its potential use as a natural added ingredient in bakery products and to fortify dairy foods and develop low-fat products [28,29]. These results are supported by those reported by Rivas et al. (2022), who reported that enzymatic treatment reduces hydration properties of dietary fibre isolated from broccoli by-products [23]. In a study by Huang et al., untreated citrus peel displayed mean functional property values of SWC at 8 mL/g, WRC at 8 g/g, and FAC at 2 g/g. The hydration property values were similar to those of the LP sample but differed from the data obtained for the treated sample (LPp), except for FAC [5].

Table 3. Functional properties of lemon peel (LP) and lemon peel treated with pectinesterase (LPp).

Functional Property	LP	LPp
SWC (mL water/g)	9.9 ± 0.07 *,[1]	2.6 ± 0.0
WRC (g water/g)	1.1 ± 0.0 *	0.8 ± 0.1
FAC (g oil/g)	9.3 ± 0.6 *	2.7 ± 0.2

[1] Values are expressed as mean ± SD (n = 3). * Indicate significant differences ($p < 0.05$) among the samples. Lemon peel (LP); lemon peel treated with pectinesterase (LPp). SWC (swelling capacity); WRC (water retention capacity); FAC (fat absorption capacity).

On the other hand, according to Zhang et al., SDF from lemon peel exhibited higher values for all three functional properties compared to the data obtained in this study for LPp [30]. These results indicate that after treatment with pectinesterase, the hydration and fat absorption properties decrease primarily due to the changes in the pectin structure discussed in the previous section.

A correlation analysis, illustrated in Figure 2, was carried out to explore the relationship between the functional properties, dietary fibre composition, and characterisation. The results indicated a positive correlation between the three analysed properties and SDF, whereas a negative correlation was observed with the presence of TDF and IDF. Regarding the correlation of functional properties with the mannose/xylose contribution index, all three properties exhibited a negative correlation. This suggests that in the analysed samples, a higher presence of soluble hemicellulose does not enhance functional properties, highlighting its association with the pectin structure. Notably, the correlation with the mannose/xylose contribution index revealed a negative impact on the functional properties, particularly with the contribution of rhamnose and uronic acids. As described earlier, longer chain lengths result in increased stiffness, thus hindering the formation of compact structures that can retain water or oils [27].

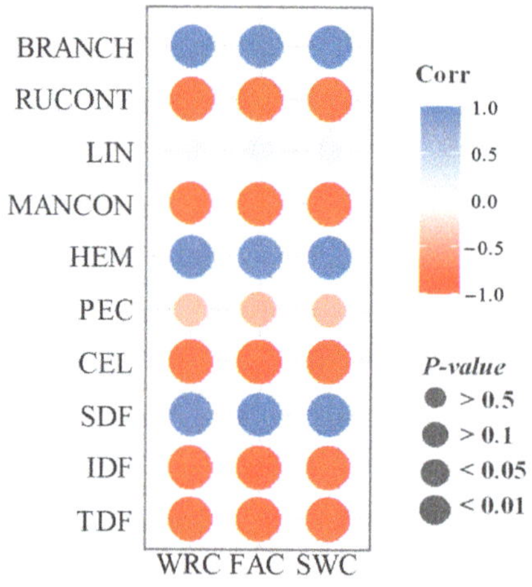

Figure 2. Correlation analysis between functional properties (FAC (fat absorption capacity), SWC (swelling capacity), and WRC (water retention capacity)) and fibre composition (TDF (total dietary fibre), IDF (insoluble dietary fibre), SDF (soluble dietary fibre), PEC (pectin content), CEL (cellulose content), HEM (hemicellulose content), MANCON (mannose contribution to hemicellulose), RUCONT (rhamnose–uronic acid contribution to pectin), LIN (linearity of pectin), and BRANCH (branching of pectin)) in the lemon peel fibre samples.

Contrary to previous findings by Belkheiri et al. [26], greater RG-I branching in the samples was linked to improved functional properties. It is essential to note that the individual characteristics of pectin are less crucial than the overall structure of pectin in determining its flexibility. In this instance, LPp molecules exhibited higher flexibility, indicating a lack of rigid structures and, consequently, reduced capacity to retain water and oil within its composition.

2.4. Total Phenolic Content

Figure 3 displays the TPC in the EPP and NEPP fractions of both samples. It is noteworthy that there were no significant differences in EPP between the samples, with a mean value of 3.7 mg GAE/g. Although an increase in EPP could be expected due to the potential enzymatic release, it should be noted that these compounds could be removed after the application of ethanol to precipitate the soluble polysaccharides. Regarding the NEPP, a notable observation is that the LP fraction exhibited the highest significant content, being 1.4 times greater than that observed for LPp. The lower presence in the LPp may be due to the aforementioned effect, which resulted in an enzymatic release of compounds.

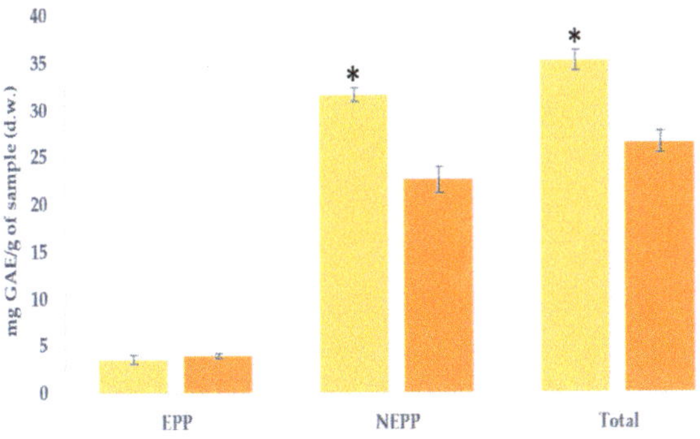

Figure 3. Total phenolic content (TPC) expressed as mg of gallic acid equivalents (GAE)/g of the sample in dry weight (d.w.) for lemon peel (■ LP) and lemon peel hydrolysed by pectinesterase (■ LPp). EPP (extractable (poly)phenols); NEPP (non-extractable (poly)phenols). Values are expressed as mean ± SD (n = 3). * Indicates significant differences ($p < 0.05$) among the samples.

Singh et al. (2020) reported a range of TPC in lemon peel from 88 mg GAE/g to 190 mg GAE/g, with values akin to those found in our sample (LP, 150.4 mg GAE/g) [3]. Other studies have cited TPC content between 65 and 72 mg GAE/g, aligning closely with our findings [7,8]. Treatment with pectinesterase resulted in reduced NEPP content, leading to a lower TPC in the LPp sample. This suggests that the applied processing may trigger the release or loss of (poly)phenols from the fibre matrix.

Although no individual (poly)phenols have been identified in this study, it should be noted that the major (poly)phenols described by other authors are eriocitrin, hesperidin, rutin, and other compounds, such as limonin, that belong to the furanolactones group [31,32]. Moreover, the literature indicates variations in (poly)phenol content based on the extraction process [33]. In this case, for LPp, where ethanol was used to precipitate soluble polysaccharides, it may also contribute to removing part of the (poly)phenols and therefore lead to a lower content compared to LP, where no ethanol was used. In addition, the drying process may also affect the (poly)phenol content, as air drying is the least preservative for these compounds, although it should be noted that in our study, where this method was used, the total values are similar to those described by other authors, as previously indicated [33].

2.5. Antioxidant Capacity

Figure 4 illustrates the antioxidant capacity of both samples, as assessed through FRAP in the EPP and NEPP extracts. Notably, significant variations were noted, indicating that the antioxidant capacity was higher in LP compared to LPp for both extracts and, consequently, in the overall content. The disparity in content was remarkable, with values 3.3 times higher for EPP, 3.4 times higher for NEPP, and 3.4 times higher for the total content in LP compared with LPp. It is important to acknowledge that these findings are in alignment with those observed for TPC. However, it is worth noting that the differing ratios may be attributed to other compounds influencing antioxidant capacity, such as carotenoids, which were not measured in the current study [34]. Additionally, there was a positive correlation observed for NEPP and the total content concerning antioxidant capacity.

It is notable that the values observed in this study appear to be lower than those reported by other researchers for lemon peel samples, ranging between 133 and 380 µmol/g. This disparity could potentially be attributed to variations in sample acquisition methods, differences in the extraction process employed for analysis, or the utilisation of different lemon varieties in the respective studies [7,8].

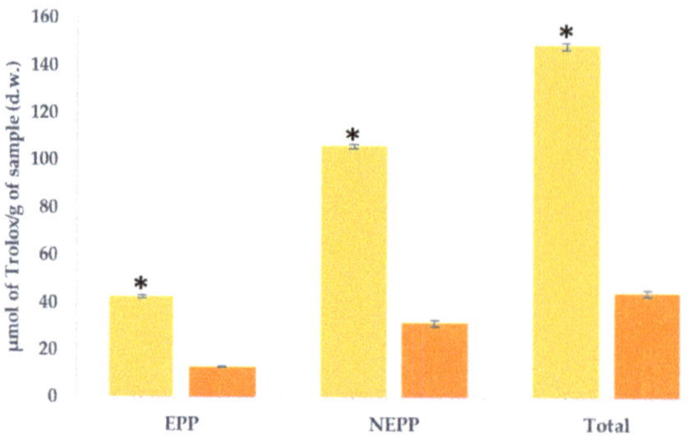

Figure 4. Antioxidant capacity expressed as μmol of Trolox/g of sample in dry weight (d.w.) for lemon peel (■ LP) and lemon peel hydrolysed by pectinesterase (■ LPp). EPP (extractable (poly)phenols); NEPP (non-extractable (poly)phenols). Values are expressed as mean ± SD (n = 3). * Indicates significant differences ($p < 0.05$) among the samples.

While other authors have reported a decline in antioxidant capacity due to hot air drying, such an effect was not evident in our study. It is important to note that the observed outcomes may not solely be attributable to this treatment or the presence of other bioactive compounds. The formation of Maillard compounds should also be considered, as they arise from heat treatment to the sugars present in the sample [35]. Although these compounds were not measured in the present study, they may potentially have been formed. As previously mentioned, the LPp sample underwent treatment with pectinesterase and precipitation with ethanol, which may contribute to the removal of part of the sugars present in this sample, thus resulting in lower production of these compounds in case they were formed.

All of the results taken together show that lemon peel is a valuable source of dietary fibre that is predominantly soluble and primarily comprises pectin; it also contains IDF due to its cellulose and hemicellulose components. It boasts good functional properties, including hydration and fat absorption properties. Moreover, the presence of (poly)phenols contribute to its antioxidant capacity.

In conclusion, upon treatment with pectinesterases, there were noticeable changes in the composition due to pectin hydrolysis. This process resulted in a reduction in the SDF percentage, an increase in the IDF content, a higher proportion of cellulose, and a diminished hemicellulose content. The alteration in the pectin structure adversely affected hydration and fat absorption properties. Simultaneously, processing led to a reduction in the (poly)phenol content, consequently the diminishing antioxidant capacity.

For all that, the treatment applied compromises further uses of the obtained ingredients and is a crucial step in the valorisation of lemon by-products. Because the valorisation of lemon peel by-products emerges as a crucial avenue, this will enable the acquisition of novel ingredients rich in dietary fibre and (poly)phenols with a notable antioxidant capacity. This approach holds promise for mitigating the environmental impact associated with substantial by-product generation at the agro-industrial level.

3. Materials and Methods

3.1. Samples

In the current study, we employed two samples derived from by-products of the citrus industry, specifically lemon peels. One sample was obtained by drying lemon peels with warm air in a chamber at 60 °C until constant weight (LP) (≈48 h), while the other sample underwent a hydrolysis process using pectinesterases (LPp), and, after the enzymatic treatment, the sample was subsequently dried using the same conditions described above. Both samples were ground, and the powders obtained were stored under refrigeration until analyses were performed.

3.2. Enzymatic Treatment

The lemon peel sample was mixed with water (1/1, *v*/*v*), and then pectin was hydrolysed by adding 1% of commercial pectinesterase (EC 3.1.1.11.) (pectin/NaCl, *w*/*v*) (Sigma, St. Louis, MO, USA). Samples were introduced to a water bath at 30 °C for 30 min at pH 7.5. To inactivate the enzyme, samples were introduced to a bath at 90 °C during 2 min, and after that they were put in cold water until reaching room temperature. Absolute ethanol was added in a proportion of 1/4, *v*/*v* to precipitate the soluble dietary fibre polysaccharides. In the following step, the solid phase was separated through centrifugation, and it was dried in a chamber at 60 °C until constant weight (≈48 h).

3.3. Dietary Fibre Quantification

Total dietary fibre (TDF), insoluble dietary fibre (IDF), and soluble dietary fibre (SDF) were determined through the enzymatic–gravimetric method described by Prosky et al. [36]. In short, an enzymatic digestion was carried out by using the Total Dietary Fibre Assay Kit (Megazyme Ltd., Wicklow, Ireland).

For analysis, 0.5 g of each sample was weighed and diluted with 40 mL of MES-TRIS Buffer. After homogenisation, 50 µL of α-amylase enzyme was added to facilitate starch hydrolysis, which was placed for 30 min at 100 °C and 60 rpm in a thermostatic bath. Subsequently, 100 µL of protease was added and maintained at 60 °C under the same conditions. Once digestion was completed, the mixture was cooled to 20 °C, and the pH was adjusted between 4.1 and 4.8 before adding 5 mL of 0.561 N HCl. Next, 200 µL of amyloglucosidase was introduced, and the mixture was incubated for 30 min at 60 °C. To separate the two phases, the Fibertec System E 1023 (Foss, Höganas, Sweden) was used. An initial filtration was performed to isolate the IDF. The volume obtained from the filtrate of the previous step was combined with a 95% ethanol solution at 65 °C and allowed to stand for one hour. Following this, the filtration procedure was repeated, and the SDF residue was collected.

Once the residues were obtained, their weight was calculated by determining the ash and protein content. This involved using a muffle furnace for the incineration of the residues at 525 °C during 24 h. Additionally, the Kjeldahl method was used to quantify both the total nitrogen and the protein [37].

3.4. Dietary Fibre Characterisation through Gas Chromatography

To determine the composition of dietary fibre polysaccharides, a process similar to that previously described for SDF and IDF was carried out, but only the TDF fraction was obtained. After enzymatic digestion, 95% ethanol was added at 65 °C, followed by centrifugation at 4500× *g* for 10 min, and the protocol described by Englyst et al. was followed [38]. Briefly, the supernatant was removed, and the residue was treated with 5 mL of 12 M H_2SO_4. It was maintained in a 35 °C bath for 30 min, with homogenisation every 10 min. Finally, the mixture was heated in a bath for 1 h at 100 °C, with homogenisation every 10 min. Once the acid hydrolysis was complete, breaking down the polysaccharide chains, sugar derivatisation was performed on the hydrolysed samples. The analysis was performed using a Gas Chromatography (GC) 7890B (Agilent, Machelen, Belgium). A mixture of neutral sugars (rhamnose, fucose, arabinose, xylose, mannose, galactose, and glucose) was used as the standard, and allose (2595-97-3, Thermo Scientific, Madrid, Spain) served as the internal standard. The results were expressed as a percentage.

Uronic acids were determined through the colorimetry method described by Scott (1979) [39] using the hydrolysed residue obtained in the previous extraction. In short, the sample was diluted with 2 M H_2SO_4 and 300 µL of 3% boric acid and 2% sodium chloride were added, followed by 5 mL of H_2SO_4. The mixture was then heated in a 70 °C bath for 40 min. After cooling, 200 µL of dimethylphenol was added, and after 15 min, the absorbance was measured at 400 nm and 450 nm to remove hexose interference. For quantification, galacturonic acid was used, and the results were expressed as a percentage.

The pectin, hemicellulose, and cellulose percentages and the pectin structure in the samples were estimated based on the calculations proposed by Houben et al. and Umaña et al. [40,41].

3.5. Functional Properties

To determine the hydration properties (SWC and WRC) of the fibre and the fat absorption capacity, three different functional properties were analysed based on the protocols previously published by Navarro-González et al. [42].

The WRC was measured by weighting 1 g of the sample, which was mixed with 30 mL of MilliQ water and then homogenised. After standing for 24 h, it was centrifuged at 3000 rpm for 20 min. The supernatant was removed, and the dry residue was weighed and kept for 18 h at 110 °C. The WRC was calculated according to the gravimetric difference.

For the SWC, 0.1 g of the sample was weighed, and 10 mL of MilliQ water was added. The mixture was mixed and left at room temperature for 18 h, followed by measuring the volume occupied by the dry residue in the graduated tube.

For the FAC, 4 g of the sample was weighed, and 24 mL of sunflower oil was added. The mixture was homogenised at 5 min intervals for 30 min. Subsequently, the mixture was centrifuged for 25 min at 3000 rpm, and the supernatant was removed. The residue with the adsorbed oil was weighed, and the calculation was performed according to the gravimetric difference.

3.6. (Poly)phenol Extraction and Total Phenolic Content Analysis

To extract (poly)phenols, two different extraction procedures described by Arranz et al. [43] were employed to yield both extractable (poly)phenols (EPP) and non-extractable (poly)phenols (NEPP). In short, to determine the EPP, 0.25 g of the sample was taken and blended with a solution of methanol and water acidified with 1% HCl (50/50, *v*/*v*) to achieve a pH close to 2. The samples were homogenised in an orbital shaker and subsequently centrifuged (4500× *g*, 10 min, room temperature) to collect the supernatant. This process was then repeated using a solution of acetone/water (70/30 *v*/*v*), and the resulting supernatant was combined with the previous one.

The pellet derived from the preceding process was used for determining the NEPP. This pellet was mixed with 10 mL of methanol/water/formic acid (79/19/1, *v*/*v*/*v*), which underwent a 20 h incubation at 85 °C, followed by centrifugation (4500× *g*, 10 min, room temperature). The collected supernatant was isolated, and the resultant residue was washed with the extraction solvent and mixed with the one previously obtained.

The total phenolic content (TPC) was analysed for both the EPP and the NEPP using the Folin–Ciocalteu method described by Singleton and Rossi (1965) [44]. This method involves an oxidation/reduction and colorimetric reaction. For the colorimetric assay, both a Folin–Ciocalteau reagent and Na_2CO_3 were added until approximately pH 10, where a blue chromophore emerged due to the reduction of phosphomolybdic-phosphotungstic complexes, leading to tungsten and molybdenum oxides. After 1 h, absorbance was measured at 750 nm using a UV–visible spectrophotometer (Evolution 300, Thermo-Scientific, Oxford, UK). Gallic acid (Riedelde Haën, Hannover, Germany) was used as the standard, and the TPC in the samples was expressed as mg of gallic acid equivalents (GAE)/g of the sample expressed in dry weight (d.w.).

3.7. Antioxidant Capacity Analysis

The antioxidant capacity analysis was determined in both fractions, EPP and NEPP, extracted following the protocol described above. For that, the ferric reducing antioxidant power (FRAP) assay was used following the protocol described by Benzie and Strain (1996) [45]. Briefly, 100 µL of both EPP and NEPP extracts were mixed with 900 µL of the FRAP reagent. Absorbance was measured at 593 nm in a UV–visible spectrophotometer (Evolution 300, Thermo-Scientific, Oxford, UK) precisely 4 min from the commencement of the reaction. The FRAP reagent composition comprises 0.3 M of acetate buffer, a 10 mM

2,4,6-tripyridyl-s-triazine (TPTZ) solution in a 40 mM HCl solution, and $FeCl_3 \cdot 6H_2O$ solution in the following proportions: 20 mL of acetate buffer, 2 mL of TPZP, and 2 mL of $FeCl_3 \cdot 6H_2O$. Trolox was used as the standard, and the results were expressed as mg Trolox equivalents (TE)/g of the sample in d.w.

3.8. Statistical Analysis

The data were processed using R studio, version 4.0.5 (R Foundation for Statistical Computing, Vienna, Austria). All assays were conducted in triplicate. Normality was determined through the Shapiro–Wilk test. The homogeneity of variances was analysed using the Bartlett test. The *t*-student test was conducted to determine significant differences at p-value < 0.05. Correlation analysis were performed using the Pearson correlation test for the relationships between the functional properties and fibre composition parameters.

Supplementary Materials: The following supporting information can be downloaded at: https://www.mdpi.com/article/10.3390/molecules29010269/s1, Figure S1: Chromatograms of (a) standard mix I for LP sample; (b) standard mix II for LPp sample; (c) LP sample; (d) LPp sample. Different peaks are 1 (rhamnose); 2 (fucose); 3 (arabinose); 4 (xylose); 5 (allose); 6 (mannose); 7 (galactose); 8 (glucose).

Author Contributions: Conceptualization, M.J.P. and V.N.-G.; methodology, M.J.P., V.N.-G. and M.S.M.; formal analysis, investigation, and data curation, M.J.P., V.N.-G. and M.S.M.; writing—original draft preparation, review, and editing, V.N.-G., R.G.-B. and M.J.P.; funding acquisition, M.J.P. All authors have read and agreed to the published version of the manuscript.

Funding: This research received no external funding.

Institutional Review Board Statement: Not applicable.

Data Availability Statement: Data are contained within the article and Supplementary Materials.

Conflicts of Interest: The authors declare no conflicts of interest.

References

1. Del Rio Osorio, L.L.; Flórez-López, E.; Grande-Tovar, C.D. The Potential of Selected Agri-Food Loss and Waste to Contribute to a Circular Economy: Applications in the Food, Cosmetic and Pharmaceutical Industries. *Molecules* **2021**, *26*, 515. [CrossRef] [PubMed]
2. Hussain, S.; Jõudu, I.; Bhat, R. Dietary Fiber from Underutilized Plant Resources—A Positive Approach for Valorization of Fruit and Vegetable Wastes. *Sustainability* **2020**, *12*, 5401. [CrossRef]
3. Singh, B.; Singh, J.P.; Kaur, A.; Singh, N. Phenolic Composition, Antioxidant Potential and Health Benefits of Citrus Peel. *Food Res. Int.* **2020**, *132*, 109114. [CrossRef] [PubMed]
4. Jiang, H.; Zhang, W.; Xu, Y.; Chen, L.; Cao, J.; Jiang, W. An Advance on Nutritional Profile, Phytochemical Profile, Nutraceutical Properties, and Potential Industrial Applications of Lemon Peels: A Comprehensive Review. *Trends Food Sci. Technol.* **2022**, *124*, 219–236. [CrossRef]
5. Huang, J.; Liao, J.; Qi, J.; Jiang, W.; Yang, X. Structural and Physicochemical Properties of Pectin-Rich Dietary Fiber Prepared from Citrus Peel. *Food Hydrocoll.* **2021**, *110*, 106140. [CrossRef]
6. Rafiq, S.; Kaul, R.; Sofi, S.A.; Bashir, N.; Nazir, F.; Ahmad Nayik, G. Citrus Peel as a Source of Functional Ingredient: A Review. *J. Saudi Soc. Agric. Sci.* **2018**, *17*, 351–358. [CrossRef]
7. Azman, N.F.I.N.; Azlan, A.; Khoo, H.E.; Razman, M.R. Antioxidant Properties of Fresh and Frozen Peels of Citrus Species. *Curr. Res. Nutr. Food Sci.* **2019**, *7*, 331–339. [CrossRef]
8. Uçak, İ.; Khalily, R. Effects of Different Solvent Extractions on the Total Phenolic Content and Antioxidant Activity of Lemon and Orange Peels. *Eurasian J. Food Sci. Technol.* **2022**, *6*, 23–28.
9. Kaya, M.; Sousa, A.G.; Crépeau, M.J.; Sørensen, S.O.; Ralet, M.C. Characterization of Citrus Pectin Samples Extracted under Different Conditions: Influence of Acid Type and PH of Extraction. *Ann. Bot.* **2014**, *114*, 1319–1326. [CrossRef]
10. Maphosa, Y.; Jideani, V.A. Dietary Fiber Extraction for Human Nutrition—A Review. *Food Rev. Int.* **2016**, *32*, 98–115. [CrossRef]
11. Barber, T.M.; Kabisch, S.; Pfeiffer, A.F.H.; Weickert, M.O. The Health Benefits of Dietary Fibre. *Nutrients* **2020**, *12*, 3209. [CrossRef]
12. Saura-Calixto, F. Antioxidant Dietary Fiber Product: A New Concept and a Potential Food Ingredient. *J. Agric. Food Chem.* **1998**, *46*, 4303–4306. [CrossRef]
13. Shahidi, F.; Chandrasekara, A. Interaction of Phenolics and Their Association with Dietary Fiber. In *Dietary Fiber Functionality in Food and Nutraceuticals: From Plant to Gut*; Wiley: Hoboken, NJ, USA, 2017; pp. 21–44. [CrossRef]

14. Pérez-Jiménez, J.; Díaz-Rubio, M.E.; Saura-Calixto, F. Non-Extractable Polyphenols, a Major Dietary Antioxidant: Occurrence, Metabolic Fate and Health Effects. *Nutr. Res. Rev.* **2013**, *26*, 118–129. [CrossRef]
15. Carboni Martins, C.; Rodrigues, R.C.; Domeneghini Mercali, G.; Rodrigues, E. New Insights into Non-Extractable Phenolic Compounds Analysis. *Food Res. Int.* **2022**, *157*, 111487. [CrossRef] [PubMed]
16. Rudrapal, M.; Khairnar, S.J.; Khan, J.; Dukhyil, A.B.; Ansari, M.A.; Alomary, M.N.; Alshabrmi, F.M.; Palai, S.; Deb, P.K.; Devi, R. Dietary Polyphenols and Their Role in Oxidative Stress-Induced Human Diseases: Insights into Protective Effects, Antioxidant Potentials and Mechanism(s) of Action. *Front. Pharmacol.* **2022**, *13*, 283. [CrossRef] [PubMed]
17. United States Department of Agriculture (USDA). FoodData Central: Lemon Peel Composition. Available online: https://fdc.nal.usda.gov/fdc-app.html#/food-details/167749/nutrients (accessed on 27 September 2023).
18. Chavan, P.; Singh, A.K.; Kaur, G. Recent Progress in the Utilization of Industrial Waste and By-Products of Citrus Fruits: A Review. *J. Food Process Eng.* **2018**, *41*, e12895. [CrossRef]
19. Czech, A.; Malik, A.; Sosnowska, B.; Domaradzki, P. Bioactive Substances, Heavy Metals, and Antioxidant Activity in Whole Fruit, Peel, and Pulp of Citrus Fruits. *Int. J. Food Sci.* **2021**, *2021*, 6662259. [CrossRef] [PubMed]
20. Wang, L.; Xu, H.; Yuan, F.; Pan, Q.; Fan, R.; Gao, Y. Physicochemical Characterization of Five Types of Citrus Dietary Fibers. *Biocatal. Agric. Biotechnol.* **2015**, *4*, 250–258. [CrossRef]
21. Qi, B.; Jiang, L.; Li, Y.; Chen, S.; Sui, X. Extract Dietary Fiber from the Soy Pods by Chemistry-Enzymatic Methods. *Procedia Eng.* **2011**, *15*, 4862–4873. [CrossRef]
22. Fuso, A.; Viscusi, P.; Larocca, S.; Sangari, F.S.; Lolli, V.; Caligiani, A. Protease-Assisted Mild Extraction of Soluble Fibre and Protein from Fruit By-Products: A Biorefinery Perspective. *Foods* **2023**, *12*, 148. [CrossRef]
23. Rivas, M.Á.; Benito, M.J.; Martín, A.; Córdoba, M.d.G.; Ruíz-Moyano, S.; Casquete, R. Improve the Functional Properties of Dietary Fibre Isolated from Broccoli By-Products by Using Different Technologies. *Innov. Food Sci. Emerg. Technol.* **2022**, *80*, 103075. [CrossRef]
24. Peng, X.; Nie, S.; Li, X.; Huang, X.; Li, Q. Characteristics of the Water- and Alkali-Soluble Hemicelluloses Fractionated by Sequential Acidification and Graded-Ethanol from Sweet Maize Stems. *Molecules* **2019**, *24*, 212. [CrossRef]
25. Mendez, D.A.; Fabra, M.J.; Martínez-Abad, A.; Martínez-Sanz, M.; Gorria, M.; López-Rubio, A. Understanding the Different Emulsification Mechanisms of Pectin: Comparison between Watermelon Rind and Two Commercial Pectin Sources. *Food Hydrocoll.* **2021**, *120*, 106957. [CrossRef]
26. Belkheiri, A.; Forouhar, A.; Ursu, A.V.; Dubessay, P.; Pierre, G.; Delattre, C.; Djelveh, G.; Abdelkafi, S.; Hamdami, N.; Michaud, P. Extraction, Characterization, and Applications of Pectins from Plant By-Products. *Appl. Sci.* **2021**, *11*, 6596. [CrossRef]
27. Shafie, M.H.; Yusof, R.; Samsudin, D.; Gan, C.Y. Averrhoa Bilimbi Pectin-Based Edible Films: Effects of the Linearity and Branching of the Pectin on the Physicochemical, Mechanical, and Barrier Properties of the Films. *Int. J. Biol. Macromol.* **2020**, *163*, 1276–1282. [CrossRef] [PubMed]
28. Cho, S.S.; Samuel, P. *Fiber Ingredients: Food Applications and Health Benefits*; CRC Press: Boca Raton, FL, USA, 2009; ISBN 9781420043853.
29. Shahidi, F.; Varatharajan, V.; Oh, W.Y.; Peng, H. Phenolic Compounds in Agri-Food by-Products, Their Bioavailability and Health Effects. *J. Food Bioact.* **2019**, *5*, 57–119. [CrossRef]
30. Zhang, Y.; Liao, J.; Qi, J. Functional and Structural Properties of Dietary Fiber from Citrus Peel Affected by the Alkali Combined with High-Speed Homogenization Treatment. *LWT* **2020**, *128*, 109397. [CrossRef]
31. Gómez-Mejía, E.; Sacristán, I.; Rosales-Conrado, N.; León-González, M.E.; Madrid, Y. Effect of Storage and Drying Treatments on Antioxidant Activity and Phenolic Composition of Lemon and Clementine Peel Extracts. *Molecules* **2023**, *28*, 1624. [CrossRef]
32. Pieracci, Y.; Pistelli, L.; Cecchi, M.; Pistelli, L.; De Leo, M. Phytochemical Characterization of Citrus-Based Products Supporting Their Antioxidant Effect and Sensory Quality. *Foods* **2022**, *11*, 1550. [CrossRef]
33. Shu, B.; Wu, G.; Wang, Z.; Wang, J.; Huang, F.; Dong, L.; Zhang, R.; Wang, Y.; Su, D. The Effect of Microwave Vacuum Drying Process on Citrus: Drying Kinetics, Physicochemical Composition and Antioxidant Activity of Dried Citrus (*Citrus reticulata* Blanco) Peel. *J. Food Meas. Charact.* **2020**, *14*, 2443–2452. [CrossRef]
34. Stahl, W.; Sies, H. Antioxidant Activity of Carotenoids. *Mol. Asp. Med.* **2003**, *24*, 345–351. [CrossRef] [PubMed]
35. Yilmaz, Y.; Toledo, R. Antioxidant Activity of Water-Soluble Maillard Reaction Products. *Food Chem.* **2005**, *93*, 273–278. [CrossRef]
36. Prosky, L.; Asp, N.G.; Schweizer, T.F.; DeVries, J.W.; Furda, I. Determination of Insoluble, Soluble, and Total Dietary Fiber in Foods and Food Products: Interlaboratory Study. *J. Assoc. Off. Anal. Chem.* **1988**, *71*, 1017–1023. [CrossRef] [PubMed]
37. AOAC. *Official Methods of Analysis of AOAC International*, 20th ed.; AOAC: Rockville, MD, USA, 2016; Volume II.
38. Englyst, H.N.; Quigley, M.E.; Hudson, G.J.; Cummings, J.H. Determination of Dietary Fibre as Non-Starch Polysaccharides by Gas-Liquid Chromatography. *Analyst* **1992**, *117*, 1707–1714. [CrossRef] [PubMed]
39. Scott, R.W. Colorimetric Determination of Hexuronic Acids in Plant Materials. *Anal. Chem.* **1979**, *51*, 936–941. [CrossRef]
40. Umaña, M.; Eim-Iznardo, V.; Roselló, M. Cinéticas de Extracción y Caracterización de Pectinas de Los Subproductos de Naranja Mediante Asistencia Acústica. Master's Thesis, Universitat de Les Illes Balears, Palma, Spain, 2016.
41. Houben, K.; Jolie, R.P.; Fraeye, I.; Van Loey, A.M.; Hendrickx, M.E. Comparative Study of the Cell Wall Composition of Broccoli, Carrot, and Tomato: Structural Characterization of the Extractable Pectins and Hemicelluloses. *Carbohydr. Res.* **2011**, *346*, 1105–1111. [CrossRef] [PubMed]

42. Navarro-González, I.; García-Valverde, V.; García-Alonso, J.; Periago, M.J. Chemical Profile, Functional and Antioxidant Properties of Tomato Peel Fiber. *Food Res. Int.* **2011**, *44*, 1528–1535. [CrossRef]
43. Arranz, S.; Saura-Calixto, F.; Shaha, S.; Kroon, P.A. High Contents of Nonextractable Polyphenols in Fruits Suggest That Polyphenol Contents of Plant Foods Have Been Underestimated. *J. Agric. Food Chem.* **2009**, *57*, 7298–7303. [CrossRef]
44. Singleton, V.L.; Rossi, J.A. Colorimetry of Total Phenolics with Phosphomolybdic-Phosphotungstic Acid Reagents. *Am. J. Enol. Vitic.* **1965**, *16*, 114–158. [CrossRef]
45. Benzie, I.F.F.; Strain, J.J. The Ferric Reducing Ability of Plasma (FRAP) as a Measure of "Antioxidant Power": The FRAP Assay. *Anal. Biochem.* **1996**, *239*, 70–76. [CrossRef]

molecules MDPI

Article

Antidiabetic, Antiglycation, and Antioxidant Activities of Ethanolic Seed Extract of *Passiflora edulis* and Piceatannol In Vitro

Flávia A. R. dos Santos [1,2], Jadriane A. Xavier [1], Felipe C. da Silva [1], J. P. Jose Merlin [2], Marília O. F. Goulart [1] and H. P. Vasantha Rupasinghe [2,*]

[1] Institute of Chemistry and Biotechnology, Federal University of Alagoas, Maceio 57072-970, Brazil; flavia.santos@iqb.ufal.br (F.A.R.d.S.); jadrianexavier@iqb.ufal.br (J.A.X.); felipe.silva@iqb.ufal.br (F.C.d.S.); mofg@qui.ufal.br (M.O.F.G.)

[2] Department of Plant, Food, and Environmental Sciences, Faculty of Agriculture, Dalhousie University, 50 Pictou Road, Truro, NS B2N 5E3, Canada; josemerlinj@dal.ca

* Correspondence: vrupasinghe@dal.ca; Tel.: +1-902-956-1917

Abstract: The objective of this work was to investigate the antidiabetic, antiglycation, and antioxidant potentials of ethanolic extract of seeds of Brazilian *Passiflora edulis* fruits (PESE), a major by-product of the juice industry, and piceatannol (PIC), one of the main phytochemicals of PESE. PESE, PIC, and acarbose (ACB) exhibited IC_{50} for alpha-amylase, 32.1 ± 2.7, 85.4 ± 0.7, and 0.4 ± 0.1 μg/mL, respectively, and IC_{50} for alpha-glucosidase, 76.2 ± 1.9, 20.4 ± 7.6, and 252 ± 4.5 μg/mL, respectively. The IC_{50} of PESE, PIC, and sitagliptin (STG) for dipeptidyl-peptidase-4 (DPP-4) was 71.1 ± 2.6, 1137 ± 120, and 0.005 ± 0.001 μg/mL, respectively. PESE and PIC inhibited the formation of advanced glycation end-products (AGE) with IC_{50} of 366 ± 1.9 and 360 ± 9.1 μg/mL for the initial stage and 51.5 ± 1.4 and 67.4 ± 4.6 μg/mL for the intermediate stage of glycation, respectively. Additionally, PESE and PIC inhibited the formation of β-amyloid fibrils in vitro up to 100%. IC_{50} values for 1,1-diphenyl-2-picrylhydrazyl radical (DPPH$^•$) scavenging activity of PESE and PIC were 20.4 ± 2.1, and 6.3 ± 1.3 μg/mL, respectively. IC_{50} values for scavenging hypochlorous acid (HOCl) were similar in PESE, PIC, and quercetin (QCT) with values of 1.7 ± 0.3, 1.2 ± 0.5, and 1.9 ± 0.3 μg/mL, respectively. PESE had no cytotoxicity to the human normal bronchial epithelial (BEAS-2B), and alpha mouse liver (AML-12) cells up to 100 and 50 μg/mL, respectively. However, 10 μg/mL of the extract was cytotoxic to non-malignant breast epithelial cells (MCF-10A). PESE and PIC were found to be capable of protecting cultured human cells from the oxidative stress caused by the carcinogen NNKOAc at 100 μM. The in vitro evidence of the inhibition of alpha-amylase, alpha-glucosidase, and DPP-4 enzymes as well as antioxidant and antiglycation activities, warrants further investigation of the antidiabetic potential of *P. edulis* seeds and PIC.

Keywords: passion fruit; piceatannol; agro-industrial residues; type 2 diabetes; cytotoxicity; seeds

Citation: dos Santos, F.A.R.; Xavier, J.A.; da Silva, F.C.; Merlin, J.P.J.; Goulart, M.O.F.; Rupasinghe, H.P.V. Antidiabetic, Antiglycation, and Antioxidant Activities of Ethanolic Seed Extract of *Passiflora edulis* and Piceatannol In Vitro. *Molecules* **2022**, 27, 4064. https://doi.org/10.3390/molecules27134064

Academic Editors: José Pinela, Maria Inês Dias, Carla Pereira and José Ignacio Alonso-Esteban

Received: 25 May 2022
Accepted: 21 June 2022
Published: 24 June 2022

1. Introduction

The *Passiflora* (Passifloraceae family) genus has more than 500 species; however, *Passiflora edulis* Sims *f. flavicarpa* O. Degenerer (yellow or sour form passion fruit) and *P. alata* (sweet form) are the two main economically important species [1]. Brazil is currently the leading producer of passion fruit in the world, with 690,364 tons of production in 2020 [2]. Several biological properties of passion fruits have been demonstrated using in vitro and in vivo studies, such as antioxidant (leaf, peel, and seed), analgesic, antidepressant, sedative and anxiolytic-like (leaf), anti-inflammatory (peel and leaf), antimicrobial (seed), antihypertensive (peel), hepatoprotective (peel and seed), and antidiabetic (peel, juice, and seed) activities [3]. In Brazil, *P. edulis*, also known as "maracuja", has been used widely in folk medicine as an anxiolytic, sedative, and analgesic agent [4].

One of the main bioactive compounds of seeds of *P. edulis* is piceatannol (PIC, 3,3′,4′,5-*trans*-tetrahydroxystilbene), an analog of resveratrol. Although the health-promoting effects of PIC have not been studied extensively as resveratrol [5], it has been shown to possess preventive effects concerning cardiovascular diseases and certain cancers [6], being anti-inflammatory [7], photoprotective [8], and as a skin hydration [9] and an antidiabetic agent [10].

Diabetes is one of the significant health problems worldwide, mainly due to its increased prevalence, irreversible complications, and high economic burden [11]. The International Diabetes Federation stated that 537 million people aged 20–79 years in the world or about 10.5% of all adults in this age group were affected by diabetes in 2021. More than 90% of diabetes cases are type 2 diabetes mellitus (T2DM), a metabolic disorder characterized by chronic hyperglycemia due to insufficient production of insulin by the pancreas or by insulin resistance [12].

The present therapeutic drugs used in the management of T2DM are sulfonylureas, biguanides, SGLT2 inhibitors, alpha-glucosidase inhibitors, dipeptidyl peptidase-4 (DPP-4) inhibitors, GLP-1 receptor agonists, thiazolidinediones, amylin analogs, and insulin [13]. However, continuous use of some FDA-approved oral drugs is accompanied by side effects; for instance, acarbose (ACB) leads to abdominal or stomach pain, flatulence, diarrhea, bloating, and cramping [14]. As such, there is a continuous need for safe, natural health products that can be used to prevent and manage T2DM.

One of the strategies to prevent postprandial hyperglycemia in T2DM is inhibiting the catalytic activity of digestive enzymes, which are responsible for catalyzing the hydrolysis of oligosaccharides into glucose, including alpha-amylase and alpha-glucosidase. These enzymes increase glucose absorption and, as a result, increase the glucose level in the bloodstream [15]. Another potential therapeutic approach that lowers blood glucose concentration in T2DM is blocking the action of DPP-4, an enzyme that degrades the hormone incretin, which is secreted into the blood after a meal to stimulate insulin secretion from pancreatic β-cells and to regulate glucose production by the liver [13].

Diabetes has also been considered a risk factor for neurodegenerative diseases such as Alzheimer's disease (AD) [16,17], which is characterized by the accumulation of extracellular insoluble senile plaque and intracellular neurofibrillary tangles. Hyperglycemia, as an inevitable consequence of increased insulin resistance, leads to the glycation of proteins and, consequently the formation of advanced glycation end products (AGEs) [18]. AGEs, which accumulate over some time and are not frequently detoxified, have been associated with amyloid-based neurodegenerative diseases [17], stabilizing protein aggregates [19]. Furthermore, the interaction of AGEs with AGE receptors (RAGE) causes inflammatory processes and oxidative stress in cells. As RAGE is not specific, it can also interact with other ligands such as β-amyloid peptide (Aβ), forming agglomerates, which lead to increased inflammation, oxidative stress, neuronal dysfunction, with consequent AD aggravation [18].

(Poly)phenol-rich plant extracts and individual (poly)phenols have been extensively studied as an alternative or complement to the current hypoglycemic medicines. In vitro studies have demonstrated that (poly)phenols interact with enzymes and other biological macromolecules. The interaction depends on the composition, molecular weight, and the position of hydroxyl groups of the bioactive compounds [20].

Daily administration of PIC caused a decrease in glucose levels in animal models. For instance, PIC supplementation reduced fasting blood glucose in both *db/db* mice and high-fat diet-fed C57BL/6J mice [21]. PIC promoted glucose uptake in the absence of insulin in cultured myotubes; enhanced AMP-activated protein kinase (AMPK) activation and glucose transporter type 4 (GLUT4) translocation to overcome insulin resistance [21]. PIC lowered the blood glucose and improved glucose tolerance in diabetic mice [22]. Therefore, PIC has therapeutic potential for the prevention and improvement of symptoms of diabetes.

However, to our best knowledge, no study has been reported in relation to the seeds of Brazilian *P. edulis* (yellow passion fruit) as an antidiabetic agent through the association of alpha-glucosidase, alpha-amylase, and DPP-4 inhibition mechanisms, not even as an antiglycant in the initial and intermediate stages of glycation and as an inhibitor of fibrillation of proteins subjected to glycation. As such, the present study investigated the potential use of bioactive chemical constituents of passion fruit seeds as a natural health product for managing chronic disorders, primarily T2DM. Total antioxidant capacity, antidiabetic, and antiglycation activities of extracts prepared from seeds of Brazilian *P. edulis* and with PIC were investigated. In addition, the cytotoxicity of passion fruit seed extracts (PESE) was studied using three cell lines.

2. Results and Discussion

2.1. Antidiabetic Activity

In vitro antidiabetic potential of PESE and PIC was investigated by determining the ability to inhibit pancreatic alpha-amylase, alpha-glucosidase, and DPP-4. In this study, ACB and STG were used for comparison purposes. All tested samples showed activity against alpha-amylase, alpha-glucosidase, and DPP-4 in a concentration-dependent manner (Table 1). For pancreatic alpha-amylase, PESE, PIC, and ACB presented IC_{50} values of 32.1 ± 2.7, 85.4 ± 0.7, and 0.4 ± 0.1 μg/mL, respectively. For yeast alpha-glucosidase, the values of IC_{50} of PESE, PIC, and ACB were 76.2 ± 1.9, 20.4 ± 7.6, and 252 ± 4.5 μg/mL, respectively. IC_{50} values of PESE and PIC were higher when compared to ACB for alpha-amylase but significantly lower ($p \leq 0.05$) for alpha-glucosidase. Interestingly, passion fruit seed presented an IC_{50} value 3-times lower than PIC in alpha-amylase but 4-fold higher in the alpha-glucosidase assay. Moreover, PIC showed antidiabetic activity in terms of alpha-glucosidase, which was 12-fold higher than ACB. For DPP-4 human recombinant enzyme, the values of IC_{50} of PESE, PIC, and STG were, respectively, 71.1 ± 2.6, 1137 ± 120, and 0.005 ± 0.001 μg/mL. The results indicate that other bioactive compounds are also involved in the inhibition of alpha-amylase, while PIC may be a key component against alpha-glucosidase.

Table 1. Antidiabetic activity in vitro of ethanolic extract of *P. edulis* seeds (PESE) and reference compounds.

Sample	Alpha-Amylase IC_{50} (μg/mL)	Alpha-Glucosidase IC_{50} (μg/mL)	DPP-4 IC_{50} (μg/mL)
PESE	32.1 ± 2.7 ^{a}	76.2 ± 1.9 ^{A}	71.1 ± 2.6 ^{x}
PIC	85.4 ± 0.7 ^{b}	20.4 ± 7.6 ^{B}	1137.5 ± 120.2 ^{y}
ACB	0.4 ± 0.1 ^{c}	251.6 ± 4.5 ^{C}	-
STG	-	-	0.005 ± 0.001 ^{z}

IC_{50} values for alpha-amylase, alpha-glucosidase, and DPP-4 are given. $^{a–c}$, $^{A–C}$, $^{x–z}$, mean ± SD followed by different letters within the same assay represent significant differences (ANOVA analysis was performed followed by the Tukey test, $p \leq 0.05$). Data are means of triplicates. Abbreviation: PIC, piceatannol; ACB, acarbose; STG, sitagliptin.

The extract of passion fruit peel also displayed higher potential towards alpha-glucosidase than alpha-amylase. The ethanolic extract of a hybrid passion fruit peel was shown to have a weak inhibitory effect in alpha-amylase with IC_{50} values of 1.8 ± 0.1 mg/mL (1800 ± 100 μg/mL), but a stronger effect in alpha-glucosidase with IC_{50} values of 0.6 ± 0.1 mg/mL (600 ± 100 μg/mL), which was 4.3-fold higher than standard drug ACB, with IC_{50} values of 0.3 ± 0.1 mg/mL (300 ± 100 μg/mL) [23].

The same pattern was described by Loizzo et al. [24], when investigating the hypoglycemic activity of ethanol extracts of seeds, peel, and pulp of ten Columbia native *Passiflora* species. Among all studied extracts, *P. ligularis* seeds + pulp had the greatest activity with IC_{50} values of 22.6 and 24.8 μg/mL against alpha-amylase and alpha-glucosidase, respectively, followed by *P. pinnatistipula* (IC_{50} values of 46.4 and 37.7 μg/mL against alpha-amylase and alpha-glucosidase, respectively). Positive control, ACB, had IC_{50} values of 50 μg/mL using starch as the substrate for alpha-amylase and 35.5 μg/mL, when using maltose as the substrate for alpha-glucosidase.

Pan et al. [25] obtained a different IC_{50} value for PIC, isolated from ethanol extract of *P. edulis* Sims seeds, which may be due to different conditions performed (substrate, enzyme concentrations, and incubation time). However, PIC derivatives exhibited potent inhibition against in vitro alpha-glucosidase with IC_{50} values ranging from 1.7 to 35.1 μM. Among them, IC_{50} of PIC was 4.3 ± 0.07 μM, while for ACB, it was 218 ± 3.14 μM.

Concerning other plant families, the leaves of *Annona muricata* are frequently used by Brazilians as a natural treatment of T2DM and its complications. Ethyl acetate fractions obtained from ethanolic extract of *A. muricata* leaves, which are rich in (poly)phenolic compounds, showed inhibition of alpha-amylase (IC_{50} 9.2 ± 2.3 μg/mL) and alpha-glucosidase (IC_{50} 413 ± 121 μg/mL), respectively, although less active than the ACB standard (IC_{50} 0.05 ± 0.01, for alpha-amylase, and 3.4 ± 0.6 μg/mL for alpha-glucosidase) [26].

Antidiabetic drugs aim mainly to achieve normoglycemia and minimize the progression of diabetes complications. Current antidiabetic drugs can interfere in one or more of the eight pathways which results in hyperglycemia. ACB, for instance, inhibits digestive enzymes, such as alpha-amylase, alpha-glucosidase, sucrase, and maltase. The inhibitor of these enzymes delays the digestion and absorption of dietary carbohydrates from the epithelium of the small intestine, and, thus, decreases the demand for insulin. Differently, STG acts to reduce the incretin effect in the small intestine by blocking the DPP-4 enzyme [13].

Due to the variety of mechanisms by which normoglycemia can be achieved, several tests can be performed to assess the antidiabetic potential of natural products. Generally, in vitro assays usually test the inhibition of digestive and DPP-4 enzymes. Our findings reveal that PESE and PIC can act as inhibitors of alpha-amylase, alpha-glucosidase, and DPP-4. Even though ACB and STG are more effective in terms of concentration in the inhibition of alpha-amylase and DPP-4, this extract and the pure compound PIC can be further investigated for use in natural antidiabetic formulations. Factors such as cost, safety, and side effects should be analyzed since the major concerns of the current drugs are the critical side effects associated with their over-use [13,15].

A recent study indicated that PIC inhibited the formation of *p*-nitrophenol by alpha-glucosidase through hydrophobic interactions and hydrogen bonding between hydroxy groups and amino acid residues in a non-competitive mechanism. The formation of the PIC-alpha-glucosidase complex may induce changes in the enzyme conformation, altering its function of digesting carbohydrates [27].

2.2. Inhibition of AGE in the Initial and Intermediate States of Glycation by the Extract

Glycation of BSA was performed in the presence and absence of PESE and PIC. Aminoguanidine (AMG) was used as a positive control. Inhibition of the initial and intermediate stages of glycation was carried out using a mixture of reducing sugars (fructose + glucose) and methylglyoxal (MGO), respectively.

The AGE formation inhibitory effect was proportional to the sample concentration. PESE and PIC were able to inhibit the formation of fluorescent AGEs for seven days of incubation at 37 °C with IC_{50} of 366 ± 1.9 and 51.5 ± 1.4 μg/mL for the BSA/fructose + glucose model, and 360 ± 9.1 and 67.4 ± 4.6 μg/mL for BSA/MGO, respectively (Table 2).

Table 2. The antiglycation activity and percentage of inhibition of formation of amyloid fibrils of ethanolic extract of *P. edulis* seeds (PESE) and reference compounds.

Sample	Fructose + Glucose		MGO	
	Antiglycation IC_{50} (μg/mL)	Inhibition of Amyloid Fibrils (%)	Antiglycation IC_{50} (μg/mL)	Inhibition of Amyloid Fibrils (%)
PESE	367 ± 1.9 ^{a}	87.4 ± 2.7 ^{b}	360 ± 9.1 ^{a}	71.9 ± 4.5 ^{b}
PIC	51.5 ± 1.4 ^{b}	100 ± 5.3 ^{a}	67.4 ± 4.6 ^{b}	100 ± 3.8 ^{a}
AMG	25.5 ± 5.0 ^{c}	35.0 ± 5.9 ^{c}	50.4 ± 1.8 ^{c}	30.3 ± 4.4 ^{c}

IC_{50} values were estimated using two models of glucose + fructose and methylglyoxal (MGO). The percentage values of inhibition of the formation of amyloid fibrils presented are the highest obtained for PESE, PIC, and AMG at concentrations of 200, 200, and 100 μg/mL, respectively. $^{a–c}$ Mean ± SD followed by different letters in the same assay represent significant differences (ANOVA analysis was performed followed by the Tukey test, $p \leq 0.05$). Data are means of triplicates. Abbreviations: PIC, piceatannol; AMG, aminoguanidine; MGO, methylglyoxal.

PIC showed higher activity when compared to PESE. The activity of PESE was only 2-fold lower than AMG in the initial stage of glycation and 1.3-fold lower in the intermediate stage. Therefore, PESE and PIC can act as antiglycation agents, probably by preventing the reaction of (di)carbonyl compounds from binding to protein and thus preventing the accumulation of AGEs.

In diabetes, AGEs are found in many organelles, associated with its complication, such as kidneys, retina, and atherosclerotic plaques. The inhibition of glycation is considered an effective strategy against the development or progression of degenerative diseases such as T2DM, atherosclerosis, and Alzheimer's disease, and their complications, as well as in the reduction of chronic inflammatory processes [28,29].

The MGO trapping assay was performed employing its derivatization with *ortho*-phenylenediamine (OPD). PESE, PIC, and the positive control AMG presented 21.9%, 65.1%, and 99.6% of MGO trapping, respectively. Despite the above-shown differences, these results indicate that PESE is capable of acting in the intermediate stage of glycation, inhibiting reactive carbonyl species such as MGO and, consequently, it may reduce the formation of AGEs. The intermediate stage of the glycation reaction leads to the formation of reactive carbonyl species, such as MGO (a model compound), which in turn reacts with amino groups of biomolecules for the formation of AGEs [30]. Moreover, although the classic pathway of the Maillard reaction is well established as a trigger for the formation of AGEs, all reactions leading to the formation of α, β-dicarbonyl compounds in the body also contribute to the formation of AGEs [18]. Thus, the trapping capacity of MGO, a dicarbonyl compound produced during glycation, is an important step against the formation of AGEs.

Zhang, Wang, and Liu [31] observed that after reaction with MGO (5 mM), robinin, procyanidins, luteolin, quercetin, chrysoeriol, kaempferol, genistein, apigenin, and rutin at 5 mM showed capture percentage values of 54.4%, 46.3%, 40.7%, 21.4%, 41.8%, 58.9%, 43.4%, 36.5%, and 49.3%, respectively, values lower than those found in this work for PIC (65.1%, 2 mM). In another study, with ethanolic extract of onion peel (0.5 mg/mL), it was observed elimination of approximately 70% of MGO (0.33 mM) after 1 h of reaction, an excellent capture potential [32], while the methanolic extracts of berries and grapes, using a similar method, showed from 20% to 50% of MGO (10 mM) capture potential. Raspberry, strawberry, blackberry, cranberry, blueberry, and noble grape at a concentration of 2.5 mg/mL showed capture potential of 20%, 30%, 45%, 50%, and 50%, respectively, after 1 h of reaction with MGO [33].

2.3. *Effect on the Formation of Amyloid Fibrils In Vitro*

Protein glycation plays an important role in triggering or facilitating the aggregation of proteins, through the formation of crosslinked β structures that in turn leads to the formation of amyloid fibrils [34]. Thus, the potential of PESE and PIC for inhibiting the formation of amyloid fibrils was evaluated in two different systems, one with BSA in the presence of glucose and fructose (BSA/Fru + Glu) and the other in the presence of MGO (BSA/MGO). The thioflavin T (ThT) fluorescence assay was used as a general indicator for the formation of amyloid fibrils [35].

The formation of β-amyloid fibrils decreased as sample concentration increased for both systems, achieving 100% of inhibition, when treated with PIC at a concentration of 200 μg/mL (Table 2). AMG, however, was able to inhibit up to 35% of the formation of aggregates from the glycation reaction, being therefore less effective in maintaining the native structure of the protein, when compared to PESE and PIC. Our findings suggest that (poly)phenols of PESE might protect the protein backbone since the treatment with PESE and PIC decreased the formation of amyloid fibrils in BSA.

2.4. *Antioxidant Activity*

Our previous study identified six major phenolic compounds of PESE, including PIC, astringin, scirpusin A, scirpusin B, isookanin-7-*O*-glucoside, and naringenin-7-*O*-

glucoside. Among them, PIC has emerged as a promising compound with important biological activities [36]. Hence, PIC was also investigated in this study, along with PESE.

The TPC of PESE corresponds to 227 mg GAE/g, dry weight (DW). DPPH$^\bullet$ and HOCl scavenging assays of PESE and PIC revealed that the antioxidant capacity of PESE is lower than that of PIC; however, this is similar to quercetin (QCT) in the DPPH$^\bullet$ scavenging assay (Table 3). HOCl scavenging activity showed that both PESE and PIC exhibit a strong antioxidant effect in a concentration-dependent manner; PIC had the highest activity and PESE was similar to QCT. Regarding the capacity of scavenging $O_2^{\bullet-}$, PIC presented the highest potential followed by QCT and PESE.

Table 3. Total phenol content (TPC) and antioxidant capacity of ethanolic extract of passion fruit seeds and piceatannol, toward HOCl and $O_2^{\bullet-}$.

Sample	TPC (mg GAE/g DW)	DPPH$^\bullet$ IC_{50} µg/mL	HOCl IC_{50} µg/mL	$O_2^{\bullet-}$ IC_{50} µg/mL
PESE	227 ± 3.9	20.4 ± 2.1 ^{a}	1.7 ± 0.3 ^{a}	38.2 ± 0.5 ^{a}
PIC	-	6.3 ± 1.3 ^{b} (25.8)	1.2 ± 0.5 ^{b} (7.8)	7.3 ± 0.02 ^{c} (30.1)
QCT	-	4.8 ± 1.0 ^{b} (15.9)	1.9 ± 0.3 ^{a} (4.0)	8.8 ± 0.3 ^{b} (29.2)

PESE, ethanolic extract of *P. edulis* seeds; PIC, piceatannol, QCT, quercetin, DW, dry weight; GAE, gallic acid equivalents; DPPH$^\bullet$, 1,1-diphenyl-2-picrylhydrazyl; HOCl, hypochlorous acid; $O_2^{\bullet-}$, superoxide anion radical; $^{a-c}$, mean ± SD followed by different letters within the same column represent significant differences (ANOVA analysis was performed followed by the Tukey test, $p \leq 0.05$). Data in parenthesis are in µM. Data are means of triplicates.

Therefore, the results suggest that PIC may be an important (poly)phenolic compound in passion fruit seeds extracts. Furthermore, these results indicate that PESE is a potential antioxidant mixture against ROS. Recently, Rotta et al. [37] obtained similar results (250 ± 20 mg GAE/g, DW) for TPC and IC_{50} (19 ± 3 µg/mL) for DPPH$^\bullet$ in extracts of *P. edulis* seeds prepared by using ethanol:water solution (70:30, *v*/*v*) at 80 °C for 4 h.

Inhibiting intracellular ROS formation would provide a therapeutic strategy to counteract oxidative stress and cell damage [38]. Several in vitro and animal models and limited human studies have revealed that supplementation with a (poly)phenol-rich diet has attenuated oxidative stress, reduced postprandial and fasting blood glucose levels, and improved insulin release and sensitivity [39]. Therefore, antioxidant phytochemicals present in PESE, such as PIC, can contribute to homeostasis.

2.5. The Extract Reduces Intracellular ROS-Induced by a Carcinogen

To evaluate the pro-oxidant and/or antioxidant effects of PESE and PIC, cultured BEAS-2B, AML-12, and MCF-10A cells were exposed to a carcinogen alone or pretreated with PESE and PIC [40]. We have demonstrated that 100 µM of 4-[(acetoxymethyl) nitrosamine]-1-(3-pyridyl)-1-butanone (NNKOAc) induces intracellular ROS production in BEAS-2B cells without reducing cell viability [41]. NNKOAc is a precursor of NNK, which is the most abundant *N*-nitrosamine and the most potent carcinogen in cigarette smoke [42]. NNKOAc generates similar cytosolic reactive electrophilic compounds as NNK, which binds to DNA, forming DNA adducts and inducing DNA damage [41].

It was observed that PESE and PIC at 50 µg/mL were able to protect all cultured cells from the oxidative stress caused by the carcinogen (Figure 1). PESE at 100 µg/mL also prevented the ROS generation in BEAS-2B cells with NNKOAc. However, they were not able to prevent oxidative stress in AML-12 and MCF-10A cells. Our results demonstrate that the PESE and PIC can mitigate ROS generation, especially under oxidative stress, which is considered the primary event in diabetes.

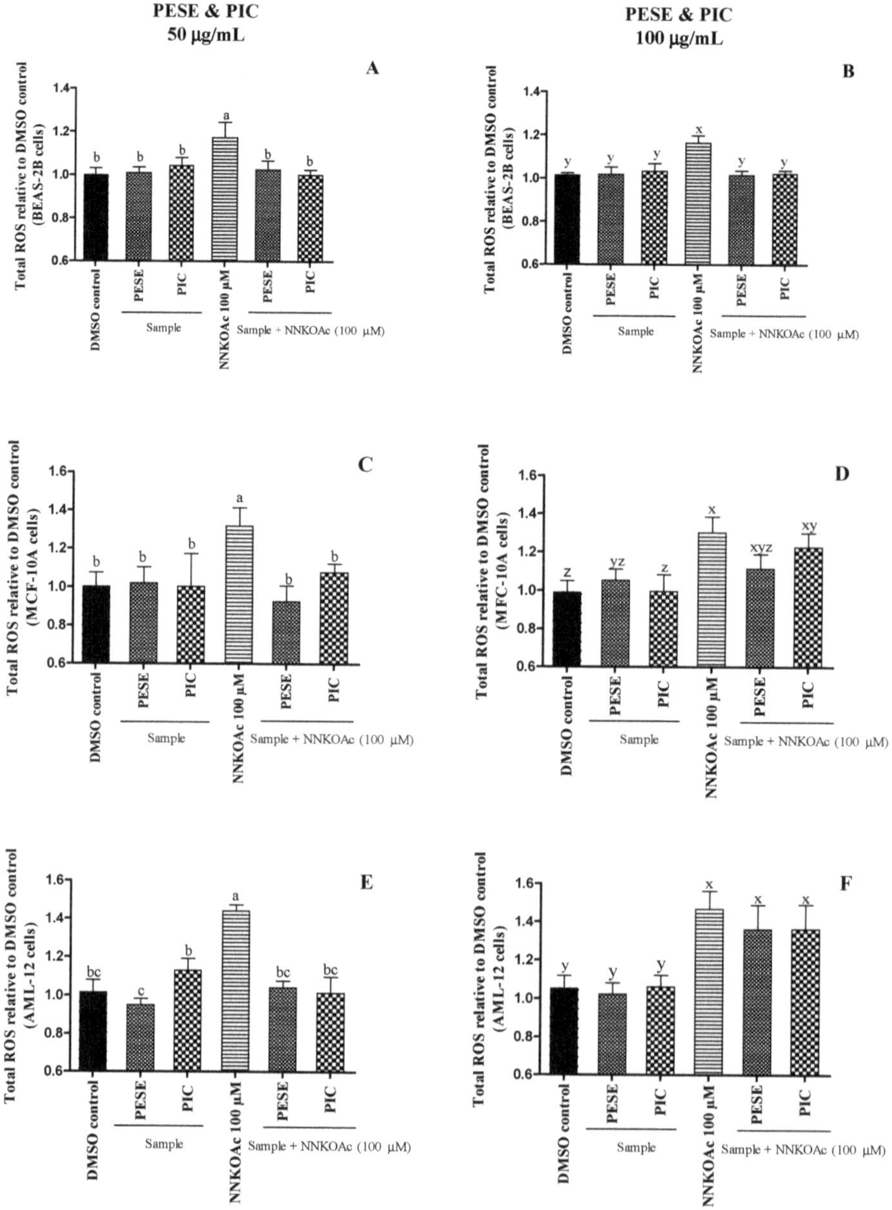

Figure 1. (**A–F**). The relative amount of reactive oxygen species (ROS) assessed on BEAS-2B, AML-12, and MCF-10A cell lines after exposure to either carcinogen alone or with pretreatment of PESE or PIC. All the treatment groups were compared to dimethyl sulfoxide (DMSO) control. a–c, x–z, mean ± SD followed by different letters represent significant differences (ANOVA analysis was performed followed by the Tukey test, $p \leq 0.05$). Data are means of triplicates. Abbreviation: AML-12, alpha mouse liver 12; BEAS-2B, normal human bronchial epithelial cells; MCF-10, non-tumorigenic epithelial cells; DMSO, dimethylsulfoxide; NNKOAC, 4-[(acetoxymethyl)nitrosamine]-1-(3-pyridyl)-1-butanone; PESE, ethanolic extract of *P. edulis* seeds; PIC, piceatannol.

It has long been recognized that hyperglycemia induces the formation of ROS, including superoxide radical anion ($O_2^{\bullet -}$), hydroxyl radical ($^{\bullet}OH$), and hydrogen peroxide (H_2O_2). Under aerobic conditions, ROS are generated during normal metabolic activities, but the imbalance between the excessive ROS generation and lack of antioxidant defense are the primary factors that might cause lipid peroxidation, protein denaturation, and DNA mutation in healthy cells. These reactive species deteriorate β-cell function and increase insulin resistance, which triggers the aggravation of T2DM [29]. Treatment with PESE and PIC at lower concentrations might protect cells against glucose toxicity through scavenging ROS.

2.6. Cytotoxicity of the Seed Extract of *P. edulis* (PESE) and PIC

The examination of the cytotoxicity of natural extracts helps in the evaluation of their safety, which is important in the development of nutraceuticals or dietary supplements. When the cell viability is greater than 90%, a sample is nontoxic at the tested dose [43]. A dose-responsive decline in the viability of BEAS-2B, AML-12, and MCF-10A cells was observed with increasing PESE and PIC concentrations ($p \leq 0.05$) (Figure 2). However, PESE did not impact the viability of BEAS-2B cells up to 100 μg/mL, while in AML-12 and MCF-10A cells, their viabilities were reduced by 25%. Up to 250 μg/mL PIC, the viability of AML-12 cells was maintained while in MCF-10A cells, viability was maintained above 80% up to 100 μg/mL PIC.

BEAS-2B cells appear to be more sensitive to PIC, with a reduction of approximately 50% of cell viability at the concentration of 100 μg/mL (409 μM). Despite the reduction in viability in some of the concentrations in the different cells studied, PESE and PIC do not reduce the cell viability in the bioactive concentrations that were effective in the inhibition of alpha-amylase, alpha-glucosidase, DPP-4, and antiglycation.

Hyperglycemia is well documented to have long-term effects on multiple organs, including the kidney, heart, brain, liver, and eyes. Diabetes also affects lung function by inducing inflammatory processes and fibrotic changes in the tissue [28]. Here, our results suggest that PESE and PIC are relatively safe for normal liver and lung cells and thus may not be harmful to organs and tissues. However, extensive toxicological assessments using experimental animal models need to be performed.

The cytotoxicity of PESE was investigated in the Vero E6 cell line and no decrease in cell viability was observed; however, at the highest concentration (100 μg/mL) there was an increase in cell viability, which according to the authors [36] may be related to a possible increase in mitochondrial proliferation or enzyme activity. The cell viability of human placental HTR-8/SVneo cells was also assessed in the presence of the extract, also with no reduction in cell viability at up to 100 μg/mL [36].

Yepes and colleagues have reported that the ethanol extract of purple passion fruit seeds at 1000 and 4000 μg/mL concentrations did not decrease the viability of normal human leukocyte cells, which is in contrast with the results of the present study [44]. Another study stated that an extract of defatted yellow passion fruit seed obtained using pressurized liquid extraction significantly decreased viability in all prostate cancer cell lines (PC-3, 22Rv1, LNCaP, and VcaP) in a dose- and time-dependent manner (10, 20, and 30 μM) [45].

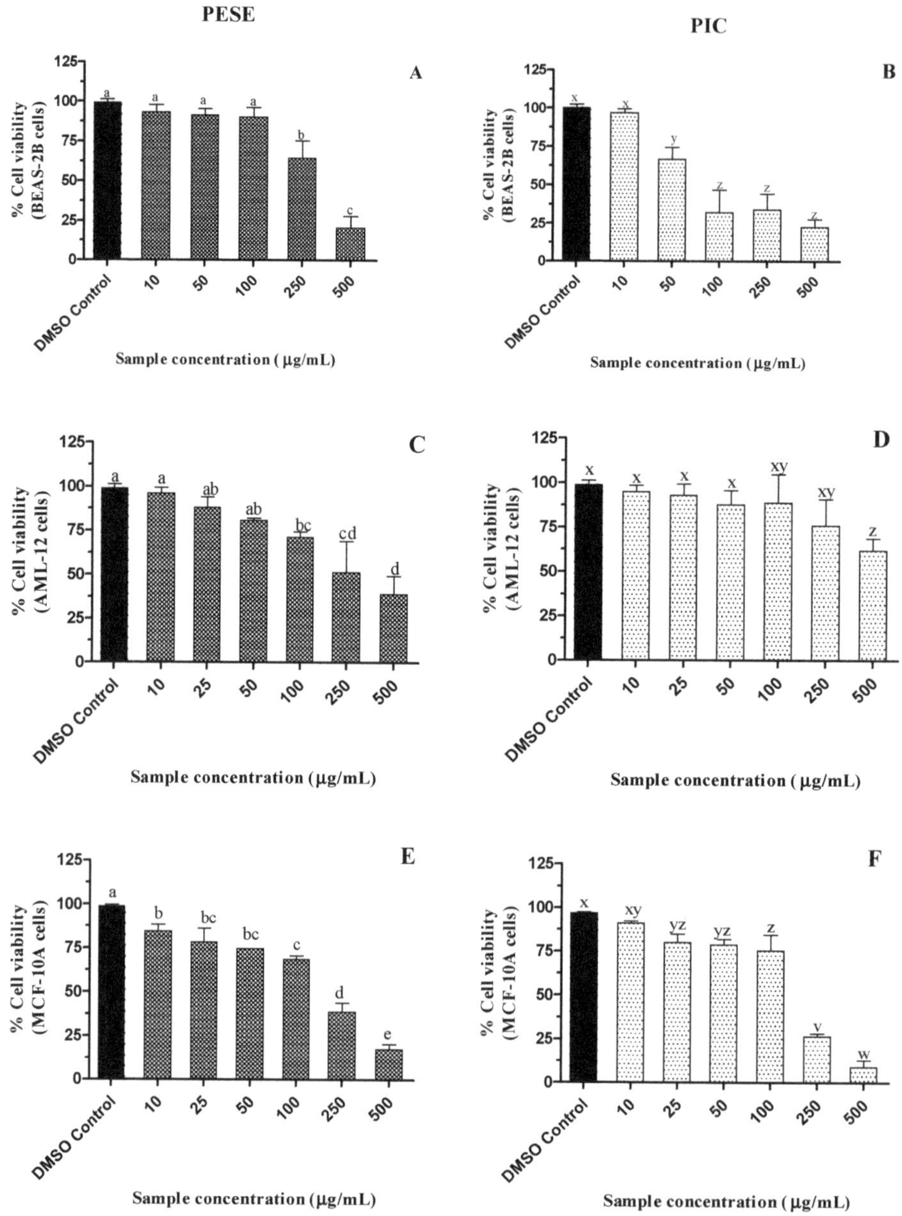

Figure 2. (**A–F**). Dose-dependent effect of ethanolic extract of passion fruit seeds on the viability of BEAS-2B, AML-12, and MCF-10A cell lines after 24 h of incubation. All the treatment groups were compared to dimethyl sulfoxide (DMSO) control. a–e, v–z, mean ± SD followed by different letters represent significant differences (ANOVA analysis was performed followed by the Tukey test, $p \leq 0.05$). Data are means of triplicates. Abbreviation: AML-12, alpha mouse liver 12; BEAS-2B, normal human bronchial epithelial cells; MCF-10A, non-tumorigenic epithelial cells; DMSO, dimethylsulfoxide; PESE, ethanolic extract of *P. edulis* seeds; PIC, piceatannol.

3. Material and Methods

3.1. Chemicals

The analytical solvents and chemicals used for antioxidant and antiglycation activities were purchased from Sigma-Aldrich (Steinheim, Germany): Folin–Ciocalteu reagent (FC), DPPH•, β-nicotinamide adenine dinucleotide (NADH), 4-nitro blue tetrazolium chloride (NBT), *N*-methylphenazonium methyl sulfate (PMS), AMG, sodium hypochlorite solution (NaOCl), dihydrorhodamine 123 (DHR), QCT, OPD, and ThT. PIC was obtained from AK Scientific (Union City, CA, USA). The analytic solvents, chemicals, and enzymes used for antidiabetic assays were obtained from Sigma-Aldrich (Oakville, ON, Canada): pancreatic alpha-amylase, yeast alpha-glucosidase, ACB, STG, 2-chloro-4-nitrophenyl-α-D-maltotrioside, 4-nitrophenyl-α-D-glucopyranoside, trisodium phosphate, *N*-(2-hydroxyethyl)piperazine-*N*′-(2-ethanesulfonic acid), 4-(2-hydroxyethyl)piperazine-1-ethanesulfonic acid (HEPES), and DPP-4. All the reagents were of analytical grade, and the stock solutions and buffers were prepared with Milli-Q purified water. LHC-9 growth medium for BEAS-2B cells was purchased from ThermoFischer 91 (Chelmsford, MA, USA). MEBMTM mammary epithelial basal medium and supplementation (MEGMTM Mammary Epithelial Cell Growth Medium Kit) for MCF10-A cells, and DMEM F-12 medium (ATCC) supplemented with insulin-transferrin-selenium (ITS-G) for AML-12 cells were purchased from Lonza (Basel, Switzerland). NNKOAc was purchased from Toronto Research Chemicals (Toronto, ON, Canada). MGO, 3-(4,5-dimethylthiazol-2-yl)-5-(3-carboxy methyl phenyl)-2-(4-sulfophenyl)-2H-tetrazolium (MTS), phenazine methosulfate (PMS), and 2′,7′-dichlorofluorescein diacetate dye (DCFH-DA) were purchased from Sigma-Aldrich (Oakville, ON, Canada). Stock solutions were prepared with DMSO, and the final concentrations were not exceeding 0.5% (*v*/*v*) in the culture treatment medium.

3.2. Plant Material

The *P. edulis* material used in this study was collected in 2019 from the company "Polpa de Frutas Santa Luzia", a local fruit processing industry located in Marechal Deodoro, Alagoas Brazil. Firstly, the remaining pulp in the seeds was removed manually. Seeds were washed in distilled water, dried in an oven at 50 °C for 48 h, and ground in a blender. The seed samples were kept in an amber flask at room temperature. Next, extraction was performed in a Soxhlet apparatus using 12 g of dry ground seeds. n-hexane (200 mL) was used to remove fat from the samples. Then, extraction was carried out with 200 mL ethanol for 6 h. Both solvents were removed by reduced pressure with a rotary evaporator. Extracts were maintained at 4 °C until further analysis. In this study, the ethanolic extract of *P. edulis* seeds (PESE) was evaluated.

3.3. Antidiabetic Activity Assays In Vitro

3.3.1. Alpha-Amylase Inhibition Assay

Plant extract screening for alpha-amylase inhibition was conducted using the previously described method [46], with some modifications. Except as otherwise indicated, solutions were prepared in 0.01 M potassium phosphate buffer containing sodium chloride (60 mM) and sodium azide (0.02% *w*/*v*), pH 6.8. The PESE and PIC solutions were prepared in a buffer containing 8% ethanol. To a 96-well clear plate, 20 μL of plant extract and 20 μL of alpha-amylase from porcine pancreas (4 U/mL in buffer) were added. After 10 min of incubation at 37 °C, 20 μL of the substrate 2-chloro-4-nitrophenyl-α-D-maltotrioside (5 mM) was added. The mixture was then incubated for 30 min at 37 °C for the reaction to take place. The reaction was terminated by adding 240 μL of trisodium phosphate solution of pH 11 (1% *w*/*v*) to stop enzyme activity. Then, the absorbance at 405 nm was recorded using the microplate reader (Infinite® 200 PRO, TECAN, Männedorf, Switzerland) to quantify the amount of 2-chloro-4-nitrophenol released. The effectiveness of the tested inhibitors was compared with ACB, a prescribed antidiabetic drug for alpha-amylase inhibition that was used as a means of comparison. The positive control was a mixture of enzyme and substrate without inhibitors. Sample control containing sample and buffer was used to eliminate color interference. Buffer was used as blank. The percentage of the alpha-amylase inhibition was calculated as I% = [(Abs_0 − Abs_1)/Abs_0] × 100, where A_0 is the absorbance of the positive control subtracted from the blank, Abs_1 is the absorbance in the presence of the extract subtracted from sample control. The IC_{50} (half-maximal inhibitory concentration) was calculated graphically, using an analytical curve by plotting the concentration versus the inhibition percentage (I%).

3.3.2. Alpha-Glucosidase Inhibition Assay

The alpha-glucosidase inhibition assay was performed using a previously described method [15] with minor modifications. First, solutions were prepared in 0.01 M potassium phosphate buffer (pH 6.8), unless otherwise stated. Various concentrations of PESE and PIC were prepared in a buffer containing 2.5% ethanol. To a 96-well clear plate, 120 μL of the samples and 20 μL of alpha-glucosidase enzyme (0.25 U/mL) were added. The plate was incubated at 37 °C for 15 min before adding 20 μL of 4-nitrophenyl-α-D-glucopyranoside (5 mM) substrate. The mixture was then incubated at 37 °C for 15 min for the reaction to take place. After incubation, the reaction was stopped by adding 80 μL of 0.2 M sodium carbonate in 0.1 M potassium phosphate buffer, pH 6.8. The absorbance at 405 nm was recorded using a microplate reader (Infinite® 200 PRO, TECAN, Männedorf, Switzerland) to quantify the amount of *p*-nitrophenol (PNP) released. ACB, an antidiabetic drug used as an alpha-glucosidase inhibitor for T2DM, was used for comparison purposes. The positive control was the mixture of enzyme and substrate without inhibitors. The sample control consisted of the mixture of sample and buffer. The buffer was used as a blank. The percentage of the alpha-glucosidase inhibition and IC_{50} was calculated as described in Section 3.3.1.

3.3.3. Dipeptidyl Dipeptidase Enzyme (DPP-4) Inhibition Assay

The DPP-4 inhibition assay was performed according to an established method [46]. Briefly, to a 96 well plate, 20 μL of the sample at different concentrations, 20 μL DPP-4 human recombinant enzyme solution (3.125 mU), and 50 μL of Gly-Pro-7-amido-4-methylcoumarin hydrobromide substrate (2.5 μM) were added. The reaction mixture was incubated for 30 min in the dark at 37 °C. Then, the fluorescent product was recorded at excitation/emission wavelengths of 350/450 nm using the microplate reader (Infinite® 200 PRO, TECAN, Männedorf, Switzerland). STG, a standard DPP-4 inhibitor, was used to compare the effectiveness of PESE and PIC. The percentage of the DPP-4 inhibition and IC_{50} was calculated as described in Section 3.3.1.

3.4. Antiglycation Activity Assays In Vitro

3.4.1. Inhibition of Advanced Glycation End Products (AGE) Formation in the Initial Stage of Glycation

The formation of AGEs was measured after incubation of a system containing protein and carbohydrates. This assay was based on previous methods [47]. The reaction system was obtained by adding 300 μL of plant extract or of the pure compound (at different concentrations), 150 μL D-fructose (200 mM), 150 μL of D-glucose (200 mM), and 300 μL of bovine serum albumin (BSA, 3 mg/mL) to a test tube. All these solutions, except plant extract, were dissolved in 0.05 M potassium phosphate buffer (pH 7.4) containing NaCl (100 mM) and NaN_3 (0.02% *w*/*v*). The plant extract was prepared in a buffer containing 60% ethanol. The mixture was incubated in the dark at 37 °C for 7 days with constant stirring. After the incubation, 200 μL of the reaction mixture was transferred to a 96-well black plate, and fluorescent AGEs were measured using a microplate reader (Infinite® 200 PRO, TECAN, Männedorf, Switzerland) at λex = 360 and λem = 440 nm. X. Aminoguanidine (AMG), a known AGE formation inhibitor, was used as a standard. The reaction mixture without inhibitors was used as a positive control. Color interference from the sample was eliminated by a sample control (buffer and sample). The fluorescence readings for the experimental treatment were blanked against extract blanks to eliminate the baseline fluorescence of the sample. The percentage of the AGE inhibition and IC_{50} was calculated as described in Section 3.3.1.

3.4.2. Inhibition of AGE Formation in the Intermediate Stage of Glycation

The BSA-MGO assay was based on those described before [33], with some modifications. This experiment is based on the formation of fluorescent AGEs from the middle stage of protein glycation. MGO (1.5 mM) and BSA (3 mg/mL) were dissolved separately in 0.05 M phosphate buffer (pH 7.4) containing NaCl (100 mM) and NaN_3 (0.02% *w*/*v*). Then, 300 µL of MGO was incubated with samples at different concentrations (dissolved in a buffer containing 60% ethanol) for an hour at 37 °C, in the dark, with constant stirring. After BSA (3 mg/mL) was added to the reaction mixture and incubation time was extended to 48 h at 37 °C in darkness. The rest of the procedure was the same as that for the BSA-glucose/fructose model that is described above. The percentage of the AGE inhibition and IC_{50} was calculated as described in Section 3.3.1.

3.4.3. Inhibition of the Formation of Amyloid Fibrils Using ThT Assay

The fibrillar state of modified BSA was determined with ThT as previously described [48]. ThT is a probe that binds directly to amyloid fibrils to give a strong fluorescent signal and thus can be used to quantify β-amyloid [35] due to protein glycation. Basically, 30 µL of glycated protein (approx. 1 mg/mL), obtained as described previously, was mixed with 100 µL ThT reagent (10 µM ThT in 100 mM phosphate buffer, pH 7.0), and fluorescence was collected at excitation/emission wavelengths of 450/490 nm in a microplate reader (Infinite® 200 PRO, TECAN, Männedorf, Switzerland). Fresh BSA solution (1 mg/mL) and PBS were used as controls.

3.4.4. Evaluation of MGO Capture Using Derivatization with OPD

MGO scavenging was tested using a described procedure [33], with slight modifications. The quantification of MGO was based on derivatization with OPD, leading to the formation of the product 2-methylquinoxyline (2-MQ). MGO and OPD were dissolved in phosphate buffer (50 mM, pH 7.4) to a concentration of 2 mM and 4 mM, respectively. AMG (2 mM) was used as a positive control, PESE and PIC were prepared with a concentration of 2 mg/mL and 2 mM, respectively. The reaction system was composed of 125 µL of the MGO solution, 125 µL of phosphate buffer (negative control), or PESE incubated at 37 °C for 1 h. After 250 µL of the OPD solution was added, the tubes were kept for 30 min for the derivatization reaction between MGO and OPD to complete. Then, the chromatographic analysis was performed. All solutions were previously filtered (microfilter pore diameter 0.45 µm). The conditions for the analysis by high-performance liquid chromatography (HPLC) were: deionized water with formic acid (0.1% *v*/*v*, solvent A) and methanol (solvent B) were used as mobile phases, the flow rate of 1.0 mL/min, and the injection volume was 20 µL. The linear gradient for elution was: starting at 5% of solvent B, 0–3 min, 5 to 50% B; 3–16 min, isocratic in 50% B; 16–17 min, 50–90% B; 17–19 min, isocratic in 90% B and 19–19.5 min, 90–5% B. The derivatization product, 2-methylquinoxyline (2-MQ), was detected at 315 nm, with a retention time of 10 min. The percentage of trapping efficiency of MGO was calculated using the following equation: % trapping MGO = [100 − (peak area after adding sample/peak area without adding sample) × 100].

3.5. Scavenging of Free Radical and Reactive Oxygen Species (ROS) Assays

3.5.1. Total Phenolic Content (TPC)

The TPC was estimated using the Folin–Ciocalteu (FC) method, as described [49], with some modifications. Briefly, 180 µL of deionized water, 300 µL of FC reagent, and 2.4 mL of 5% sodium carbonate (*w*/*v*) were added to 120 µL of diluted samples. After incubation in a water bath at 40 °C in the dark for 20 min, the absorbance of the resulting mixture was measured at 760 nm using a UV–Vis spectrophotometer (Agilent 8453). The results were expressed as milligrams of gallic acid equivalents (mg GAE) per gram of dry extract.

3.5.2. Radical Scavenging Assay DPPH$^{\bullet}$

Antioxidant activity of PESE was determined using the DPPH$^{\bullet}$ method [50]. Briefly, aliquots of 0.30 mL of sample dissolved in ethanol (5–25 μg/mL) were mixed with 2.70 mL of DPPH$^{\bullet}$ solution (40 μg/mL in methanol). After incubation in the dark for 30 min, the absorbance was read at 516 nm, using a UV–Vis spectrophotometer (Agilent Technologies, Santa Clara, CA, USA). Results were expressed as the half-maximal inhibitory concentration (IC_{50}) in μg/mL.

3.5.3. Hypochlorous Acid (HOCl) Scavenging Activity

HOCl scavenging activity of PESE and PIC was determined [51]. Briefly, a freshly prepared HOCl solution, by adjusting the pH of a 1% solution of NaOCl to 6.2 with dropwise addition of diluted H_2SO_4, was diluted to 30 μM, using 100 mM phosphate buffer pH 7.4. For the analysis, in a 96-well plate, the following reagents were added at the indicated final concentrations: 150 μL of buffer solution (100 mM, pH 7.4), 50 μL of PESE (1, 5, 10, 25, 50, 100, 200, and 300 μg/mL), 50 μL of DHR 123 (5 μM) and 50 μL of HOCl (5 μM). QCT was used as means of comparison. Fluorescence assays were performed in a microplate reader (Infinite® 200 PRO, TECAN, Männedorf, Switzerland), at 37 °C, at wavelengths of 505 ± 10 nm and 530 ± 10 nm, for excitation and emission, respectively. The results were expressed as IC_{50} (μg/mL) of extract solution.

3.5.4. Superoxide Anion Radical Scavenging Activity

The potential of the samples to scavenger superoxide anion radicals was determined [51], with minor modifications. In a 96-well plate, the following solutions were added to the final concentrations indicated: 50 μL of PESE (1 to 300 μg/mL), 50 μL of NADH (166 μM), 150 μL of NBT (43.3 μM), and 50 μL of PMS (2.7 μM). Phosphate buffer (19 mM, pH 7.4) was used to dissolve NADH, NBT, and PMS. QCT was used as means of comparison. The experiment was conducted at 37 °C in a microplate reader (Infinite® 200 PRO, TECAN, Männedorf, Switzerland), and the absorbance was measured at 560 nm. The results were expressed as IC_{50} of extract solution in μg/mL.

3.5.5. Measurement of Intracellular ROS Level

The generation of intracellular ROS in BEAS-2B, AML-12, and MCF-10A cells after treatments was measured as described previously [40]. Briefly, DCFH-DA dye was readily taken up by cells and is subsequently hydrolyzed to DCFH, which can be oxidized to a measurable fluorescent product dichlorofluorescein (DCF). The cells, pre-treated with PESE or PIC for 3 h, were exposed to the carcinogen NNKOAc, for 3 h, or alone in different experimental groups. Cells with DMSO media (0.5%) served as the vehicle control. After treatments, DCFH-DA was added to the cell culture plates at a final concentration of 5 μM followed by 40 min incubation at dark. The fluorescence degradation was then measured at an excitation and emission wavelength of 485 nm and 535 nm, using a microplate reader (Infinite® 200 PRO, TECAN, Männedorf, Switzerland).

3.6. Cell Cultures and Cell Viability Assay

BEAS-2B, MCF10-A, and AML-12, cells were purchased from the American Tissue Type Culture Collection (ATCC; CRL-9609, CRL-10317, and CRL-2254) and cultured with a specific medium. BEAS-2B cells were cultured with LHC-9 media at 37 °C in an incubator with 5% CO_2. Culture flasks (polystyrene T75) were pre-coated with a mixture of 0.01 mg/mL fibronectin, 0.03 mg/mL bovine collagen type I, and 0.01 mg/mL bovine serum albumin were dissolved in LHC-9 medium overnight. MCF10-A cells were cultured in mammary epithelial basal medium (MEBM) culture medium supplemented with the Mammary Epithelial Cell Growth Medium Kit (Lonza), and AML-12 cells in DMEM-F12 medium (ATCC) were supplemented with insulin-transferrin-selenium (ITS, GIBCO), 10% fetal bovine serum, and 1% antibiotic (penicillin-streptomycin, Gibco) at 37 °C in a 5% CO_2

incubator. Cells were grown to around 70% confluence on the culture flask, and passages (<10) were employed for all experimental conditions.

MTS Cell Titer 96™ aqueous cell proliferation assay was used to determine cell viability [41]. BEAS-2B, AML-12, and MCF-10A cells were treated with PESE and PIC at different concentrations to determine the non-cytotoxic dose. For that, 1×10^4 cells were plated on a 96-well plate with a growth medium of 100 μL/well. After 24 h, cells were treated with PESE or PIC for an additional 24 h. Fifteen microliters of MTS reagent (with PMS) was then added to each well and incubated for 3 h in the dark. Absorbance was recorded at 490 nm using a microplate reader (Infinite® 200 PRO, TECAN, Männedorf, Switzerland). For each experiment, cells with DMSO media served as control, and cells with only culture medium and MTS reagent served as the blank.

3.7. Statistical Analysis

All analyses were performed in triplicate (n = 3) with three independent studies and using Graph-Pad Prism software (GraphPad Software Inc., San Diego, CA, USA). Data were presented as the mean ± standard deviation (SD), and analyses of variance, one-way ANOVA, followed by Tukey test and $p \leq 0.05$ were considered significant between experimental groups.

4. Conclusions

Passion fruit seeds, by-products of the juice industry, have the potential for use as a low-cost antioxidant and bioactive source for developing nutraceuticals and dietary supplements, for managing blood glucose levels, and consequently, reducing the progression of complications of T2DM. In this work, PESE showed a potent antidiabetic, antiglycant, and antioxidant potential without being toxic to BEAS-2B cells at bioactive concentrations. Furthermore, it was observed that PESE could protect some of the tested cultured cell lines from the oxidative stress caused by the presence of the chemical carcinogen, NNKOAc. Although it is still early to suggest the use of ethanolic extract from passion fruit seeds in the treatment of T2DM, this work demonstrates its potential dietary use to manage T2DM and glycation-associated complications. Further investigations on the efficacy and safety of PESE, using pre-clinical experimental animal models are encouraged and are underway.

Author Contributions: Conceptualization, M.O.F.G., J.A.X. and H.P.V.R.; validation, M.O.F.G., J.A.X., H.P.V.R. and F.A.R.d.S.; formal analysis, F.A.R.d.S.; investigation, F.A.R.d.S., F.C.d.S. and J.P.J.M.; writing—original draft preparation, F.A.R.d.S.; writing—review and editing, M.O.F.G., J.A.X. and H.P.V.R.; visualization, M.O.F.G.; and supervision, M.O.F.G., J.A.X. and H.P.V.R. All authors have read and agreed to the published version of the manuscript.

Funding: The authors are grateful for the financial support provided by the Global Affairs Canada Student Exchange Program and the International Office of Dalhousie University and by the Brazilian research funding agencies Conselho Nacional de Desenvolvimento Científico e Tecnológico (CNPq) for financial support (435704/2018-4) and fellowships (MOFG), Coordenação de Aperfeiçoamento de Pessoal de Nível Superior-Brazil (CAPES)—Finance Code 001, CAPES/RENORBIO/PROAP (23038.011373/2017-31), INCT-Bioanalítica (process 465389/2014-7), and Fundação de Apoio à Pesquisa de Alagoas (FAPEAL).

Institutional Review Board Statement: Biosafety protocol (2013-10) of Dalhousie University.

Informed Consent Statement: Not applicable.

Data Availability Statement: Not applicable.

Acknowledgments: The authors are grateful for the support from Killam Chair funds of HPVR.

Conflicts of Interest: The authors declare no conflict of interest.

Sample Availability: Samples of passion seeds are available from M.O.F.G. and J.A.X.

References

1. Da Silva Francischini, D.; Lopes, A.P.; Segatto, M.L.; Stahl, A.M.; Zuin, V.G. Development and Application of Green and Sustainable Analytical Methods for Flavonoid Extraction from Passiflora Waste. *BMC Chem.* **2020**, *14*, 56. [CrossRef] [PubMed]
2. IBGE. Produção de Maracujá. Available online: https://www.ibge.gov.br/explica/producao-agropecuaria/maracuja/br (accessed on 10 December 2020).
3. He, X.; Luan, F.; Yang, Y.; Wang, Z.; Zhao, Z.; Fang, J.; Wang, M.; Zuo, M.; Li, Y. Passiflora Edulis: An Insight into Current Researches on Phytochemistry and Pharmacology. *Front. Pharmacol.* **2020**, *11*, 617. [CrossRef] [PubMed]
4. Kuete, V.; Taiwe, G.S. Passiflora edulis. In *Medicinal Spices and Vegetables from Africa: Therapeutic Potential Against Metabolic, Inflamatory, Infectious and Systemic Diseases*; Elsevier Science & Technology: Amsterdam, The Netherlands, 2017; pp. 513–523. ISBN 978-0-12-809286-6.
5. Krambeck, K.; Oliveira, A.; Santos, D.; Pintado, M.M.; Silva, J.B.; Lobo, J.M.S.; Amaral, M.H. Identification and Quantification of Stilbenes (Piceatannol and Resveratrol) in *Passiflora edulis* by-Products. *Pharmaceuticals* **2020**, *13*, 73. [CrossRef] [PubMed]
6. Song, H.; Jung, J.I.; Cho, H.J.; Her, S.; Kwon, S.H.; Yu, R.; Kang, Y.H.; Lee, K.W.; Park, J.H.Y. Inhibition of Tumor Progression by Oral Piceatannol in Mouse 4T1 Mammary Cancer Is Associated with Decreased Angiogenesis and Macrophage Infiltration. *J. Nutr. Biochem.* **2015**, *26*, 1368–1378. [CrossRef]
7. Kershaw, J.; Kim, K.H. The Therapeutic Potential of Piceatannol, a Natural Stilbene, in Metabolic Diseases: A Review. *J. Med. Food* **2017**, *20*, 427–438. [CrossRef]
8. Nagapan, T.S.; Ghazali, A.R.; Basri, D.F.; Lim, W.N. Photoprotective Effect of Stilbenes and Its Derivatives against Ultraviolet Radiation-Induced Skin Disorders. *Biomed. Pharmacol. J.* **2018**, *11*, 1199–1208. [CrossRef]
9. Maruki-Uchida, H.; Morita, M.; Yonei, Y.; Sai, M. Effect of Passion Fruit Seed Extract Rich in Piceatannol on the Skin of Women: A Randomized, Placebo-Controlled, Double-Blind Trial. *J. Nutr. Sci. Vitaminol.* **2018**, *64*, 75–80. [CrossRef]
10. Surh, Y.-J.; Na, H.-K. Anti-Inflammatory Nutraceuticals and Chronic Diseases. In *Advances in Experimental Medicine and Biology*; Gupta, S.C., Prasad, S., Aggarwal, B.B., Eds.; Springer International Publishing: Berlin/Heidelberg, Germany, 2016; Volume 928, pp. 185–212. ISBN 9783319413327.
11. Gregg, E.; Buckley, J.; Ali, M.; Davies, J.; Flood, D.; Griffiths, B.; Lim, L.-L.; Manne-Goehler, J.; Pearson-Stuttard, J.; Shaw, J. *Improving Health Outcomes of People with Diabetes Mellitus: Target Setting to Reduce the Global Burden of Diabetes Mellitus by 2030*; World Health Organization: Geneva, Switzerland, 2021; Volume 8.
12. International Diabetes Federation. *IDF Diabetes Atlas*, 10th ed.; International Diabetes Federation: Brussels, Belgium, 2021; Volume 102, ISBN 9782930229980.
13. Tupas, G.D.; Otero, M.C.B.; Ebhohimen, I.E.; Egbuna, C.; Aslam, M. Antidiabetic Lead Compounds and Targets for Drug Development. In *Phytochemicals as Lead Compounds for New Drug Discovery*; Egbuna, C., Kumar, S., Ifemeje, J., Ezzat, S., Kaliyaperumal, S., Eds.; Elsevier: Amsterdam, The Netherlands, 2019; pp. 127–141. ISBN 9780128178904.
14. Marín-Peñalver, J.J.; Martín-Timón, I.; Sevillano-Collantes, C.; del Cañizo-Gómez, F.J. Update on the Treatment of Type 2 Diabetes Mellitus. *World J. Diabetes* **2016**, *7*, 354–395. [CrossRef]
15. Sekhon-Loodu, S.; Rupasinghe, H.P.V. Evaluation of Antioxidant, Antidiabetic and Antiobesity Potential of Selected Traditional Medicinal Plants. *Front. Nutr.* **2019**, *6*, 53. [CrossRef]
16. Jayaraj, R.L.; Azimullah, S.; Beiram, R. Diabetes as a Risk Factor for Alzheimer's Disease in the Middle East and Its Shared Pathological Mediators. *Saudi J. Biol. Sci.* **2020**, *27*, 736–750. [CrossRef]
17. Kong, Y.; Wang, F.; Wang, J.; Liu, C.; Zhou, Y.; Xu, Z.; Zhang, C.; Sun, B.; Guan, Y. Pathological Mechanisms Linking Diabetes Mellitus and Alzheimer's Disease: The Receptor for Advanced Glycation End Products (RAGE). *Front. Aging Neurosci.* **2020**, *12*, 217. [CrossRef] [PubMed]
18. Torres, N.M.P.O.; Xavier, J.A.; Goulart, M.O.F.; Alves, R.B.; Freitas, R.P. The Chemistry of Advanced Glycation End-Products. *Rev. Virtual Quim.* **2018**, *10*, 375–392. [CrossRef]
19. Zaman, M.; Khan, A.N.; Wahiduzzaman; Zakariya, S.M.; Khan, R.H. Protein Misfolding, Aggregation and Mechanism of Amyloid Cytotoxicity: An Overview and Therapeutic Strategies to Inhibit Aggregation. *Int. J. Biol. Macromol.* **2019**, *134*, 1022–1037. [CrossRef] [PubMed]
20. Sun, L.; Miao, M. Dietary Polyphenols Modulate Starch Digestion and Glycaemic Level: A Review. *Crit. Rev. Food Sci. Nutr.* **2020**, *60*, 541–555. [CrossRef]
21. Uchida-Maruki, H.; Inagaki, H.; Ito, R.; Kurita, I.; Sai, M.; Ito, T. Piceatannol Lowers the Blood Glucose Level in Diabetic Mice. *Biol. Pharm. Bull.* **2015**, *38*, 629–633. [CrossRef]
22. Zhao, M.; Gao, P.; Tao, L.; Wen, J.; Wang, L.; Yi, Y.; Chen, Y.; Wang, J.; Xu, X.; Zhang, J.; et al. Piceatannol Attenuates Streptozotocin-Induced Type 1 Diabetes in Mice. *Biocell* **2020**, *44*, 353–361. [CrossRef]
23. Cao, Q.; Teng, J.; Wei, B.; Huang, L.; Xia, N. Phenolic Compounds, Bioactivity, and Bioaccessibility of Ethanol Extracts from Passion Fruit Peel Based on Simulated Gastrointestinal Digestion. *Food Chem.* **2021**, *356*, 129682. [CrossRef]
24. Loizzo, M.R.; Lucci, P.; Núñez, O.; Tundis, R.; Balzano, M.; Frega, N.G.; Conte, L.; Moret, S.; Filatova, D.; Moyano, E.; et al. Native Colombian Fruits and Their By-Products: Phenolic Profile, Antioxidant Activity and Hypoglycaemic Potential. *Foods* **2019**, *8*, 89. [CrossRef]

25. Pan, Z.H.; Ning, D.S.; Fu, Y.X.; Li, D.P.; Zou, Z.Q.; Xie, Y.C.; Yu, L.L.; Li, L.C. Preparative Isolation of Piceatannol Derivatives from Passion Fruit (*Passiflora edulis*) Seeds by High-Speed Countercurrent Chromatography Combined with High-Performance Liquid Chromatography and Screening for α-Glucosidase Inhibitory Activities. *J. Agric. Food Chem.* **2020**, *68*, 1555–1562. [CrossRef]
26. Justino, A.B.; Miranda, N.C.; Franco, R.R.; Martins, M.M.; da Silva, N.M.; Espindola, F.S. *Annona muricata* Linn. Leaf as a Source of Antioxidant Compounds with In Vitro Antidiabetic and Inhibitory Potential against α-Amylase, α-Glucosidase, Lipase, Non-Enzymatic Glycation and Lipid Peroxidation. *Biomed. Pharmacother.* **2018**, *100*, 83–92. [CrossRef]
27. Jiang, L.; Wang, Z.; Wang, X.; Wang, S.; Cao, J.; Liu, Y. Exploring the Inhibitory Mechanism of Piceatannol on α-Glucosidase Relevant to Diabetes Mellitus. *RSC Adv.* **2020**, *10*, 4529–4537. [CrossRef] [PubMed]
28. Talakatta, G.; Sarikhani, M.; Muhamed, J.; Dhanya, K.; Somashekar, B.S.; Mahesh, P.A.; Sundaresan, N.; Ravindra, P.V. Diabetes Induces Fibrotic Changes in the Lung through the Activation of TGF-β Signaling Pathways. *Sci. Rep.* **2018**, *8*, 11920. [CrossRef] [PubMed]
29. Kaneto, H.; Katakami, N.; Matsuhisa, M.; Matsuoka, T.A. Role of Reactive Oxygen Species in the Progression of Type 2 Diabetes and Atherosclerosis. *Mediat. Inflamm.* **2010**, *2010*, 453892. [CrossRef] [PubMed]
30. Rabbani, N.; Xue, M.; Thornalley, P.J. Dicarbonyl Stress, Protein Glycation and the Unfolded Protein Response. *Glycoconj. J.* **2021**, *38*, 331–340. [CrossRef]
31. Zhang, D.; Wang, Y.; Liu, H. Corn Silk Extract Inhibit the Formation of Nε-Carboxymethyllysine by Scavenging Glyoxal/Methyl Glyoxal in a Casein Glucose-Fatty Acid Model System. *Food Chem.* **2020**, *309*, 125708. [CrossRef]
32. Kim, J.H.; Kim, M., II; Syed, A.S.; Jung, K.; Kim, C.Y. Rapid Identification of Methylglyoxal Trapping Constituents from Onion Peels by Pre-Column Incubation Method. *Nat. Prod. Sci.* **2017**, *23*, 247–252. [CrossRef]
33. Wang, W.; Yagiz, Y.; Buran, T.J.; Nunes, C.D.N.; Gu, L. Phytochemicals from Berries and Grapes Inhibited the Formation of Advanced Glycation End-Products by Scavenging Reactive Carbonyls. *Food Res. Int.* **2011**, *44*, 2666–2673. [CrossRef]
34. Taghavi, F.; Habibi-Rezaei, M.; Amani, M.; Saboury, A.A.; Moosavi-Movahedi, A.A. The Status of Glycation in Protein Aggregation. *Int. J. Biol. Macromol.* **2017**, *100*, 67–74. [CrossRef]
35. Xue, C.; Lin, T.Y.; Chang, D.; Guo, Z. Thioflavin T as an Amyloid Dye: Fibril Quantification, Optimal Concentration and Effect on Aggregation. *R. Soc. Open Sci.* **2017**, *4*, 160696. [CrossRef]
36. Xavier, J.A.; Santos, J.C.; Nova, M.A.V.; Gonçalves, C.M.; Borbely, K.S.C.; Pires, K.S.N.; dos Santos, F.A.R.; Valentim, I.B.; Barbosa, J.H.P.; da Silva, F.C.; et al. Anti-Zika Virus Efects, Placenta Protection and Chemical Composition of Passiflora edulis Seeds Ethanolic Extract. *J. Braz. Chem. Soc.* **2022**, *33*, 701–714. [CrossRef]
37. Rotta, E.M.; Giroux, H.J.; Lamothe, S.; Bélanger, D.; Sabik, H.; Visentainer, J.V.; Britten, M. Use of Passion Fruit Seed Extract (*Passiflora edulis* Sims) to Prevent Lipid Oxidation in Dairy Beverages during Storage and Simulated Digestion. *LWT* **2020**, *123*, 109088. [CrossRef]
38. Tsao, R. Chemistry and Biochemistry of Dietary Polyphenols. *Nutrients* **2010**, *2*, 1231–1246. [CrossRef] [PubMed]
39. Aryaeian, N.; Sedehi, S.K.; Arablou, T. Polyphenols and Their Effects on Diabetes Management: A Review. *Med. J. Islam. Repub. Iran* **2017**, *31*, 886–892. [CrossRef] [PubMed]
40. Merlin, J.P.J.; Dellaire, G.; Murphy, K.; Rupasinghe, H.P.V. Vitamin-Containing Antioxidant Formulation Reduces Carcinogen-Induced Dna Damage through Atr/Chk1 Signaling in Bronchial Epithelial Cells in Vitro. *Biomedicines* **2021**, *9*, 1665. [CrossRef]
41. Amararathna, M.; Hoskin, D.W.; Rupasinghe, H.P.V. Anthocyanin-Rich Haskap (*Lonicera caerulea* L.) Berry Extracts Reduce Nitrosamine-Induced DNA Damage in Human Normal Lung Epithelial Cells in Vitro. *Food Chem. Toxicol.* **2020**, *141*, 111404. [CrossRef]
42. Clague, J.; Shao, L.; Lin, J.; Chang, S.; Zhu, Y.; Wang, W.; Wood, C.G.; Wu, X. Sensitivity to NNKOAc Is Associated with Renal Cancer Risk. *Carcinogenesis* **2009**, *30*, 706–710. [CrossRef]
43. Sjögren, G.; Sletten, G.; Dahl, J.E. Cytotoxicity of Dental Alloys, Metals, and Ceramics Assessed by Millipore Filter, Agar Overlay, and MTT Tests. *J. Prosthet. Dent.* **2000**, *84*, 229–236. [CrossRef]
44. Yepes, A.; Ochoa-Bautista, D.; Murillo-Arango, W.; Quintero-Saumeth, J.; Bravo, K.; Osorio, E. Purple Passion Fruit Seeds (*Passiflora Edulis f. Edulis* Sims) as a Promising Source of Skin Anti-Aging Agents: Enzymatic, Antioxidant and Multi-Level Computational Studies. *Arab. J. Chem.* **2021**, *14*, 102905. [CrossRef]
45. Kido, L.A.; Hahm, E.-R.; Kim, S.-H.; Baseggio, A.M.; Cagnon, V.H.A.; Singh, S.V.; Maróstica, M.R. Prevention of Prostate Cancer in Transgenic Adenocarcinoma of the Mouse Prostate Mice by Yellow Passion Fruit Extract and Antiproliferative Effects of Its Bioactive Compound Piceatannol. *J. Cancer Prev.* **2020**, *25*, 87–99. [CrossRef]
46. de Silva, A.B.K.H.; Rupasinghe, H.P.V. Polyphenols Composition and Anti-Diabetic Properties in Vitro of Haskap (*Lonicera caerulea* L.) Berries in Relation to Cultivar and Harvesting Date. *J. Food Compos. Anal.* **2020**, *88*, 103402. [CrossRef]
47. de Melo, I.S.V.; dos Santos, A.F.; de Lemos, T.L.G.; Goulart, M.O.F.; Santana, A.E.G. Oncocalyxone a Functions as an Anti-Glycation Agent In Vitro. *PLoS ONE* **2015**, *10*, e0131222. [CrossRef]
48. Moosavi-Movahedi, A.A.; Ghamari, F.; Ghaffari, S.M.; Salami, M.; Farivar, F.; Moosavi-Movahedi, F.; Johari, A.; Aminin, A.L.N. Natural Peptide Anti-Glycation Effect in the Presence of Aloe vera Phenolic Components on Human Serum Albumin. *RSC Adv.* **2015**, *5*, 248–254. [CrossRef]
49. Cicco, N.; Lanorte, M.T.; Paraggio, M.; Viggiano, M.; Lattanzio, V. A Reproducible, Rapid and Inexpensive Folin-Ciocalteu Micro-Method in Determining Phenolics of Plant Methanol Extracts. *Microchem. J.* **2009**, *91*, 107–110. [CrossRef]

50. de Almeida Xavier, J.; Valentim, I.B.; Camatari, F.O.S.; de Almeida, A.M.M.; Goulart, H.F.; de Souza Ferro, J.N.; de Oliveira Barreto, E.; Cavalcanti, B.C.; Bottoli, C.B.G.; Goulart, M.O.F. Polyphenol Profile by Uhplc-Ms/Ms, Anti-Glycation, Antioxidant and Cytotoxic Activities of Several Samples of Propolis from the Northeastern Semi-Arid Region of Brazil. *Pharm. Biol.* **2017**, *55*, 1884–1893. [CrossRef]
51. Lucas, M.; Freitas, M.; Xavier, J.A.; Moura, F.A.; Goulart, M.O.F.; Ribeiro, D.; Fernandes, E. The Scavenging Effect of Curcumin, Piperine and Their Combination against Physiological Relevant Reactive Pro-Oxidant Species Using In Vitro Non-Cellular and Cellular Models. *Chem. Pap.* **2021**, *75*, 5269–5277. [CrossRef]

molecules

MDPI

Article

Simultaneously Determined Antioxidant and Pro-Oxidant Activity of Randomly Selected Plant Secondary Metabolites and Plant Extracts

Tibor Maliar [1,*], Mária Maliarová [1], Marcela Blažková [2,3], Marek Kunštek [2,3], Ľubica Uváčková [3], Jana Viskupičová [4], Andrea Purdešová [1] and Patrik Beňovič [1]

[1] Department of Chemistry and Environmnetal Sciences, Faculty of Natural Sciences, University of Ss. Cyril and Methodius in Trnava, Nám. J. Herdu 2, 917 01 Trnava, Slovakia; maria.maliarova@ucm.sk (M.M.); andrea.purdesova@ucm.sk (A.P.); benovic2@ucm.sk (P.B.)
[2] National Agricultural and Food Centre, Hlohovecká 2, 951 41 Lužianky, Slovakia; marcela.blazkova@nppc.sk (M.B.); marek.kunstek@nppc.sk (M.K.)
[3] Department of Biology and Biotechnology, Faculty of Natural Sciences, University of Ss. Cyril and Methodius in Trnava, Nám. J. Herdu 2, 917 01 Trnava, Slovakia; lubica.uvackova@ucm.sk
[4] Centre of Experimental Medicine SAS, Institute of Experimental Pharmacology and Toxicology, Slovak Academy of Sciences, Dúbravská cesta 9, 841 04 Bratislava, Slovakia; janaviskupicova@gmail.com
* Correspondence: tibor.maliar@ucm.sk

Abstract: Oxidative stress is a well-known phenomenon arising from physiological and nonphysiological factors, defined by the balance between antioxidants and pro-oxidants. While the presence and uptake of antioxidants are crucial, the pro-oxidant effects have not received sufficient research attention. Several methods for assessing pro-oxidant activity, utilizing various mechanisms, have been published. In this paper, we introduce a methodology for the simultaneous determination of antioxidant and pro-oxidant activity on a single microplate in situ, assuming that the FRAP method can measure both antioxidant and pro-oxidant activity due to the generation of pro-oxidant Fe^{2+} ions in the Fenton reaction. Systematic research using this rapid screening method may help to distinguish between compounds with dominant antioxidant efficacy and those with dominant pro-oxidant effects. Our preliminary study has revealed a dominant pro-oxidant effect for compounds with a higher number of oxygen heteroatoms, especially sp2 hybridized compounds (such as those containing keto groups), such as flavonoids and plant extracts rich in these structural types. Conversely, catechins, carotenoids, and surprisingly, extracts from birch leaves and chestnut leaves have demonstrated dominant antioxidant activity over pro-oxidant. These initial findings have sparked significant interest in the systematic evaluation of a more extensive collection of compounds and plant extracts using the developed method.

Keywords: oxidative stress; antioxidant activity; pro-oxidant activity; compounds; plant extract

Citation: Maliar, T.; Maliarová, M.; Blažková, M.; Kunštek, M.; Uváčková, Ľ.; Viskupičová, J.; Purdešová, A.; Beňovič, P. Simultaneously Determined Antioxidant and Pro-Oxidant Activity of Randomly Selected Plant Secondary Metabolites and Plant Extracts. *Molecules* **2023**, *28*, 6890. https://doi.org/10.3390/molecules28196890

Academic Editors: José Pinela, Maria Inês Dias, Carla Pereira and José Ignacio Alonso-Esteban

Received: 9 August 2023
Revised: 22 September 2023
Accepted: 27 September 2023
Published: 30 September 2023

1. Introduction

The commercial database SciFinder, one of the most widely utilized databases, reveals a significant lack of scientific papers on pro-oxidants. Keyword searches for "ANTIOXIDANT/PROOXIDANT/ANTIOXIDANTS and PROOXIDANTS" in titles yielded 55,177 records for antioxidants, 542 records for pro-oxidants, and only 227 records for papers mentioning both antioxidants and pro-oxidants. The database categorizes these results into manuscripts, patents, reviews, clinical studies, books, conference contributions, etc. A similar situation arises when searching for antioxidant/pro-oxidant activity in plant extracts. To provide a systematic overview of this field, it is essential to describe and explain the FRAP method applied in this paper. FRAP stands for "Ferric Reducing Antioxidant Power". This assay measures antioxidant power by reducing ferric-tripyridyltriazine (Fe^{3+}-TPTZ) to an intense blue-colored ferrous-tripyridyltriazine complex (Fe^{2+}-TPTZ) at a low pH, with an absorption maximum of 593 nm [1]. Trolox is commonly used as a positive control, and results can be expressed as μM of Trolox equivalents or μM Fe^{2+} based on a standard curve. The FRAP assay is a proven method for assessing the antioxidant capacity of foods and legume seeds, which is closely related to their polyphenol contents [1]. In 1996, Benzie and Strain developed the FRAP assay to estimate the ferric-reducing power of human plasma [2]. Dragsted et al. [3] adapted the FRAP assay for use with a microtiter plate reader in a 96-well format. Their team successfully determined the FRAP values of plant extracts and agrarian crop extracts using a slightly modified method on microtiter plates [4–8]. In the paper by Wojtunik-Kulesza [9], it was highlighted that the FRAP assay requires specific conditions, including an acidic medium (pH 3.6) to facilitate iron solubility and a temperature of 37 °C. The low pH reduces the ionization potential, promoting electron transfer and increasing the redox potential, which affects the dominant reaction mechanism [10,11]. However, it is essential to note that the FRAP assay, similar to other assays, has its limitations. The redox potential of the Fe^{3+}/Fe^{2+} pair plays a crucial role, as compounds with a lower redox potential may yield falsely high Fe^{3+} reduction results. Additionally, FRAP assay results depend on the timescale of the analysis [11].

In the field of antioxidant and pro-oxidant activity methods, numerous papers and concepts have been introduced. In brief, reduction of a chemical refers to a gain of electrons, while oxidation refers to a loss of electrons [12]. A reductant or reducing agent donates electrons, causing another reactant to be reduced, while an oxidant or oxidizing agent accepts electrons, leading to the oxidation of another reactant. These terms have specific meanings in the context of a biological system [12]. Based on this perspective, assays can be divided into two categories: antioxidant capacity assays involving oxidants that are not necessarily pro-oxidants, and antioxidant capacity assays involving oxidants that are pro-oxidants, as published in [1,13]. The FRAP assay falls under the first category, as it involves an oxidant, Fe^{3+}. However, it is essential to note that Fe^{3+} is not necessarily a pro-oxidant. On the other hand, Fe^{2+}, produced from the reduction in Fe^{3+} in the FRAP assay, could act as a pro-oxidant due to its reaction with H_2O_2. Nevertheless, neither Fe^{2+} nor Fe^{3+} directly cause oxidative damage to lipids, proteins, or nucleic acids.

The explanation of Fenton reactions and the significance of Fe^{2+} and Fe^{3+} ions in these reactions is crucial for understanding the subsequent discussions. The Fenton system uses ferrous ions (Fe^{2+}) to react with hydrogen peroxide (H_2O_2), producing hydroxyl radicals ($OH^\bullet$) and hydroxide ions (OH^-) (Equation (1)). Fenton-like reactions involve a two-step system, generating hydroperoxyl radicals ($HO_2^\bullet$) in the subsequent reactions (Equations (2) and (3)).

$$H_2O_2 + Fe^{2+} \rightarrow Fe^{3+} + OH^\bullet + OH^- \quad (1)$$

The hydroxyl radical is a very reactive oxidant capable of rapidly reacting with surrounding molecules.

$$H_2O_2 + Fe^{3+} \rightarrow Fe\cdots OOH_2{}^+ + H^+ \quad (2)$$

$$Fe\cdots OOH^{2+} \rightarrow Fe^{2+} + HO_2{}^\bullet \quad (3)$$

The significance of Fenton-like reactions lies not only in the production of the $HO_2^\bullet$ radical but also in the conversion of ferric ions (Fe^{3+}) to ferrous ions (Fe^{2+}), which can initiate another Fenton reaction cycle. The kinetics of Fenton oxidation are complex and can be described by a combined pseudo-first-order kinetic model, while Fenton-like reactions follow simpler, pseudo-first-order kinetics [14]. However, several studies have shown that the rate of decomposition of H_2O_2 and the rate of oxidation of organic solutes are much slower using Fe^{3+}/H_2O_2 than Fe^{2+}/H_2O_2 as a source of radicals [15]. In Fenton-like reactions, ferric ions react with H_2O_2 to produce ferrous ions at a very slow rate ($k = 0.001–0.01\ M^{-1}\ s^{-1}$) [16]. Macáková et al. [17] published a paper describing that iron reduction potentiates hydroxyl radical formation only in flavonols. Flavonoids, substantial components of the human diet, are generally considered beneficial. However, they may possess possible pro-oxidative effects based on their reducing potential. The study revealed that a substantial reduction of ferric ions occurred under acidic conditions, particularly with flavonols and flavanols containing the catecholic ring B. This paper, along with others, sheds light on the interactions between flavonoids and iron, bringing another dimension to the understanding of these reactions. The findings showed that flavonols such as morin and rutin exhibited progressive pro-oxidant effects, while 7-hydroxyflavone and hesperetin were the only flavonoids with dose-dependent inhibition of hydroxyl radical production [17].

As far as the authors are aware, there has not been a paper published describing the FRAP assay as a method for the simultaneous determination of both antioxidant and pro-oxidant activities of tested samples based on the conversion of Fe^{3+} to Fe^{2+}. The ferrous ion is currently recognized as the more pro-oxidative form of iron. However, there are similar methods based on the same principle, involving the colorization of some transition metals. Many of these methods employ both the ferrous and ferric ions in the complex for spectrophotometric determination and analysis. Particularly, several methods for determining pro-oxidant activity have been published, such as the method of reducing power (RP) using potassium ferricyanide [18], the Ferric-ferrozine assay for total antioxidant capacity using ferrozine as a ligand [19], the CUPRAC method using neocuproine as a ligand [20], or the copper reducing activity index (CRAI) assay using sodium diethyldithiocarbamate as a ligand (DDTC) coupled with TBARS determination of pro-oxidative fragments of linoleic acid [21].

In addition to the methods mentioned earlier, several techniques are available for quantifying the pro-oxidant effect on DNA, proteins, or lipids [22], or determining the pro-oxidant effects on cell morphology in vitro [23,24].

The simultaneous detection of both antioxidant and pro-oxidant activities has been published. Some papers used peroxidase (myeloglobin/H_2O_2)-generated $ABTS^{+\bullet}$ [2,2′-azinobis-(3-ethylbenzthiazoline-6-sulfonic acid)] radical cation [25], the β-carotene bleaching assay [25], and the crocin bleaching assay [26]. In this paper, we present a simultaneous determination of antioxidant activity using the DPPH method and antioxidant pro-oxidant activity of compounds and extract samples using the FRAP method, modified on a microplate.

Antioxidative active compounds and well-known antioxidants, including vitamins, may also act as pro-oxidants, depending on their concentration. For instance, vitamin C is a potent antioxidant, but it can exhibit pro-oxidant behavior depending on the dosage [27]. The pro-oxidant effect of vitamin C can also manifest when it interacts with iron, reducing Fe^{3+} to Fe^{2+}, or with copper, reducing Cu^{2+} to Cu^{+} [28,29]. The supplementation of vitamin C may result in a reduced normal biological response to free radicals and create an environment that is more susceptible to oxidation, potentially leading to mild oxidative stress due to its pro-oxidative properties [27]. Similarly, alpha-tocopherol, is another well-known potent antioxidant that can act as a pro-oxidant in high concentrations. This occurs due to its reaction to reactive oxygen species (ROS), where it remains in its reactive form without the availability of ascorbic acid [30,31].

Similarly, the same effect has been observed in another well-known category of antioxidants—flavonoids. Even flavonoids have been reported to act as pro-oxidants in systems containing transition metals. Flavonoids, such as quercetin and kaempferol, induce DNA damage and lipid peroxidation in the presence of transition metals [32]. Flavonoids can potentially act as pro-oxidants through several mechanisms, including direct interaction with oxygen via the Fl-$O^{\bullet}$ radical [33], inhibition of mitochondrial respiration, causing a substrate-independent cyanide-insensitive respiratory burst in isolated mitochondria, associated with the production of ROS [34], and oxidation by peroxidases, resulting in the formation of intracellular phenoxyl radicals by myeloperoxidase [35]. Finally, flavonoids can act as pro-oxidants by oxidizing low-molecular antioxidants [36].

A similar situation was observed for other phenolics in general. Phenolics can also display pro-oxidant effects, especially in systems containing redox-active metals. The presence of iron or copper catalyzes their redox cycling and may lead to the formation of phenolic radicals, which damage lipids and DNA [37,38].

The main goal of this paper was to investigate the simultaneous determination of antioxidant/Fe^{3+}-reducing activity, producing pro-oxidant Fe^{2+} ions in randomly selected compounds and plant extracts on a microplate. The objective was to observe significant differences among the tested samples and initiate a systematic, rapid evaluation for further research.

2. Results and Discussion

In this study, we assessed the antioxidant activity of 30 randomly selected compounds and 18 plant extracts using the DPPH method, as well as their ability to reduce Fe^{3+} ions using the FRAP method, with TROLOX as a standard. The pro-oxidant antioxidant balance index (PABI) was calculated from the ratio $FRAP_{50}/DPPH_{50}$ according to Equation (4). The results for the tested compounds, along with their CAS numbers, $DPPH_{50}$ and $FRAP_{50}$ values, correlation coefficients, and the targeted PABI, are presented in Table 1. These compounds were randomly selected from our ongoing parallel studies. Several compounds were excluded from the dataset shown in Table 1 because their values were outside the testing interval.

Table 1. Compounds tested, including CAS numbers, $DPPH_{50}$, $FRAP_{50}$ values, correlation coefficients (r^2), and pro-oxidant antioxidant balance index (PABI).

STANDARD/ Compound	CAS	DPPH$_{50}$ (μM)	r^2	FRAP$_{50}$ (μM)	r^2	PABI
TROLOX	53188-07-1	115.01 ± 1.5	0.9895	171.14 ± 7.9	0.906	1.49
Quercetin	117-39-5	356.03 ± 1.8	0.954	156.26 ± 4.1	0.963	0.44
Rutin	153-18-4	400.81 ± 7.9	0.970	>4086	-	-
Baicalein	491-67-8	49.16 ± 1.4	0.958	186.19 ± 4.9	0.945	3.79
Morin	480-16-0	78.87 ± 2.2	0.915	176.85 ± 3.4	0.956	2.24
7,8-Dihydroxyflavone	38183-03-8	63.17 ± 1.4	0.939	253.91 ± 2.9	0.952	4.02
Hesperidin	520-26-3	523.24 ± 14.5	0.976	>4086	-	-
Diosmin	520-27-4	>4086	-	>4086	-	-
Apigenin-7-glucoside	578-74-5	>4086	-	>4086	-	-
(−)-Epicatechin	490-46-0	22.49 ± 0.5	0.950	279 ± 0.9	0.997	12.40
(+)-Catechin	154-23-4	192.63 ± 5.5	0.997	4479.36 ± 10.8	0.944	23.25
L-Ascorbic acid	50-81-7	164.06 ± 9.9	0.938	337.2 ± 0.8	0.938	2.06
Tannic acid	1401-55-4	111.82 ± 2.3	0.982	49.51 ± 0.5	0.974	0.44
Crocin	42553-65-1	771.1 ± 9.4	0.975	884.32 ± 17.2	0.960	1.15
β-Carotene	7235-40-7	2086 ± 29.5	0.971	>4086	-	-
Purpurin	81-54-9	941.3 ± 28.6	0.993	2504.61 ± 30.8	0.955	2.66
Silibinin	22888-70-6	3336.48 ± 107.1	0.974	1365.1 ± 22.7	0.929	0.41
Olivetol	500-66-3	>4086	0.939	>4086	-	-
Gallic acid	149-91-7	50.72 ± 0.8	0.931	408.97 ± 10.07	0.978	8.06
Caffeic acid	331-39-5	177.25 ± 8.8	0.937	500.39 ± 8.6	0.969	2.82
Protocatechuic acid	99-50-3	166.31 ± 8.1	0.982	>4086	-	-
Avenanthramide A	108605-70-5	448.6 ± 11.4	0.968	1730.67 ± 16.8	0.965	3.86
Avenanthramide B	108605-69-2	1566.2 ± 60.5	0.863	2778.45 ± 43.3	0.946	1.77
Avenanthramide C	116764-15-9	431.14 ± 20.6	0.989	1468.64 ± 14.3	0.971	3.41

The obtained results clearly indicate that the majority of the tested compounds exhibited both DPPH$^{\bullet}$ scavenging ability and Fe^{3+} reducing activity, thereby producing pro-oxidative Fe^{2+} ions. However, certain compounds such as rutin, β-carotene, and protocatechuic acid demonstrated antioxidant activity according to the DPPH method, yet their $FRAP_{50}$ values exceeded the testing interval (1–4096 μM), suggesting lower pro-oxidant activity in comparison to their antioxidant activity. Based on the perspective of antioxidant activity determined by the DPPH method, the most potent compounds were the well-known flavonoids quercetin, morin, baicalein, 7,8-dihydroxyflavone, (−)-epicatechin, and polyphenolic acid-gallic acid, which exhibited higher antioxidant activity than the standard TROLOX. In the second category, compounds with $DPPH_{50}$ values falling within the range of 115–500 μM were less active than the standard. This group included compounds such as rutin, hesperidin, (+)-catechin, L-ascorbic acid, tannic acid, caffeic acid, protocatechuic acid, and avenanthramide A and C.

When comparing antioxidant activity and the production of pro-oxidant Fe^{2+} ions using the PABI parameter, we can divide tested compounds into three categories: compounds with dominant pro-oxidant activity (PABI < 1), compounds with roughly balanced pro-oxidant/antioxidant activity (PABI from 1 to 2), and compounds with dominant antioxidant effect (PABI > 2). Compounds with PABI values less than 1 were considered pro-oxidants. These compounds included quercetin, tannic acid, and silibinin. On the other hand, compounds with PABI values within the interval of 1–3, representing a balanced pro-oxidant/antioxidant profile, included the standard TROLOX, morin, L-ascorbic acid, crocin, an anthraquinone purpurin, caffeic acid, and avenanthramide B. Remarkably, compounds with significantly high PABI values (over 5) included gallic acid (PABI = 8.06), and most notably, both tested catechins/(−)-epicatechin PABI = 12.4, (+)-catechin PABI = 23.25). Overall, both antioxidant and pro-oxidant activities varied throughout the entire testing concentration range (1–4096 μM), with PABI values spanning the interval of 0.41–23.25. These findings provide insight into the diversity of PABI values within the randomly selected compound collection, demonstrating a variation of over 50 times in this study. It is important to note that this ratio could change with an expansion of the tested compound collection. From a structural perspective, our findings support the thesis presented in the Introduction section [17], which suggested that a substantial reduction in ferric ions occurs under acidic conditions, particularly with flavonols and flavanols containing the catechol moiety in the B ring. Pro-oxidant activity refers to the ability to interact with transition metals, forming coordination complexes. These chemical compounds consist of a central atom or ion, which is usually metallic and is called the coordination center, surrounded by bound molecules or ions known as ligands or complexing agents. The coordination of Fe ions can occur with biomolecules possessing free ionic pairs in relatively close proximity, such as the catechol in the B ring or the keto group of flavonols and flavanols with an OH group in the vicinal position (position 3) on the C ring or near position (position 5) on the A ring. This may explain the higher pro-oxidant activity of flavonols and flavanols compared to catechins, which lack the sp2 hybridized keto group in position 4 of the C ring.

Our study's findings indicate that flavonoids, particularly quercetin and flavolignan silibinin, exhibit more pronounced pro-oxidant properties due to specific structural features. These include the presence of a C2–C3 double bond in the C ring, the catechol conformation of vicinal hydroxyl groups, or the hydroxyl group in combination with the keto group at position 4 [33]. This observation aligns with existing literature [39,40], including other sources that suggest quercetin's pro-oxidant effect contributes to its potential anticancer efficacy [41]. There are some records on the pro-oxidant activity of silibinin [42]. Contrary to our findings, previous publications have highlighted the pro-oxidant activity of catechins, particularly in the context of their anticancer effects [43,44]. It is important to acknowledge that our results are solely derived from in vitro studies, and this could partially explain the discrepancy. The saturated C-ring of catechins, lacking double bonds, may mitigate their pro-oxidant potential. A similar situation might be observed in gallic acid [45].

The second part of the study focused on screening randomly selected plant extracts prepared using an extraction process provided in the Material and Methods section. The study encompassed medicinal plants (n = 18), which are currently being explored. The achieved results are presented in Table 2.

Table 2. Randomly selected medical plant extracts, including $DPPH_{50}$, $FRAP_{50}$ values (expressed in mg of extract dried matter per milliliter), correlation coefficients (r^2), and pro-oxidant antioxidant balance index (PABI).

Plant Species, Latin Name and Bot. Classifier	Plant Part	$DPPH_{50}$ (mg dm/mL)	r^2	$FRAP_{50}$ (mg dm/mL)	r^2	PABI
Sessile oak, *Quercus petraea*, (Matt.) Liebl.	leaves	2.54 ± 0.2	0.949	1.81 ± 0.22	0.980	0.71
Silver birch, *Betula pendula*, Roth.	leaves	1.6 ± 0.2	0.998	2.37 ± 0.3	0.986	1.48
Horse chestnut, *Aesculus hippocastanum*, L.	leaves	1.82 ± 0.1	0.979	9.09 ± 0.04	0.934	4.98
Old man's beard, *Clematis vitalba*, L.	bark	7.72 ± 0.3	0.913	32.33 ± 0.2	0.944	4.18
Rapeseed, *Brassica napus*, L.	grains	145.15 ± 0.1	0.962	11.64 ± 0.3	0.961	0.08
Rhubarb, *Rheum rhabarbarum*, L.	root	1.39 ± 0.2	0.962	0.58 ± 0.1	0.933	0.42
Pedunculate oak, *Quercus robur*, L.	bark	0.5 ± 0.1	0.973	0.3 ± 0.1	0.939	0.61
Black elderberry, *Sambuicus nigra*, L.	flower	1.09 ± 0.1	0.968	1.61 ± 0.2	0.966	1.47
Woundwort, *Prunella vulgaris*, L.	flower	1.57 ± 0.1	0.965	1.27 ± 0. 1	0.962	0.81
Green tea, *Camelia sinensis*, (L.) Kuntze.	flower	1.93 ± 0.1	0.923	0.21 ± 0.1	0.993	0.11
Liquorice, *Glycyrrhiza glabra*, L.	root	5.62 ± 0.2	0.955	3.41 ± 0.2	0.967	0.61
Common wormwood, *Artemisia absinthium*, L.	leaves	1.73 ± 0.1	0.984	1.25 ± 0.1	0.963	0.73
Thistle, *Silybum marianum*, (L.) Gaertn.	grain	7.22 ± 0.4	0.991	1.61 ± 0.1	0.970	0.22
Chamomile, *Matricaria chamomilla*, L.	flower	5.06 ± 0.3	0.984	5.03 ± 0.2	0.981	0.99
Ginger, *Zingiber officinale*, Roscoe.	root	4.62 ± 0.2	0.994	5.43 ± 0.3	0.991	1.18
Turmeric, *Curcuma longa*, L.	root	17.52 ± 0.5	0.988	11.44 ± 0.4	0.982	0.65
Sage, *Salvia officinalis*, L.	leaves	0.27 ± 0.1	0.975	1.73 ± 0.1	0.959	6.27
Grape wine, *Vitis vinifera*, L.	frost dried grapes	3.52 ± 0.1	0.976	40.63 ± 1.8	0.961	11.53

Based on the results, there are some differences among medical plant extracts, although they are not as pronounced as those seen with single compounds. These variations can be attributed to the "buffer" effect of the complex mixtures found in these extracts, which could consist of numerous compounds. The antioxidant activity of the strongest samples, such as oak, birch leaves, bark, green tea leaves, common wormwood leaves, and rhubarb roots, is well documented in the literature. In our experiments, these samples exhibited $DPPH_{50}$ values under 2 mg of extract-dried matter per milliliter. While there is less information in the literature regarding the ability to reduce Fe^{3+} ions and produce pro-oxidant Fe^{2+} ions, it is particularly interesting to interpret the results based on PABI values. Surprisingly, most of the tested extract samples (oak, rapeseed, rhubarb, wormwood, green tea, licorice, common wormwood, thistle, chamomile, or turmeric) had PABI values under 1, indicating a predominant pro-oxidant effect over antioxidants. In contrast, extract samples prepared from silver birch, horse chestnut, old man's beard, black elderberry, ginger, sage, and notably grape wine showed a dominant antioxidant activity over pro-oxidant. One of the most promising findings is related to sage, which exhibited the most potent antioxidant activity among the samples with a high PABI index.

In the current body of literature, only a single paper focuses on the potential pro-oxidant effects of polyphenols from sage (and rosemary) [46], and only a handful of papers address grape wine [32,47], despite the abundance of publications discussing its antioxidant effects. Compared to our results, only a limited number of publications have described the simultaneous determination of antioxidant and pro-oxidant activities, preferably in a single format.

3. Materials and Methods

3.1. Chemicals and Solvents

2,2-Diphenyl-1-picrylhydrazyl radical ($DPPH^{\bullet}$), 2,2-Diphenyl-1-picrylhydrazin (DPPH), 2,4,6-Tris(2-pyridyl)-s-triazine (TPTZ), $FeCl_2 \cdot 4H_2O$, $FeCl_3 \cdot 6H_2O$, TROLOX, quercetin, rutin, baicalein, morin, 7,8-dihydroxyflavone, (−)-epicatechin, (+)-catechin, L-ascorbic acid, tannic acid, crocin, -caroten, purpurin, silibinin, olivetol, gallic acid, caffeic acid, protocatechuic acid, avenanthramides A, B, and C, flavone, ergocalciferol, lipoic acid, biotin, thiamin, indole-3-carbinol, astaxanthin, uric acid, sodium acetate, ethanol, and acetic acid were purchased from Merck /Sigma/ (St. Louis, MO, USA).

3.2. Preparation of the Extract Samples

Randomly selected plant material from 18 plant species (refer to Table 2) was chosen based on their bioactivity in our current research. A total of 2 g of dried plant matter was disintegrated into small pieces (under 5 mm particle size). The plant matter was then extracted in screwed-up tubes with 20 mL of a 50% ethanol solution in the dark at room temperature for 24 h. Afterward, the extract was filtered and stored in Eppendorf tubes at 4 °C in the dark.

3.3. Principles of Microplate Methods

The DPPH method was employed for measuring antioxidant activity, while the FRAP method was used to measure both antioxidant and pro-oxidant activity. Both methods were performed on microplates that were previously modified by our research team [4–8], ensuring equal concentrations of key reagents (DPPH, TPTZ, and FeCl3) at 0.4 mM. Briefly, a 0.4 mM $DPPH^{\bullet}$ radical solution was prepared in ethanol. For the FRAP assay, separate solutions A and B were prepared, with solution A containing 0.0187 g of TPTZ in 10 mL of ethanol and solution B containing 0.338 g of sodium acetate in 88.3 mL of water and 1.748 mL of acetic acid. These solutions were mixed freshly before each experiment. Additionally, 1.2 mM solutions of $FeCl_2 \cdot 4H_2O$ and $FeCl_3 \cdot 6H_2O$ were prepared freshly prior to the experiments. The microplate template used for the assays is presented in Figure 1.

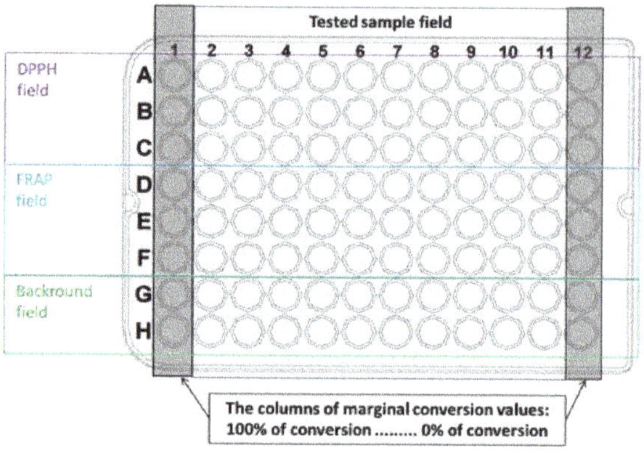

Figure 1. Microplate template. Organization of the microplate consisting of a tested sample field (wells A2–H12) and columns dedicated to standard, representing 100% conversion (wells A1–F1) and 0% conversion (wells A12–F12) for the assays.

3.4. *Definition of the Conversion Intervals*

In the first column (column 1), a 100% conversion standard for the DPPH method was applied using 2,2-diphenyl-1-picrylhydrazine (DPPH), and for the FRAP method, a solution of $FeCl_2{\cdot}4H_2O$ was used instead of the sample. In contrast, in the last column (column 12), 0% conversion standard for the DPPH method was applied using 2,2-Diphenyl-1-picrylhydrazyl radical (DPPH$^{\bullet}$), and for the FRAP method, a solution of $FeCl_3{\cdot}6H_2O$ was used instead of the sample.

3.5. *Application of Tested Sample on Microplate*

First, dilution of tested samples (compounds, extracts) in the microplate testing field was realized by both following dilution modes.

Mode 1, dilution to 1/2 concentration (100 μL of tested samples into column 2, 50 μL of ethanol into columns 3–10), transfer of 50 μL in the direction of columns 2–11 using one 8-channel micropipette. Starting from stock solution (for compounds) with a concentration of 8.192 mM, we achieve final concentrations in μM: 4096, 2048, 1024, 512, 256, 128, 64, 32, 16, and 8 μM. The extract samples were tested in 50% ethanol solutions as they were prepared.

Mode 2, dilution to 2/3 concentration (150 μL of tested samples into column 2, 50 μL of ethanol into columns 3–10), transfer of 100 μL in direction of columns 2–11 using one 8-channel micropipette, and removal of volume 100 μL from the last column. Starting from stock solution (for compounds) with optimal concentration determined by mode 1.

3.6. *Application of Tested Sample on Microplate*

Firstly, the dilution of tested samples (compounds and extracts) was realized on the microplate testing field by following both dilution modes. Mode 1, dilution to 1/2 concentration (100 μL of tested samples into column 2, 50 μL of ethanol into columns 3–10), transfer of 50 μL in the direction of columns 2–11 using one 8-channel micropipette. Starting from stock solution (for compounds) with a concentration of 8.192 mM, we achieve final concentrations in μM: 4 096, 2 048, 1 024, 512, 256, 128, 64, 32, 16, and 8 μM. The extract samples were tested as 50% ethanol solutions.

Mode 2, dilution to 2/3 concentration (150 μL of tested samples into column 2, 50 μL of ethanol into columns 3–10), transfer of 100 μL in the direction of columns 2–11 using one 8-channel micropipette, and removal of volume 100 μL from the last column. Starting from stock solution (for compounds) with optimal concentration determined by mode 1.

3.7. *Preparation of the Microplate*

Phase I—Application of the conversion standards:

- adding 50 μL of a 0.4-mM solution of 2,2-diphenyl-1-picrylhydrazine in ethanol into the wells A1–C1;
- adding 50 μL of a 1.2-mM water solution of $FeCl_2 \cdot 4H_2O$ into the wells D1–F1;
- adding 50 μL of a 0.4-mM solution of $DPPH^{\bullet}$ into the wells A12–C12;
- adding 50 μL of a 1.2-mM water solution of $FeCl_3 \cdot 6H_2O$ into the wells D12–F12.

Phase II—Application of $FeCl_3$ solution and microplate filling:

- adding 50 μL of a 1.2-mM solution of $FeCl_3 \cdot 6H_2O$ into the field of three rows D2–F11,
- adding 150 μL of 50% (v/v) ethanol into the "background" wells in rows G and H (G2–H11).

3.8. Preparation of Starting Reagents

DPPH reagents were prepared as a 0.4-mM solution in ethanol freshly before the experiment. Similarly, FRAP reagent was prepared freshly before the experiment accordingly: 10 mL of a 6-mM TPTZ solution in ethanol and 90 mL of a sodium acetate-acetic buffer (0.338 g of sodium acetate dissolved in 88.5 mL of water, followed by the addition of 1.478 mL of acetic acid).

3.9. Starting and Adapting the Microplate before Measurement

The microplate was initialized by adding 150 μL of 0.4 mM DPPH reagent to rows A, B, and C, achieving a final DPPH concentration of 0.3 mM in the reaction mixture. Similarly, 100 μL of FRAP reagent was added to rows D, E, and F to achieve a final concentration of 0.3 mM for TPTZ and Fe^{3+} ions in the reaction mixture. Subsequently, 100 μL was removed from the whole microplate to achieve an optical density (OD) within the limit of the Lambert–Beer law.

3.10. Microplate Incubation and Measurement

The microplate was incubated for 10 min at room temperature, followed by measurements at 520 nm and 630 nm for DPPH and FRAP, respectively. The optical density (OD) data of the samples were adjusted by subtracting the background data and transformed into percentile values of conversion measurement using 0% and 100% conversion data. $DPPH_{50}$ and $FRAP_{50}$ parameters correspond to a concentration of a compound responsible for 50% conversion of either DPPH radical or Fe^{3+} to Fe^{2+} (expressed in μM). They were calculated from the following function plot: percentage of conversion = f(concentration). For extract samples, the concentration of dry matter weight of the extracted sample was used. The pro-oxidant antioxidant balance index /PABI/ was calculated according to Equation (4), using the micromolar expression for compounds and the weight of dried matter expression for extract samples.

$$\text{Pro-oxidant Antioxidant Balance Index /PABI/} = FRAP_{50}/DPPH_{50} \quad (4)$$

3.11. Statistical Analysis of Data

Each experiment was repeated three times with eight replicates. Results were presented as the mean ± standard deviation (SD). The correlation coefficient was calculated using the Spearman method. A difference was considered statistically significant when * $p < 0.1$.

4. Conclusions

In summary, this study aimed at developing a rapid screening microplate method for simultaneous detection of the antioxidant and pro-oxidant activities of a diverse range of compounds and medical plant extracts. This study provides a preliminary overview, shedding light on the potential range of compounds spanning from those with strong antioxidant effects to those with pronounced pro-oxidant effects. Subsequent in-depth investigations are needed to build upon these initial insights.

Our upcoming research will focus on elucidating the structural characteristics that contribute to the prevalence of pro-oxidant effects. We plan to employ molecular mechanics and semi-empirical calculations using suitable software tools to delve into this aspect. Such analysis may be particularly beneficial for compounds commonly found in food, cosmetics, and pharmaceutical additives produced via plant suspension cultures. This will allow us to deepen our understanding of these compounds' effects and implications.

Author Contributions: Conceptualization, T.M., M.M. and A.P.; methodology, T.M., M.B. and P.B.; software, T.M. and M.K:; validation, Ľ.U., M.M. and J.V.; formal analysis, M.K., A.P. and P.B.; investigation, T.M. and J.V.; resources, M.B.; data curation, T.M., P.B.; writing—original draft preparation, T.M.; writing—review and editing, J.V., Ľ.U. and M.M.; supervision, T.M.; project administration, T.M.; funding acquisition, T.M. All authors have read and agreed to the published version of the manuscript.

Funding: This research was funded by Grants SRDA-20-0413 and VEGA 2/0113/21.

Institutional Review Board Statement: Not applicable.

Informed Consent Statement: Not applicable.

Data Availability Statement: The data presented in this study are available on request from the corresponding author.

Conflicts of Interest: The authors declare no conflict of interest.

Sample Availability: Not applicable.

References

1. Njoya, E.M. Chapter 31—Medicinal plants, antioxidant potential, and cancer. In *Cancer*, 2nd ed.; Preedy, V.R., Patel, V.B., Eds.; Academic Press: Cambridge, MA, USA, 2021; pp. 349–357.
2. Benzie, F.F.; Strain, J.J. The ferric reducing ability of plasma (FRAP) as a measure of "Antioxidant power": The FRAP Assay. *Anal. Biochem.* **1996**, *239*, 70–76. [CrossRef]
3. Dragsted, L.O.; Pedersen, A.; Hermetter, A.; Basu, S.; Hansen, M.; Haren, G.R.; Kall, M.; Breinholt, V.; Castenmiller, J.J.M.; Stagsted, J.; et al. The 6-a-day study: Effects of fruit and vegetables on markers of oxidative stress and antioxidative defense in healthy nonsmokers. *Am. J. Clin. Nutr.* **2004**, *79*, 1060–1072. [CrossRef]
4. Maliarová, M.; Maliar, T.; Krošlák, E.; Sokol, J.; Nemeček, P.; Nechvátal, P. Antioxidant and proteinase inhibition activity of main oat avenanthramides. *J. Food Nutr. Res.* **2017**, *54*, 346–353.
5. Krošlák, E.; Maliar, T.; Nemeček, P.; Viskupičová, J.; Maliarová, M.; Havrlentová, M.; Kraic, J. Antioxidant and proteinase inhibitory activities of selected poppy (*Papaver somniferum* L.) genotypes. *Chem. Biodivers.* **2017**, *14*, e1700176. [CrossRef]
6. Maliar, T.; Nemeček, P.; Ürgeová, E.; Maliarová, M.; Nesvadba, V.; Krofta, K.; Vulganová, K.; Krošlák, E.; Kraic, J. Secondary metabolites, antioxidant and anti-proteinase activities of methanolic extracts from cones of hop (*Humulus lupulus* L.) cultivars. *Chem. Pap.* **2016**, *71*, 41–48. [CrossRef]
7. Vulganová, K.; Maliar, T.; Maliarová, M.; Nemeček, P.; Viskupičová, J.; Balážová, A.; Sokol, J. Biologically valuable components, antioxidant activity and proteinase inhibition activity of leaf and callus extracts of *Salvia* sp. *Nova Biotechnol. Et. Chim.* **2019**, *18*, 25–36. [CrossRef]
8. Kulichová, K.; Sokol, J.; Nemeček, P.; Maliarová, M.; Maliar, T.; Havrlentová, M.; Kraic, J. Phenolic compounds and biological activities of rye (*Secale cereale* L.) grains. *Open Chem.* **2019**, *7*, 988–999. [CrossRef]
9. Wojtunik-Kulesza, K.A. Approach to Optimization of FRAP Methodology for Studies Based on Selected Monoterpenes. *Molecules* **2020**, *12*, 5267. [CrossRef]
10. Gulcin, I. Antioxidants and antioxidant methods. An updated overview. *Arch. Toxicol.* **2020**, *94*, 651–715. [CrossRef]
11. Hagerman, A.E.; Riedl, K.M.; Jones, G.A.; Sovik, K.N.; Ritchard, N.T.; Hartzfeld, P.W.; Reichel, T.L. High molecular weight plant phenolics (tannins) as biological antioxidants. *J. Agric. Food Chem.* **1998**, *46*, 1887–1892. [CrossRef]
12. Ronald, L.; Prior; Cao, G. In vivo total antioxidant capacity: Comparison of different analytical methods. *Free Radic. Biol. Med.* **1999**, *27*, 1173–1181.
13. Benzie, I.F.F.; Strain, J.J. Ferric reducing/antioxidant power assay: Direct measure of total antioxidant activity of biological fluids and modified version for simultaneous measurement of total antioxidant power and ascorbic acid concentration. *Methods Enzymol.* **1999**, *299*, 15–27.
14. Shaobin, W. A Comparative study of Fenton and Fenton-like reaction kinetics in decolourisation of wastewater. *Dyes Pigment.* **2008**, *76*, 714–720.
15. Gallard, H.; De Laat, J. Kinetic modelling of Fe(III)/H_2O_2 oxidation reactions in dilute aqueous solution using atrazine as a model organic compound. *Water Res.* **2000**, *34*, 3107–3116. [CrossRef]
16. Neyens, E.; Baeyens, J. A review of classic Fenton's peroxidation as an advanced oxidation technique. *J. Hazard. Mater.* **2003**, *98*, 33–50. [CrossRef] [PubMed]

17. Macáková, K.; Mladenka, P.; Filipsky, T.; Říha, M.; Jahodár, L.; Trejtnar, F.; Bovicelli, P.; Silvestri, I.P.; Hrdina, R.; Saso, L. Iron reduction potentiates hydroxyl radical formation only in flavonols. *Food Chem.* **2012**, *135*, 2584–2592. [CrossRef]
18. Seck, I.; Hosu, A.; Cimpoiu, C.; Ndoye, S.F.; Ba, L.A.; Sall, C.; Seck, M. Phytochemicals content, screening, and antioxidant/pro-oxidant activities of *Carapa procera* (barks) (Meliaceae). *S. Afr. J. Bot.* **2021**, *137*, 369–376. [CrossRef]
19. Berker, K.I.; Güçlü, K.; Demirata, B.; Apa, R. A novel antioxidant assay of ferric reducing capacity measurement using ferrozine as the colour forming complexation reagent. *Anal. Methods* **2010**, *2*, 1770–1778. [CrossRef]
20. Apak, R.; Güçlü, G.; Demirata, B.; Ozyürek, M.; Celik, S.E.; Bektaşoğlu, B.; Berker, K.I.; Ozyurt, D. Comparative evaluation of various total antioxidant capacity assays applied to phenolic compounds with the CUPRAC assay. *Molecules* **2007**, *12*, 1496–1547. [CrossRef]
21. Rufian-Henares, J.A.; Delgado-Andrade, C.; Morales, F.J. Assessing the antioxidant and pro-oxidant activity of phenolic compounds by means of their copper reducing activity. *Eur. Food Res. Technol.* **2006**, *223*, 225–231. [CrossRef]
22. Damien Dorman, H.J.; Hiltunen, R. Antioxidant and pro-oxidant in vitro evaluation of water-soluble food-related botanical extracts. *Food Chem.* **2011**, *129*, 1612–1618. [CrossRef]
23. Costa, A.S.G.; Alves, R.C.; Vinha, A.F.; Costa, E.; Costa, C.S.G.; Nunes, M.A.; Almeida, A.A.; Santos-Silva, A.; Beatriz, M.; Oliveira, P.P. Nutritional, chemical and antioxidant/pro-oxidant profiles of silverskin a coffee roasting by-product. *Food Chem.* **2018**, *267*, 28–35. [CrossRef]
24. An, J.; Liu, J.; Liang, Y.; Ma, Y.; Chen, C.; Cheng, Y.; Peng, P.; Zhou, N.; Zhang, R.; Addy, M.; et al. Characterization, bioavailability, and protective effects of phenolic-rich extracts from almond hulls against pro-oxidant induced toxicity in Caco-2 cells. *Food Chem.* **2020**, *322*, 126742. [CrossRef] [PubMed]
25. Prieto, M.A.; Rodríguez-Amado, I.; Vázquez, J.A.; Murado, M.A. β-Carotene Assay Revisited. Application to characterize and quantify antioxidant and prooxidant activities in a microplate. *J. Agric. Food Chem.* **2012**, *60*, 8983–8993. [CrossRef] [PubMed]
26. Prieto, M.A.; Vázquez, J.A.; Murado, M.A. Crocin bleaching antioxidant assay revisited: Application to microplate to analyse antioxidant and pro-oxidant activities. *Food Chem.* **2015**, *167*, 299–310. [CrossRef] [PubMed]
27. Sotler, R.; Poljšak, B.; Dahmane, R.; Jukić, T.; Jukić, D.P.; Rotim, C.; Trebše, P.; Starc, A. Prooxidant activities of antioxidants and their impact on health. *Acta Clin. Croat.* **2019**, *58*, 726–736. [CrossRef]
28. Urbański, N.K.; Beresewicz, A. Generation of •OH initiated by interaction of Fe^{2+} and Cu^{+} with dioxygen; comparison with the Fenton chemistry. *Acta Biochim. Pol.* **2000**, *47*, 951–962. [CrossRef]
29. Asplund, K.U.; Jansson, P.J.; Lindqvist, C.; Nordström, T. Measurement of ascorbic acid (vitamin C) induced hydroxyl radical generation in household drinking water. *Free Radic. Res.* **2002**, *36*, 1271–1276. [CrossRef] [PubMed]
30. Carlisle, D.L.; Pritchard, D.E.; Singh, J.; Owens, B.M.; Blankenship, L.J.; Orenstein, J.M.; Patierno, S.R. Apoptosis and P53 induction in human lung fibroblasts exposed to chromium (VI): Effect of ascorbate and tocopherol. *Toxicol. Sci.* **2000**, *55*, 60–68. [CrossRef]
31. Blumberg, J.; Block, G. The alpha-tocopherol beta-carotene cancer prevention in Finland. *Nutr. Rev.* **1994**, *52*, 242–245. [CrossRef]
32. Halliwell, B. Are polyphenols antioxidants or pro-oxidants? What we learn from cells culture and in vivo studies? *Arch. Biochem. Biophys.* **2008**, *476*, 107–112. [CrossRef] [PubMed]
33. Amić, D.; Davidovič-Amić, D.; Bešlo, D.; Rastija, V.; Lučić, B.; Trinajstić, N. SAR and QSAR of the antioxidant activity of flavonoids. *Curr. Med. Chem.* **2007**, *14*, 827–845. [CrossRef]
34. Hodnick, W.F.; Duval, D.L.; Pardini, R.S. Inhibition of mitochondrial respiration and cyanide-stimulated generation of reactive oxygen species by selected flavonoids. *Biochem. Pharmacol.* **1994**, *47*, 573–580. [CrossRef]
35. Goldman, R.; Claycamp, R.; Sweetland, G.H.; Sedlov, M.A.; Tyurin, A.V.; Kisin, E.R.; Tyurina, Y.Y.; Ritov, V.B.; Wenger, S.L.; Grant, S.G.; et al. Myeloperoxidase-catalyzed redox-cycling of phenol promotes lipid peroxidation and thiol oxidation in HL-60 cells. *Free Radic. Biol. Med.* **1999**, *27*, 1050–1063. [CrossRef] [PubMed]
36. Galati, G.; Chan, T.; Wu, B.; O'Brien, P.J. Glutathione-dependent generation of reactive oxygen species by the peroxidase-catalyzed redox cycling of flavonoids. *Chem. Res. Toxicol.* **1999**, *12*, 521–525. [CrossRef]
37. Terpinc, P.; Polak, T.; Šegatin, N.; Hanzlowsky, A.; Ulrih, N.P.; Abramovic, H. Antioxidant properties of 4-vinyl derivatives of hydroxycinnamic acids. *Food Chem.* **2011**, *128*, 62–68. [CrossRef]
38. Yordi, E.G.; Pérez, E.M.; Matos, M.J.; Villares, E.U. Antioxidant and Pro-Oxidant Effects of Polyphenolic Compounds and Structure-Activity Relationship Evidence. In *Nutrition, Well-Being and Health*; Bouayed, J., Bohm, T., Eds.; InTech: Rijeka, Croatia, 2012; Volume 2, pp. 23–48.
39. Laughton, M.J.; Barry Halliwell, B.; Evans, P.J.; Robin, J.; Hoult, S. Antioxidant and pro-oxidant actions of the plant phenolics quercetin, gossypol and myricetin: Effects on lipid peroxidation, hydroxyl radical generation and bleomycin-dependent damage to DNA. *Biochem. Pharm.* **1989**, *38*, 2859–2865. [CrossRef] [PubMed]
40. Utrera, M.; Estévez, M. Impact of trolox, quercetin, genistein and gallic acid on the oxidative damage to myofibrillar proteins: The carbonylation pathway. *Food Chem.* **2013**, *141*, 4000–4009. [CrossRef] [PubMed]
41. Jain, A.K.; Thanki, K.; Jain, S. Novel self-nanoemulsifying formulation of quercetin: Implications of pro-oxidant activity on the anticancer efficacy. *Nanomed. Nanotechnol. Biol. Med.* **2014**, *10*, 959–969. [CrossRef]
42. Vieira do Carmo, M.A.; Granato, D.; Azevedo, L. Chapter Seven—Antioxidant/pro-oxidant and antiproliferative activities of phenolic-rich foods and extracts: A cell-based point of view. In *Advances in Food and Nutrition Research*; Granato, D., Ed.; Academic Press: Cambridge, MA, USA, 2021; Volume 98, pp. 253–280.

43. Lambert, J.D.; Elias, R.J. The antioxidant and pro-oxidant activities of green tea polyphenols: A role in cancer prevention. *Arch. Biochem. Biophys.* **2010**, *501*, 65–72. [CrossRef]
44. Naihao Lu, N.; Puqing Chen, P.; Qin Yang, Q.; Peng, Y.-Y. Anti- and pro-oxidant effects of (+)-catechin on hemoglobin-induced protein oxidative damage. *Toxicol. Vitr.* **2011**, *25*, 833–838.
45. Yen, G.C.; Duh, P.D.; Tsai, H.L. Antioxidant and pro-oxidant properties of ascorbic acid and gallic acid. *Food Chem.* **2002**, *79*, 307–313. [CrossRef]
46. De Haan, L.; Lee-Hilz, Y.; Wemmenhoven, E.; Aarts, J.; Reitjens, I. EpRE-mediated gene induction by the pro-oxidant mechanism of polyphenolic anti-oxidants from rosemary and sage. *Toxicol. Lett.* **2007**, *172*, S42. [CrossRef]
47. Laggner, H.; Hermann, M.; Sturm, B.; Gmeiner, B.M.K.; Kapiotis, S. Sulfite facilitates LDL lipid oxidation by transition metal ions: A pro-oxidant in wine? *FEBS Lett.* **2005**, *579*, 6486–6492. [CrossRef] [PubMed]

molecules

MDPI

Article

A Standardized Botanical Composition Mitigated Acute Inflammatory Lung Injury and Reduced Mortality through Extracellular HMGB1 Reduction

Mesfin Yimam [1,*,†], Teresa Horm [1], Alexandria O'Neal [1], Ping Jiao [1], Mei Hong [1], Lidia Brownell [1], Qi Jia [1], Mosi Lin [2,†], Alex Gauthier [2,†], Jiaqi Wu [2], Kranti Venkat Mateti [2], Xiaojian Yang [2], Katelyn Dial [2], Sidorela Zefi [2] and Lin L. Mantell [2]

1 Unigen Inc., 2121 South State Street, Suite #400, Tacoma, WA 98405, USA; teresamarie522@gmail.com (T.H.); alexhoneal@gmail.com (A.O.); pjiao@unigen.net (P.J.); meih@unigen.net (M.H.); lbrownell@unigen.net (L.B.); qjia@unigen.net (Q.J.)

2 College of Pharmacy and Health Sciences, St John's University, Queens, NY 11439, USA; linm@stjohns.edu (M.L.); alexgauthier4@gmail.com (A.G.); wuj@stjohns.edu (J.W.); kranthi.mateti16@my.stjohns.edu (K.V.M.); xiaojing.yang@stjude.org (X.Y.); kd800@georgetown.edu (K.D.); zefi19@my.stjohns.edu (S.Z.); mantelll@stjohns.edu (L.L.M.)

* Correspondence: myimam@unigen.net

† These authors contributed equally to this work.

Abstract: HMGB1 is a key late inflammatory mediator upregulated during air-pollution-induced oxidative stress. Extracellular HMGB1 accumulation in the airways and lungs plays a significant role in the pathogenesis of inflammatory lung injury. Decreasing extracellular HMBG1 levels may restore innate immune cell functions to protect the lungs from harmful injuries. Current therapies for air-pollution-induced respiratory problems are inadequate. Dietary antioxidants from natural sources could serve as a frontline defense against air-pollution-induced oxidative stress and lung damage. Here, a standardized botanical antioxidant composition from *Scutellaria baicalensis* and *Acacia catechu* was evaluated for its efficacy in attenuating acute inflammatory lung injury and sepsis. Murine models of disorders, including hyperoxia-exposed, bacterial-challenged acute lung injury, LPS-induced sepsis, and LPS-induced acute inflammatory lung injury models were utilized. The effect of the botanical composition on phagocytic activity and HMGB1 release was assessed using hyperoxia-stressed cultured macrophages. Analyses, such as hematoxylin-eosin (HE) staining for lung tissue damage evaluation, ELISA for inflammatory cytokines and chemokines, Western blot analysis for proteins, including extracellular HMGB1, and bacterial counts in the lungs and airways, were performed. Statistically significant decreases in mortality (50%), proinflammatory cytokines (TNF-α, IL-1β, IL-6) and chemokines (CINC-3) in serum and bronchoalveolar lavage fluid (BALF), and increased bacterial clearance from airways and lungs; reduced airway total protein, and decreased extracellular HMGB1 were observed in in vivo studies. A statistically significant 75.9% reduction in the level of extracellular HMGB1 and an increase in phagocytosis were observed in cultured macrophages. The compilations of data in this report strongly suggest that the botanical composition could be indicated for oxidative-stress-induced lung damage protection, possibly through attenuation of increased extracellular HMGB1 accumulation.

Keywords: HMGB1; inflammatory lung injury; oxidative stress; sepsis

Citation: Yimam, M.; Horm, T.; O'Neal, A.; Jiao, P.; Hong, M.; Brownell, L.; Jia, Q.; Lin, M.; Gauthier, A.; Wu, J.; et al. A Standardized Botanical Composition Mitigated Acute Inflammatory Lung Injury and Reduced Mortality through Extracellular HMGB1 Reduction. *Molecules* **2023**, *28*, 6560. https://doi.org/10.3390/molecules28186560

Academic Editor: George Grant

Received: 29 July 2023
Revised: 4 September 2023
Accepted: 7 September 2023
Published: 11 September 2023

1. Introduction

The respiratory system is an easy target for constant endogenous and exogenous harmful attacks. Among the most pressing global challenges, inhalation of environmental fine particulate matter (PM2.5) from air pollution is a crucial risk factor that is known to cause significant oxidative and inflammatory damage to the lungs. Air pollution, including from wildfires, causes poor air quality problems and threatens the general public health.

Recently, in 2022, researchers from the University of Southern California found a 5% increase in mortality risk on extreme PM2.5-only days compared with nonextreme days, highlighting the severe consequence of air pollution [1]. The mortality risk increased to 21% when heat was coupled with air pollution in this same survey. These recurrent exposures to air pollution are known to compromise the innate pulmonary immune response, leaving the lung susceptible to bacterial and viral infections as well as air-pollution-induced oxidative damage [2,3]. Older adults and people with weakened immune systems are more prone to and at greater risk of catastrophic health outcomes. Current conventional pharmaceutical intervention for this global health challenge is inadequate. Daily supplementation of antioxidants in the form of dietary polyphenols could potentially be considered as frontline defense and/or adjunct for air-pollution-induced oxidative stress damage of the lung.

Extracellular HMGB1 is among the key inflammatory mediators elevated by exposure to pollution. It is integral to oxidative-stress-associated downstream effects, surrendering the lung to injury and impeding its function [4,5]. It is a potent, late systemic inflammatory mediator known to serve as an alarmin to surrounding tissues, signaling the loss of cellular homeostasis and triggering a subsequent cytokine storm. The extracellular HMGB1 secreted passively by oxidative stressed cells or actively by activated innate immune cells is highly pro-inflammatory, with a crucial role in sepsis. High levels of HMGB1 in pulmonary tissues have been reported to be associated with a compromised immune response. An increased level of extracellular HMGB1 decreases alveolar macrophage phagocytic activity, subjecting sensitive lung tissue to pathogen invasion or pollution-related damage [6–8]. Antioxidants known to decrease the release of extracellular HMBG1 or inhibit its activity may restore the normal function of innate immune cells (i.e., alveolar macrophage phagocytic activity). For instance, the inhibition of HMGB1 expression in acute lung injury in murine models has been shown to reduce inflammation and tissue damage in addition to sustaining macrophage phagocytic activity, preserving immune activity and respiratory tract function [7,9–12].

Delivery of natural polyphenols in dietary supplements provides a unique advantage in oxidative stress management compared to the administration of simple antioxidant vitamins, as natural polyphenols have a greater structural diversity with a possibility of additional benefit than antioxidation [13]. In this regard, significantly diverse antioxidation properties have been reported for *Scutellaria baicalensis* and *Acacia catechu* extracts or their active constituents that make up the UP446 composition. For example, baicalin, a flavonoid from the root of *S. baicalensis*, has been found to have antioxidant properties that were significantly better than ascorbic acid and butylated hydroxytoluene (BHT) when evaluated on diphenylpicrylhydrazyl radical (DPPH) scavenging activity by iron-chelating assays [14]. It was also reported that baicalin and its aglycon, baicalein, scavenged hydroxyl radicals, DPPH radicals and alkyl radicals in a dose-dependent manner while effectively inhibiting lipid peroxidation of rat brain cortex mitochondria induced by ferrous ascorbic acid, AAPH or NADPH [15]. *S. baicalensis* root extract was also found to protect lipid peroxidation in lung tissue after free-radical-induced injury using linoleic acid hydroperoxide (LHP) [16]. Similarly, record numbers of health benefits of *A. catechu* and its active catechins have been reported through a reduction in ROS. For instance, *A. catechu* heartwood extract, characterized for its high catechin content, was assessed for its neuroprotection activity in both human neuroblastoma cells and rat brain tissues following hydrogen peroxide exposure. The extract was found to reduce ROS formation and protect the mitochondria from oxidative-stress-induced damage in both species [17]. Catechins exhibited strong properties of neutralizing reactive oxygen and nitrogen species in various oxidative-stress-induced lipid peroxidation assays in vivo and in vitro providing protection from oxidative-stress-induced damage [18,19].

UP446 is a standardized composition consisting primarily of a free B-ring flavonoid, baicalin, from *S. baicalensis* and a flavan, catechin, from the heartwoods of *A. catechu* as detailed in the materials and methods section. The composition is a dual cyclooxygenase (COX) and lipoxygenase (LOX) enzyme inhibitor which was found to decrease mRNA expression and protein levels of the proinflammatory cytokines, such as interleukin (IL)-1β, IL-6, and tumor necrosis factor (TNF)-α in preclinical studies [20,21]. Recently, it has been reported that the composition was considered beneficial in mounting a robust humoral response (elevated total IgA and IgG levels) in healthy participants following influenza vaccination paired with strong antioxidation capacity (increased glutathione peroxidase) in a randomized double-blind placebo control clinical trial [22]. These preclinical and clinically proven anti-inflammatory, antioxidant and immune support properties of the composition may have significant contributions to a healthy respiratory system. The current studies were designed to explore additional mechanisms by which the composition could provide protection to the lung and/or the respiratory system in general.

The botanicals constituting the composition tested in this report have been studied separately or together and indicated for various uses in respiratory system support. *S. baicalensis* was recorded in the classical Chinese medical literature (Shen Nong Ben Cao) from the Eastern Han dynasty (circa 200 C.E. or 2200 years ago). A recent list of the top 30 herbs in Traditional Chinese Medicine (TCM) for treating respiratory infections based on the analysis of two TCM databases (the World Traditional Medicine Patent database (WTM) and the Saphron TCM database) had *Radix Scutellariae* as the second most utilized herb, with a 38% frequency in all TCM compositions for treatment of respiratory tract infections [23].

Bioflavonoids, mainly baicalin and its aglycon, baicalein, have been identified as the bioactive components of *Radix Scutellariae*, with biological functions related to antioxidation, anti-inflammation, reduction in the allergic response, and antibacterial activity [24]. Broad spectrum anti-viral activity of these bioflavonoids for commonly isolated causative agents of respiratory tract infection has been reported from in vitro and in vivo studies. For instance, baicalin and baicalein exhibited potent antiviral activity through the inhibition of proteins that viruses need to bind to and bud from host cells–activities that are essential for infection [25]. In mice infected with Influenza A H1N1 virus (swine flu), an extract from *Radix Scutellariae* modulated their inflammatory response to reduce disease severity, decreased lung tissue damage, and ultimately increased their survival rate [26]. Recently, it was also found that the ethanol extract of *S. baicalensis* inhibits SARS-CoV-2 3CLpro in vitro with an IC_{50} of 8.52 μg/mL and inhibits the replication of SARS-CoV-2 virus in Vero cells with an EC_{50} of 0.74 μg/mL [27]. Among the major components of *S. baicalensis*, baicalein strongly inhibits SARS-CoV-2 3CLpro activity in vitro with an IC_{50} of 0.39 μM and inhibits replication of SARS-CoV-2 in Vero cells with an EC_{50} of 2.9 μM. The study demonstrates that the ethanol extract of *S. baicalensis* limited coronavirus replication in live cells at a concentration that could be achieved following oral administration for a clinically meaningful outcome [27].

Similarly, *Acacia catechu* has been used in ayurvedic medicine for centuries, with significant science-based reports suggesting its application in respiratory system support. *Acacia catechu* has been found to increase the number of antibody-producing cells, increase macrophage phagocytic activity, and inhibit the release of pro-inflammatory cytokines [28]; it has been found to be beneficial in ameliorating chemically induced oxidative stress, inflammation, and apoptosis in the lungs of mice [29], and it has been shown to increase immune modulation effects on both cell-mediated and humoral immunity in vivo [30].

Recently, Feng et al. have shown that a standardized blend of *Scutellaria baicalensis* and *Acacia catechu* may have a therapeutic advantage for COVID-19 through a "multi-compound and multi-target" approach to directly inhibit the virus, improve immune function, and reduce the inflammatory response associated with COVID-19. The author's suggestion was based on a systems pharmacology methodology integrated by ADME screening, target prediction, network analysis, gene ontology (GO), Kyoto Encyclopedia of Genes and Genomes (KEGG) enrichment analysis, molecular docking, and molecular dynamic simulations [31].

Given these research-based findings for *S. baicalensis* and *A. catechu*, in the present study, we evaluated the effect of an antioxidant botanical composition (UP446) on the level of HMGB1 release from hyperoxia-stressed macrophages in vitro and assessed its impact on survival in a lipopolysaccharide (LPS)-induced sepsis and acute inflammatory lung injury model. We further expanded our evaluations and tested the composition of hyperoxia-exposed and bacteria (*Pseudomonas aeruginosa*)-challenged acute lung injury in vivo for its application in respiratory system support.

2. Results

2.1. Effect of UP446 on LPS-Induced Mortality Rate

Three hours following intraperitoneal injection of LPS, mice started to show early signs of endotoxemia. Exploratory behavior of mice was progressively reduced and was accompanied by ruffled fur (piloerection), decreased mobility, lethargy, and diarrhea. While these signs and symptoms seemed to be present in all the treatment groups, the magnitude of severity was more pronounced in the vehicle-treated group.

Two mice from the vehicle-treated group were found deceased 24 h after LPS injection. The survival rate was determined for this group and was found as 62.5% at this time point (Figure 1). Mice treated with UP446 had a 100% survival rate 24 h after LPS injection. Survival rates of 87.5% and 50% were observed for mice treated with UP446, and vehicle, respectively, 34 h after LPS injection. Perhaps the most significant observation for UP446-treated mice was observed 48 h after LPS injection. At this time point, there was only a 12.5% survival rate for the vehicle-treated mice, while UP446-treated mice showed a 75% survival rate. On the third day (72 h after LPS injection), the survival rates for the groups were 62.5% and 12.5% for UP446 and vehicle, respectively. All mice in the vehicle control group were deceased after 82 h of LPS injection, leaving a 0% survival rate for this group.

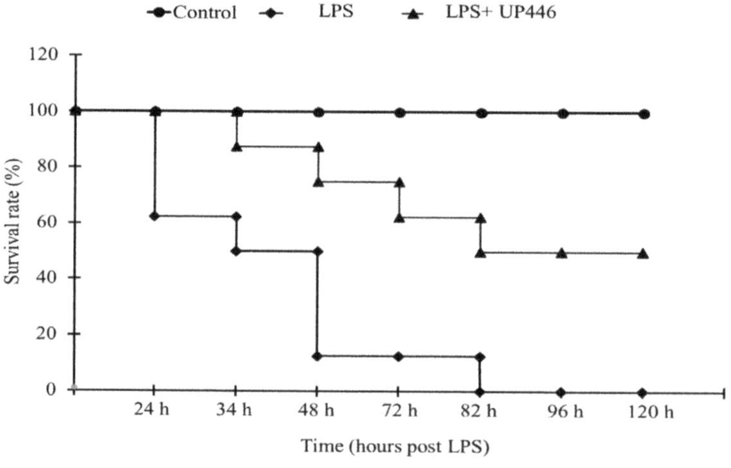

Figure 1. Effect of UP446 on the survival rate of mice with LPS-induced endotoxemia. Eight weeks old, male CD-1 mice (n = 8) were used in this study. While the control mice were injected with only PBS, the model mice (LPS) were pretreated with composition—UP446 for 7 days before lethal dose intraperitoneal injection of LPS at 25 mg/kg with a 10 mL/kg PBS volume an hour after the last dose. Animals were observed hourly and monitored for 5 days after LPS injection. The survival rate was calculated as: 100 − [(deceased mice/total number of mice) × 100]%.

On the other hand, mice treated with UP446 showed a 50% survival rate and remained the same 96 h and 120 h after LPS injection. This survival rate was statistically significant ($p = 0.001$) when compared to the vehicle-treated animals (Figure 1). Surviving animals showed progressive improvements in their well-being. Mice appeared physically better and gradually resumed normal behaviors.

2.2. *Effect of UP446 on LPS-Induced Acute Inflammatory Lung Injury*

The severity of lung damage as a result of intratracheal LPS was assessed using H&E-stained lung tissue (Figure 2A). Rats in the vehicle-treated group showed statistically significant increases in the severity of pulmonary edema (2.5-fold increase) (Figure 2B) and lung damage (3.5-fold increase) (Figure 2C). Daily oral treatment of rats for a week with the high dose of UP446 at 250 mg/kg resulted in a statistically significant 20.8% reduction in overall lung damage severity when compared to vehicle-treated, LPS-induced, acute lung injury rats (Figure 2C). Similarly, a strong trend in the reduction in pulmonary edema (23.3% reduction, $p = 0.08$) was observed for the high dose of UP446 when compared to the vehicle-treated rats (Figure 2B). The low dose UP446 group saw minimal changes in the histopathology evaluation relative to the vehicle-treated diseased rats.

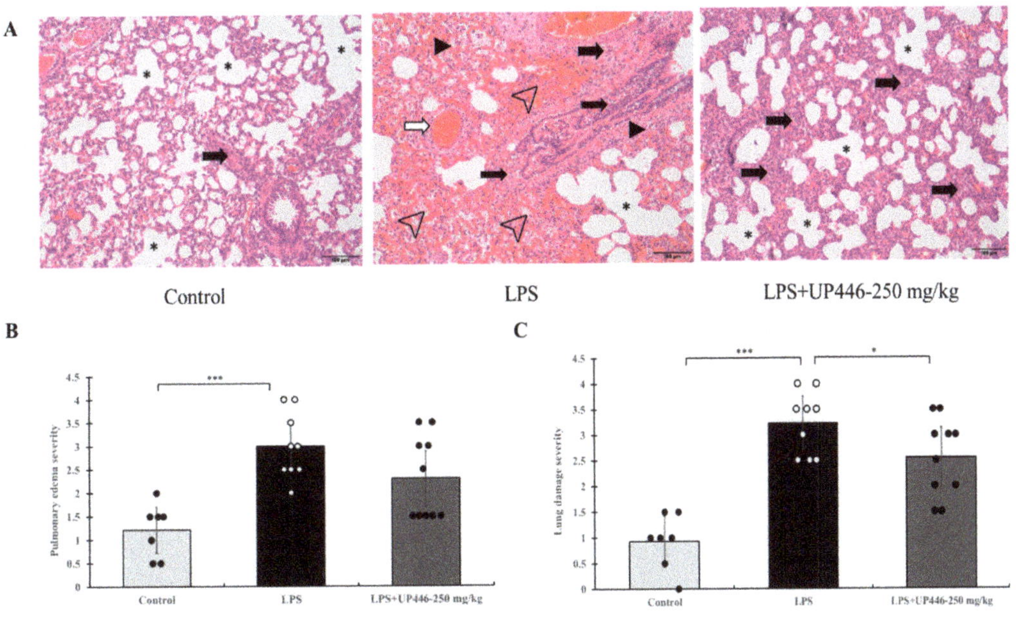

Figure 2. Effects of UP446 on lung histopathology from LPS-induced acute inflammatory lung injury model in rats. Male SD rats (n = 10) at the age of 9 weeks old were treated with the composition—UP446 orally for 7 days before the treatment with LPS. On the 8th day, an hour after oral treatment, LPS was instilled intratracheal (i.t.) at 10 mg/kg dissolved with PBS. The control rats (n = 7) received only PBS. Treatment groups include G1= control, G2 = LPS, G3 = LPS + UP446-125 mg/kg and G4 = LPS + UP446-250 mg/kg. G1 and G2 received the carrier vehicle (i.e., 0.5% CMC) during treatment period. G1, G2 and G4 were used for histopathological analysis. Rats were sacrificed 24 h post intratracheal LPS administration. At necropsy, the left lobe was dissected out, fixed in formalin and submitted to Nationwide Histology for histopathology analysis by a certified pathologist. Tissue and slide preparation were done per company protocol. (**A**) lung histopathology Magnification 100×. dark arrowhead: interstitial edema; open arrowhead: hemorrhagic alveolar sac; open arrow: endothelial damage and hemorrhage; dark arrow: infiltration of inflammatory cells mainly neutrophils and monocytes/macrophages; asterisk: alveolar space. (**B**) Pulmonary edema: alveolar, duct and bronchial, alveolar wall and Interstitial edema, congestion, hemorrhagic perivascular, alveolar sac edema, fibrin exudate, hemorrhagic alveolar sac, alveolar duct thicken dt Hyalin membrane type I loss, apoptotic cells, specific parameter scores 0–4. (**C**) Severity: Normal, minimum-mild, moderate, and extreme. Focal, regional, reginal extensive coalescing, diffuse, score 0–4. * $p \leq 0.05$, *** $p \leq 0.00001$.

Statistically significant elevations in proinflammatory cytokine and inflammatory protein levels were observed for vehicle-treated rats challenged with LPS. These increases were significantly reduced when rats were treated with UP446 (Figure 3A–E). Statistically significant and dose-correlated reductions were observed for rats treated with UP446 at 250 mg/kg and 125 mg/kg orally. These reductions were calculated against the vehicle control and were found to be 90.7% and 69.8% for TNF-α (Figure 3A) and 81.2% and 61.8% for IL-1β (Figure 3B) when UP446 was administered at 250 mg/kg and 125 mg/kg, respectively. While the highest dose (250 mg/kg) resulted in a 74.6% reduction in the level of BAL IL-6, the lower dose showed a 58.3% reduction (Figure 3C). In this study, LPS rats treated with the high dose of UP446 showed statistically significant 42.4% reductions in CRP compared to the vehicle-treated disease model (Figure 3D).

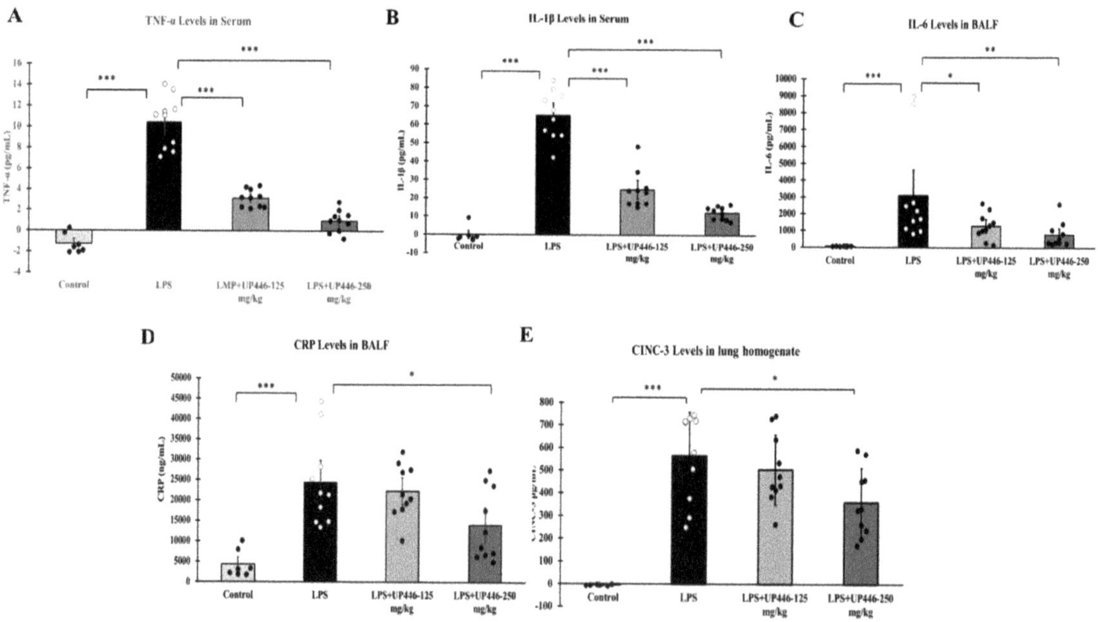

Figure 3. Effect of UP446 on inflammatory cytokines, and chemokines in LPS-induced ALI in rats. Male SD rats (n = 10) at the age of 9 weeks old were treated with the composition—UP446 at 125 mg/kg and 250 mg/kg orally for 7 days before the treatmpleent with LPS. On the 8th day, an hour after oral treatment, LPS was instilled intratracheal (i.t.) at 10 mg/kg dissolved with PBS. The control rats (n = 7) received only PBS. Treatment groups include G1 = control, G2 = LPS, G3 = LPS + UP446-125 mg/kg and G4 = LPS + UP446-250 mg/kg. G1 and G2 received the carrier vehicle (i.e., 0.5% CMC) during treatment period. Rats were sacrificed, serum for TNF-α (**A**) and IL-1β (**B**) and bronchoalveolar lavage (BAL) for IL-6 (**C**) and CRP (**D**) was collected 24 h post intratracheal LPS administration. The right lobe was homogenized for MIP-2/CINC-3 (**E**) activity analysis. * $p \leq 0.05$; ** $p \leq 0.001$; *** $p \leq 0.00001$.

The daily oral treatment of UP446 at 250 mg/kg for a week caused a statistically significant reduction in cytokine-induced neutrophil chemoattractant 3 (CINC-3) in LPS-induced acute lung injury animals (Figure 3E). The level of CINC-3 in the normal control rats receiving only the PBS intratracheally was near zero. In contrast, intratracheal LPS-induced acute lung injury rats treated with the carrier vehicle showed an average lung homogenate level of CINC-3 at 563.7 ± 172.9 pg/mL. This level was reduced to an average value of 360.8 ± 110.7 pg/mL for the 250 mg/kg UP446 treated rats. This 36% reduction in CINC-3 level for the rats treated with 250 mg/kg of UP446 was statistically significant when compared to the vehicle-treated disease model. The lower dose UP446 group only had a marginal 10.5% reduction in lung homogenate CINC-3 level in comparison to the vehicle-treated rats.

2.3. *Effect of UP446 on Hyperoxia-Exposed, Bacteria-Challenged Mice*

Pre-exposure to hyperoxia (O_2) caused significantly more severe acute lung injury, indicated by the amount of protein secretion and lung edema in these mice compared to the mice that remained at room air (RA) (Figure 4A). The composition UP446 significantly reduced this effect by 58.1%. The reduction in the total protein content in lung lavage fluids of mice in the UP446-treated group was statistically significant compared to that of mice treated with hyperoxia and vehicle control (O_2).

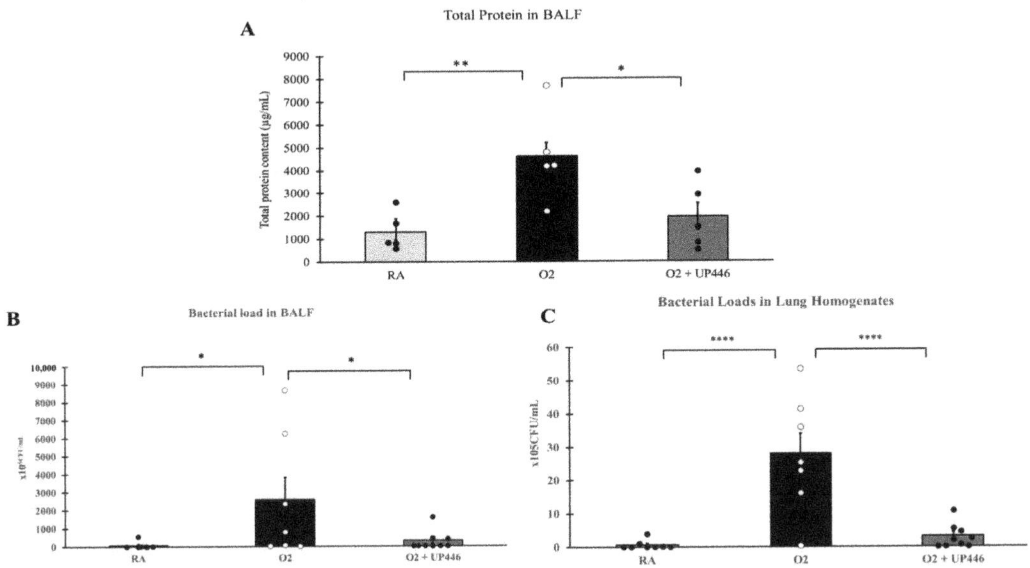

Figure 4. Effect of UP446 on BALF total protein (**A**), BALF bacterial load (**B**) and lung homogenate bacterial load (**C**) in oxidative stress/bacterial challenged mice. Following treatment with the composition UP446 (250 mg/kg) orally for seven days, mice were exposed to >90% oxygen for 48 h and continued oral treatment for 2 more days before being inoculated with *Pseudomonas aeruginosa* (PA). Mice were euthanized 24 h after bacteria inoculation; lungs were lavaged, and the lavage fluid was used to determine the total protein content (**A**) and bacterial load (**B**). Bacterial load was also determined from the lung homogenate (**C**). * $p \leq 0.05$, ** $p \leq 0.001$; **** $p \leq 0.0001$.

Bacterial loads in the airways (Figure 4B) and lung homogenates (Figure 4C) were elevated 36- and 44-fold, respectively, by preexposure of the mice to hyperoxia (O_2), compared to those of mice that remained at room air (RA). Mice treated with UP446 had significantly lower (88.3% and 88.8% reductions) bacterial loads in their airways and lung homogenates, respectively, compared to mice exposed to hyperoxia and treated with vehicle alone. These differences in the bacterial loads in airways and lung homogenates were statistically significant compared to that of mice treated with hyperoxia and vehicle control (O_2).

2.4. *Effect of the Composition on HMGB1 Release*

Mice with hyperoxia-induced lung injury that were challenged with *Pseudomonas aeruginosa* (PA) bacteria showed a 5-fold increase in extracellular HMGB1 in the lung lavage compared to the room air (RA) mice challenged with PA. Pretreating animals with UP446 showed a 71.6% reduction in the level of extracellular HMGB1 in the lung lavage compared to vehicle-treated mice exposed to hyperoxia and PA infection (Figure 5A). These reductions were statistically significant for both the positive control and UP446 groups.

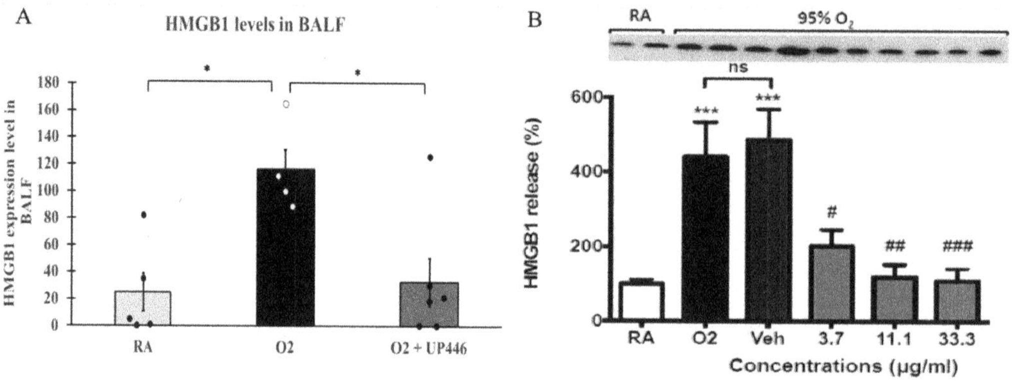

Figure 5. (**A**) Effect of UP446 on HMGB1 in oxidative stress/bacterial challenged mice. Following treatment with the composition UP446 (250 mg/kg) orally for seven days, mice were exposed to >90% oxygen for 48 h and continued oral treatment for 2 more days before being inoculated with *Pseudomonas aeruginosa* (PA). Mice were euthanized 24 h after bacteria inoculation; lungs were lavaged, and the lavage fluid was used to determine HMGB1 expression levels. (**B**) Hyperoxia-induced HMGB1 release in RAW 264.7 cells: RAW 264.7 cells either remained at room air (21% O_2) or were exposed to 95% O_2 for 24 h in the presence of a composition at indicated concentrations. HMGB1 levels in the media were determined by Western blot analysis. Each value represents the mean ± SEM of 2 independent experiments, in duplicates. "ns": no significance. * $p \leq 0.05$, *** $p < 0.001$ compared to room air control (AR). # $p < 0.05$, ## $p < 0.01$, ### $p < 0.001$ compared to vehicle control.

Compared to the room air control group (21% O_2) (RA), HMGB1 release in the hyperoxia control group (95% O_2) (O_2) was significantly increased. The vehicle, DMSO, did not significantly alter HMGB1 release compared to the hyperoxia control group. In contrast, treatment with the composition UP446 resulted in dose-correlated, statistically significant reductions (75.9–89.7%) in the level of HMGB1 when tested at 3.7 μg/mL, 11.1 μg/mL and 33.3 μg/mL (Figure 5B).

2.5. *Effect of the Composition UP446 on Macrophage Phagocytic Activity*

Cultured macrophages (RAW264.7) were subjected to hyperoxia for 24 h in the presence of either different concentrations of the composition UP446 or the vehicle alone. As indicated in the images, hyperoxia exposure significantly compromised macrophage phagocytic activity. The composition UP446 at doses as low as 3.7 μg/mL significantly enhanced macrophage function. Increasing the concentration further to 11.1 and 33.3 μg/mL did not seem to enhance the phagocytic activity of macrophages from what was already observed for the lowest concentration, though a slight increase was seen for the higher concentration tested. Optimum activation of macrophages for their phagocytic activity was already observed at the minimum concentration tested. The phagocytosis activity from each UP446 concentration was statistically significant (Figure 6). These results suggest that the composition UP446 enhances alveolar macrophage phagocytic activity, protecting lung functions under oxidative stress.

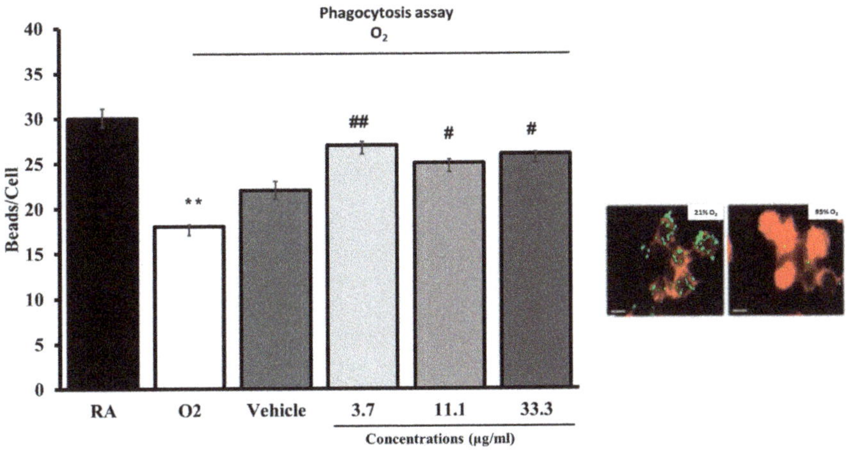

Figure 6. Hyperoxia-compromised macrophage phagocytic function. RAW 264.7 cells either remained at room air (21% O_2) or were exposed to 95% O_2 for 24 h in the presence of a standardized composition (specifics and conc.). Cells were then incubated with FITC-labeled latex mini-beads for 1 h and stained with phalloidin and DAPI to visualize the actin cytoskeleton and nuclei, respectively. For quantification of phagocytic activity, at least 200 cells per group were counted and the number of beads per cell was represented as a percentage of the 21% O_2 (0 μg/mL) control group. Each value represents the mean ± SEM of 2 independent experiments for each group, in duplicates. Significance is compared to the 95% O_2 (0 μg/mL) control group. ** $p < 0.001$ compared to room air control (AR). # $p < 0.05$, ## $p < 0.01$compared to vehicle control.

3. Materials and Methods

3.1. The Botanical Composition

A proprietary botanical composition UP446 is a mixture of standardized extracts from roots of *S. baicalensis* and heartwoods of *A. catechu* with a baicalin content not less than 60% and aa catechin content not less than 10% in the composition (Figure 7). Other minor flavonoids, such as wogonin 7-glucuronide and baicalein, etc., account for about 15% of the total weight. Moisture, ash, fat, and fiber constitute the remaining weight. A detailed method for the preparation of the two major flavonoids, baicalin and catechin, from the roots of *S. baicalensis* and the heartwoods of *A. catechu*, respectively, was disclosed in a US patent [32].

3.2. Lipopolysaccharide (LPS)-Induced Sepsis in Mice

Purpose-bred CD-1 mice purchased from Charles River Laboratories at the age of 8 weeks were used for this study. Mice were acclimated for a week before being assigned to study groups. Groups included: G1 = Normal control, G2 = vehicle control (0.5% carboxymethyl cellulose (CMC)), and G3 = UP446 (250 mg/kg). Eight mice were allocated to each group. Mice were pretreated with the composition (UP446) for 7 days before receiving a lethal dose intraperitoneal injection of LPS (*E. coli*, 055:B5; Sigma, St. Louis, MO, USA; Lot# 081275) at 25 mg/kg with a 10 mL/kg PBS volume. LPS was dissolved in phosphate-buffered saline (PBS; Lifeline, Lot # 07641). Animals were observed hourly and monitored for 5 days after LPS injection. The survival rate compared LPS + vehicle (0.5% CMC; Spectrum, New Brunswick, NJ, USA; Lot # 1IJ0127) and LPS + UP446. Mice were kept in a temperature-controlled room and were provided with food and water ad libitum.

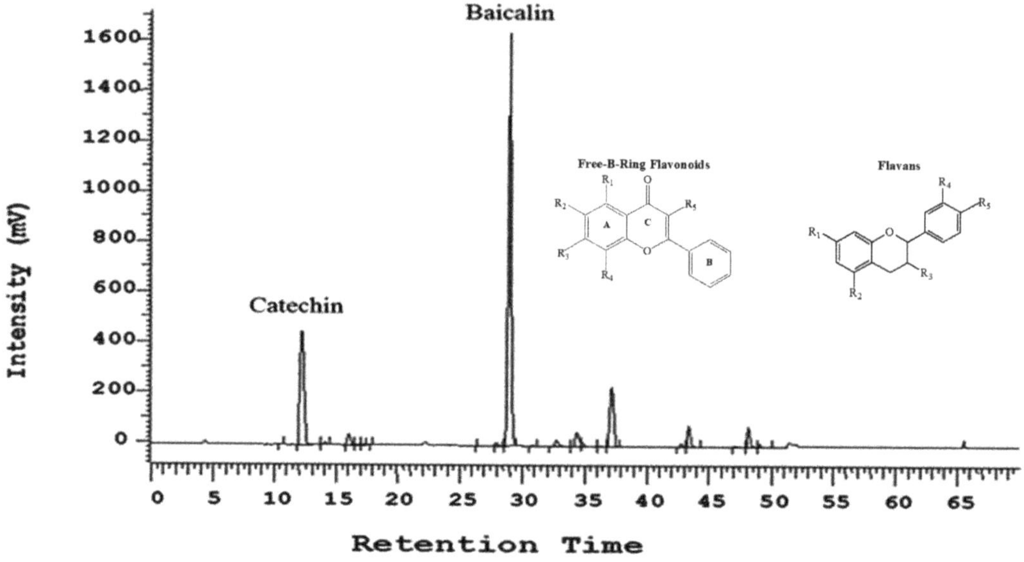

Figure 7. HPLC chromatogram and chemical structure of free B-ring flavonoid (baicalin) and flavan (catechin). The flavonoids were detected using a UV detector at 275 nm and identified based on retention time by comparison with known flavonoid standards. Free B-Ring Flavonoids: Baicalin: R1 = R2 = OH, R3 = Glucuronide, R4 = R5 = H; Baicalein: R1 = R2 = R3 = OH, R4 = R5 = H; Oroxylin A: R1 = R3 = OH, R2 = OMe, R4 = R5 = H; Chrysin: R1 = R3 = OH, R2 = R4 = R5 = H; Wogonin: R1 = R3 = OH, R4 = OMe, R3 = R5 = H; Wogonin 7-Glucuronide: R1 = OH, R3 = Glucuronide, R4 = OMe, R2 = R5 = H. Flavans: Catechin (+): R1 = R2 = R3 = R4 = R5 = OH; Epicatechin (−): R1 = R2 = R3 = R4 = R5 = OH.

3.3. Lipopolysaccharide (LPS)-Induced Acute Inflammatory Lung Injury in Rats

The study was designed to evaluate the direct impact of the composition UP446 in alleviating LPS-induced acute lung injury, with daily oral administration at 250 mg/kg (high dose) and 125 mg/kg (low dose). Animals were pretreated with the test materials for 7 days before model induction with LPS. On the 8th day, an hour after oral treatment, LPS was delivered intratracheally (i.t.) at 10 mg/kg, dissolved in 0.1 mL/100 g PBS, to each rat. The normal control rats received the same volume i.t. of PBS only. Groups included G1 = Normal control without LPS, G2 = vehicle control (0.5% CMC) with LPS, G3 = UP446 high dose (250 mg/kg) with LPS and G4 = UP446 low dose (125 mg/kg) with LPS. Ten rats were allocated to each group.

Surviving animals were sacrificed 24 h after intratracheal LPS administration. At necropsy, bronchoalveolar lavage (BAL) was collected by intratracheal injection of 1.5 mL PBS into the right lobe of the lung, followed by gentle aspiration at least 3 times. Recovered fluid was pooled, centrifuged at 1500 rpm for 10 min at 4 °C, and was used to measure cytokines (e.g., IL-6) and pulmonary protein level. This same right lobe was collected for tissue homogenization from each rat for MIP-2/CINC-3 protein quantification. The left lobe was fixed with formalin and submitted for histopathology evaluation to Nationwide Histology for analysis by a certified pathologist. Serum collected at necropsy was used to measure cytokines, such as TNF-α and IL-1β.

3.4. The Effects of the Botanical Composition on Hyperoxia-Exposed-, Bacterial-Challenged, Acute Inflammatory Lung Injury In Vivo

The effect of the botanical composition UP446 on bacterial clearance and HMGB1 expression level was evaluated in vivo. Male C57BL/6 mice (6 to 10 weeks old; The Jackson

Laboratory, Bar Harbor, ME, USA) were used in this study, which was approved by the Institutional Animal Care and Use Committee at St. John's University. The mice were housed in a pathogen-free environment, maintained at 22 °C and 50% relative humidity with a 12 h light/dark cycle. All mice had ad libitum access to standard rodent food and water. Mice were randomized to G1 = Normal control without bacterial inoculation, G2 = vehicle-treated disease model, and G3 = Disease model + UP446 (250 mg/kg). Following treatments with the vehicle, or the composition UP446 at 250 mg/kg orally for seven days, mice were exposed to >90% oxygen for 48 h and continued oral treatment for 2 more days before being inoculated with 0.1×10^8 colony-forming units (CFUs) of *Pseudomonas aeruginosa* (PA). Mice were placed in microisolator cages (Allentown Caging Equipment, Allentown, NJ, USA) that were kept in a plexiglass chamber (Bio-Spherix, Lacona, NY, USA) for hyperoxia exposure. An oxygen analyzer (MSA; Ohio Medical Corporation, Gurnee, IL, USA) was used to monitor the O_2 concentration in the chamber. Mice were euthanized 24 h after bacteria inoculation and bronchoalveolar lavage (BAL) fluid was collected. The lungs were gently lavaged twice with 1 mL of a sterile, nonpyrogenic phosphate-buffered saline (PBS) solution (Mediatech, Herndon, VA, USA), containing a cocktail of protease and phosphatase inhibitors (Pierce, ThermoFisher Scientific, Waltham, MA, USA). BAL samples were centrifuged at 200× *g* at 4 °C for 5 min, and the resultant supernatants were immediately used for quantitative bacteriology. Lung tissues were immediately collected into 1 mL cold PBS containing a protease and phosphatase inhibitor cocktail (Pierce, ThermoFisher Scientific) followed by homogenization by a dounce tissue homogenizer [6].

3.5. Hyperoxia-Induced HMGB1 Release in Macrophages

The effect of the composition in reducing the level of HMGB1 release was tested in murine immune cells in vitro. RAW264.7 cells either remained in room air (21% O_2) or were exposed to 95% O_2 for 24 h in the presence of the composition UP446 at concentrations of 0, 3.7, 11.1 and 33.3 µg/mL. HMGB1 levels in the cell culture media were determined by Western blot analysis described below in the assay section.

3.6. Phagocytosis Activity of Macrophages

The effect of the composition on the phagocytic activity of macrophages was tested in vitro. RAW264.7 cells either remained in room air (21% O_2) or were exposed to 95% O_2 for 24 h in the presence of a standardized composition (UP446) at concentrations of 3.7, 11.1, 33.3 and 100 µg/mL. Cells were then incubated with FITC-labeled latex mini-beads for one hour and stained with phalloidin and DAPI to visualize the actin cytoskeleton and nuclei, respectively. For quantification of phagocytic activity, at least 200 cells per group were counted, and the number of beads per cell was represented as a percentage of the 21% O_2 (0 µg/mL) control group.

3.7. Assays

3.7.1. Cytokines and Chemokines

The presence of TNF-α/IL-1β/IL-6 in undiluted rat serum was measured using the rat TNF-α/IL-1β/IL-6 Quantikine ELISA kit from R&D Systems (product#: RTA00, RLB00 and R6000B, respectively) as follows: undiluted serum was added to a microplate coated with TNF-α/IL-1β/IL-6 antibody. After 2 h at room temperature, TNF-α/IL-1β/IL-6 in serum was bound to the plate and the plate was thoroughly washed. Enzyme-conjugated TNF-α/IL-1β/IL-6 antibody was added to the plate and allowed to bind for 2 h at room temperature. The washing was repeated, and enzyme substrate was added to the plate. After developing for 30 min at room temperature, a stop solution was added, and the absorbance was read at 450 nm. The concentration of TNF-α/IL-1β/IL-6 was calculated based on the absorbance readings of a TNF-α/IL-1β/IL-6 standard curve.

The presence of CRP in rat bronchoalveolar lavage (BAL) diluted 1:1000 was measured using the C-Reactive Protein (PTX1) Rat ELISA kit from Abcam (product#: ab108827) as follows: 1:1000 diluted BAL was added to a microplate coated with CRP antibody. After

2 h on a plate shaker at room temperature, CRP in BAL was bound to the plate and the plate was thoroughly washed. Biotinylated C-Reactive Protein Antibody was added to the plate and allowed to bind for 1 h on a plate shaker at room temperature. The washing was repeated, and Streptavidin-Peroxidase Conjugate was added to the plate. After incubating for 30 min at room temperature, washing was repeated, and chromogen substrate was added. After developing for 10 min at room temperature, a stop solution was added, and the absorbance was read at 450 nm. The concentration of CRP was calculated based on the absorbance readings of a CRP standard curve.

Cytokine-induced neutrophil chemoattactant 3 (CINC-3)/macrophage inflammatory protein 2 (MIP-2) ELISA was carried out as follows: 50 μL of each rat lung homogenate sample (10 per group for vehicle, UP446 low dose, UP446 high dose, 7 per group for control) and 50 μL of assay diluent buffer was added to the wells of a 96-well microplate coated with monoclonal CINC-3 antibody and allowed to bind for 2 h. The plate was subjected to 5 washes before an enzyme-linked polyclonal CINC-3 antibody was added and allowed to bind for 2 h. The wells were washed another 5 times before a substrate solution was added to the wells and the enzymatic reaction was allowed to commence for 30 min at room temperature protected from light. The enzymatic reaction produced a blue dye that changed to yellow with the addition of the stop solution. The absorbance of each well was read at 450 nm (with a 580 nm correction) and compared to a standard curve of CINC-3 in order to approximate the amount of CINC-3 in each rat lung homogenate sample.

3.7.2. Bacterial Counts

Bacterial counts in the lung homogenate and bronchoalveolar lavage were quantitated in serially diluted Luria–Bertani broth using a colony formation unit assay plated onto Pseudomonas Isolation Agar (Difco, Sparks, MD, USA) at 37 °C for 18 h.

3.7.3. Western Blot Analysis

Cells were washed three times with PBS and lysed using a cell lysis buffer (Cell Signaling Technology, Danvers, MA, USA) supplemented with Halt Protease and Phosphatase Inhibitor Cocktail (78440, ThermoFischer, Waltham, MA, USA) for intracellular protein analysis. The total protein content of cell lysate was determined by using the Bicinchoninic Acid (BCA) assay kit (23225, ThermoFisher, Waltham, MA, USA), as per the manufacturer's instructions. Samples were loaded onto 12% or 15% SDS-polyacrylamide gels (Bio-Rad, Hercules, CA, USA) and transferred to Immobilon-P membranes (Millipore, Bedford, MA, USA). Nonspecific binding sites on the membrane were blocked by incubating the membrane with 5% nonfat dry milk (Bio-Rad, Hercules, CA, USA) in Tris-buffered saline, containing 0.1% Tween 20 (TBST), for 1 h at room temperature. Next, the membranes were washed three times with TBST, and incubated overnight at 4 °C with anti-HO-1 (1:1000, #ab13248, Abcam, Cambridge, UK) and anti-pan-actin (1:1000, #8456, Cell Signaling) antibodies, diluted in 5% nonfat dry milk in TBST. After three washes with TBST, the membranes were incubated with goat anti-rabbit horseradish peroxidase-coupled secondary antibody (1:5000; GE Healthcare, Chicago, IL, USA) for 1 h at room temperature. Subsequently, membranes were again washed three times with TBST, and the immunoreactive proteins were visualized using the SuperSignal West Pico Plus Chemiluminescent Substrate (ThermoFisher, Waltham, MA, USA), as per the manufacturer's instructions. Images were obtained using the Bio-Rad ChemiDoc XRS imaging system (Bio-Rad, Hercules, CA, USA). The immunoreactive bands were quantified using ImageJ software (version 2.0.0).

3.7.4. Statistical Analysis

Data were analyzed using Sigmaplot (Version 11.0, San Jose, CA, USA). The results are represented as mean ± one SD. Statistical significance between groups was calculated by means of single factor analysis of variance followed by a paired *t*-test. *p*-values less than or equal to 0.05 ($p \leq 0.05$) were considered statistically significant. When the normality test failed, for nonparametric analysis, data were subjected to Mann–Whitney sum ranks for *t*-test and Kruskal–Wallis one-way analysis of variance on ranks for ANOVA. When the treatment group sizes were unequal, Dunn's test was used for pairwise comparisons and comparisons against the placebo group following rank-based ANOVA.

4. Discussion

Poor air quality from recent wildfires and other sources of air pollution poses a major public health threat worldwide. Antioxidants could potentially be considered as frontline defense and/or adjunct for air-pollution-induced oxidative stress damage of the lung. Historically, herbal medicines have been considered as an alternative to pharmaceuticals because of their relative safety; however, scientifically sound, research-backed, safe and efficacious nutritional supplements from natural sources for respiratory system support (i.e., mitigation of pollution-induced oxidative stress damage) are very limited. *Scutellaria baicalensis* and *Acacia catechu*, known antioxidants, have long been used in Traditional Chinese Medicine (TCM) and ayurvedic preparations, respectively, for centuries for varieties of human ailments. For instance, Radix Scutellaria has been reported as the second most utilized herb, with a 38% frequency in all TCM compositions for the treatment of respiratory tract infections [20]. Findings depicted in this report may support the frequent historical usage of these botanicals for respiratory system support.

Recent discoveries in pulmonary pathophysiology have shown extracellular HMGB1 as an alarmin to trigger profound inflammatory and immune responses following oxidative-stress-induced lung injury [11]. In the current report, we evaluated UP446 (an antioxidant composition) in multiple pre-clinical in vivo and in vitro models suggestive of its respiratory-system-support-related mode of action with the direct or indirect involvement of extracellular HMGB1.

The hypothesis that the composition could be involved in reducing the level of extracellular HMGB1 was first tested in a sepsis model indirectly. The composition UP446 effectively prevented the development of sepsis in a Lipopolysaccharide (LPS)-induced sepsis model. Mortality was significantly reduced (i.e., 50%) as a result of UP446 supplementation. Since mortality was the end point measurement, we did not measure the level of serum HMGB1 in this study; however, previously, it was reported that there is a correlation between high serum levels of HMGB1 and sepsis, resulting in an unbalanced host immune response, culminating in the death of the mice [33]. As such, the increased survival rate in UP446-treated animals in the current study could possibly be linked to the reduction in extracellular HMGB1 and mitigation of its downstream inflammatory effects.

We proceeded to assess the impact of the composition on acute inflammatory lung injury using an LPS-induced lung injury in rats. Intratracheal instillation of lipopolysaccharide (LPS) into the lungs of animals triggers an intense inflammatory response. It causes acute inflammatory lung injury that mimics the pathology in humans, characterized by pathological changes, such as diffuse alveolar damage, accompanied by profound increases in proinflammatory cytokines, infiltration of polymorphonuclear cells, increased alveolar capillary membrane permeability, and accumulation of protein-rich fluid in the alveolar space, leading to edema [34–36]. These pathological changes may be a direct consequence of increased accumulation of HMGB1 in the airways and the lungs, causing severe inflammatory injury. It has been reported that extracellular HMGB1 triggers the production of potent proinflammatory cytokines and chemokines, such as TNF-α, IL-1β, CINC-3, and IL-6, increased endothelial/epithelial barrier permeability, infiltration of polymorphonuclear cells, and formation of lung edema, causing lung injury and loss of lung function [12,37]. As such, major cytokines and chemotactic factors involved in acute inflammatory response in the lung have significant clinical relevance in cytokine storm intervention and alleviating the severity of acute respiratory distress syndrome (ARDS). In the current study, the composition UP446 mitigated LPS-induced acute inflammatory lung injury, possibly through inhibition of HMGB1 though our assay failed to show differences in the airway HMGB1 level in this model, suggesting the presence of additional pathways for the composition.

Nevertheless, as evidenced by the histopathology data, a statistically significant reduction in cellular and structural damage was observed as a result of oral supplementation of the composition UP446. Marked reductions in pulmonary edema further strengthened the lung protection capacity of the composition, possibly through a reduction in extracellular HMGB1 and hence improving the phagocytic activity of alveolar macrophages. Effective alleviation of acute inflammatory lung injury by the composition UP446 was further demonstrated by a reduction in key proinflammatory cytokines and chemokines (TNF-α, IL-1β and IL-6 and CINC-3) as well as the inflammatory protein CRP. Substantiating our findings, it was reported that the blend of *Scutellaria* and *Acacia* was found to decrease lung wet-to-dry weight ratio, mitigate lung histopathological changes, and reduce the release of inflammatory mediators, such as TNF-α and IL-1β, in BAL in a rat model of acute lung injury (ALI) [38]. Similarly, baicalein, a phenolic flavonoid from the root of *Scutellaria baicalensis* has been shown to significantly mitigate LPS-induced lung edema and attenuate the levels of IL-1β, TNF-α, IL-6, CINC-3 in broncho alveolar lavage fluid, with marked improvement of lung histopathological symptoms, in an LPS-induced rat model of acute inflammatory lung injury [39]. The major constituent of UP446, baicalin, has also been found to inhibit the cytoplasmic translocation of HMGB1 induced by lipopolysaccharide in vitro. In vivo, baicalin decreased serum HMGB1, TNF-α, IL-1β and IL-6 while improving survival and tissue injury of septic mice [40].

We further carried out an additional preclinical in vivo study to better understand the mode of action of the composition UP446 in a model that mimics mechanical ventilation-induced oxidative stress and secondary pulmonary bacterial infection. It has been previously shown that exposure to hyperoxia can compromise the host defense against bacterial infections, resulting in higher bacterial loads in lung tissues and airways and increased total protein content in lung lavage fluids upon microbial infection [6]. In the current study, we observed these phenomena in the vehicle-treated disease model with significant increases in bacterial load and protein exudate, whereas supplementation with UP446 showed significant reductions in both protein and bacterial load, suggesting the standardized botanical composition's ability to improve host defense mechanisms, possibly through the reduction in extracellular HMGB1.

Accumulation of extracellular HMGB1 in the airways compromises innate immunity, leading to the impaired ability of alveolar macrophages to clear invading pathogens, exacerbating inflammatory lung injury. HMGB1-triggered cytokine storm can subsequently cause severe acute respiratory tract and lung injury that could ultimately cause death [6,38]. The involvement of HMGB1 in the pathogenesis of acute lung injury has been demonstrated in mouse models whereby the intratracheal administration of anti-HMGB1 antibodies mitigated the development of lung injury and increased survival rates [37]. To determine whether UP446-attenuated acute lung injury in mice exposed to hyperoxia and PA is due to its impact on the accumulation of extracellular HMGB1 in the airways, the levels of HMGB1 were measured in the lung lavage fluids. Prolonged exposure of the mice to hyperoxia and microbial infection increased the accumulation of HMGB1 in the airways by about 5-fold. Pretreating animals with UP446 showed a 71.6% reduction in the level of HMGB1 expression compared to vehicle-treated mice exposed to hyperoxia and PA infection. These data suggest that the composition UP446 can reduce the accumulation of airway HMGB1 in mice exposed to hyperoxia and PA infection. These findings suggest the enhanced ability of UP446 to improve host defense mechanisms against oxidative stress and microbial infection involving the respiratory system. Previously, the antioxidant activity of the composition through increasing endogenous antioxidant enzyme has been demonstrated in a human clinical study. In this study, healthy participants who were supplemented with the composition for 56 days orally showed a statistically significant increase in the serum level of glutathione peroxidase (GSH-Px) both at the interim (day 28) and the end of study (day 56) compared to baseline [20].

5. Conclusions

Reducing the level of extracellular HMGB1 induced by oxidative stress and/or respiratory tract infection could have a significant impact on protecting the lungs from injury and preserving normal physiological respiratory tract functions. The compilations of data in this report strongly suggest that the antioxidant composition UP446 could be indicated for lung protection through mitigation of HMGB1. Further confirmatory studies focused on the impact of the composition on this molecular target in humans is suggested.

Author Contributions: Conceptualization, M.Y., L.B., Q.J. and L.L.M.; Data curation, M.Y., T.H., L.L.M., M.L., A.G. and Q.J.; Formal analysis, M.Y., L.L.M., M.L., A.G., T.H. and Q.J.; Funding acquisition, Q.J. and L.L.M.; Investigation, M.Y., T.H., A.O., P.J., M.H., M.L., A.G., J.W., K.V.M., X.Y., K.D. and S.Z.; Methodology, M.Y., T.H., A.O., P.J., M.H., M.L., A.G., J.W., K.V.M., X.Y., K.D. and S.Z.; Project administration, M.Y., L.L.M. and Q.J.; Resources, L.B., Q.J. and L.L.M.; Supervision, M.Y., Q.J. and L.L.M.; Validation, T.H., P.J., M.H., L.L.M., M.L., A.G. and Q.J.; Visualization, L.B., L.L.M., M.L., A.G. and Q.J.; Writing—original draft, M.Y., L.L.M., M.L., A.G. and T.H.; Writing—review and editing, M.Y., L.B., T.H., L.L.M., M.L., A.G. and Q.J. All authors have read and agreed to the published version of the manuscript.

Funding: The authors would like to extend their utmost gratitude to Bill Lee (owner of Econet/Unigen) and St. Johns Univestsity for their fincnaial contribution.

Institutional Review Board Statement: All animal experiments were conducted according to the institutional guidelines congruent with guide for the care and use of laboratory animals with IACUC#s IA-P02-090120 and IA-P03-090120.

Informed Consent Statement: Not applicable.

Data Availability Statement: Conclusions were made based on data depicted in this report. Raw data could be requested through corresponding authors.

Acknowledgments: The authors would like to express their gratitude to Jessica Cross for her support in animal husbandry and assisting the in vivo study. We would like to thank Maleka Stewart, Mathieu David and Miranda Bogdanowicz, and Joanna Woo for the technical assistance and insightful discussion.

Conflicts of Interest: M.Y., P.J., M.H., L.B. and Q.J. are current Unigen employees, therefore they have competing financial interests. Studies carried out at St. John's University were sponsored by Unigen, though Unigen had no role in the design of the study, in the collection, analyses, or interpretation of data.

Sample Availability: Not applicable.

References

1. Rahman, M.; McConnell, R.; Schlaerth, H.; Ko, J.; Silva, S.; Lurmann, F.W.; Palinkas, L.; Johnston, J.; Hurlburt, M.; Yin, H.; et al. The Effects of Coexposure to Extremes of Heat and Particulate Air Pollution on Mortality in California: Implications for Climate Change. *Am. J. Respir. Crit. Care Med.* **2022**, *206*, 1117–1127. [CrossRef]
2. Ma, J.H.; Song, S.H.; Guo, M.; Zhou, J.; Liu, F.; Peng, L.; Fu, Z.R. Long-term exposure to PM2.5 lowers influenza virus resistance via down-regulating pulmonary macrophage Kdm6a and mediates histones modification in IL-6 and IFN-β promoter regions. *Biochem. Biophys. Res. Commun.* **2017**, *493*, 1122–1128. [CrossRef]
3. Sigaud, S.; Goldsmith, C.A.; Zhou, H.; Yang, Z.; Fedulov, A.; Imrich, A.; Kobzik, L. Air pollution particles diminish bacterial clearance in the primed lungs of mice. *Toxicol. Appl. Pharmacol.* **2007**, *223*, 1–9. [CrossRef]
4. Hou, W.; Zhang, H.; Jiang, M.; Wu, Y.; Li, T.; Cong, L.; Duan, J. Gu-Ben-Zhi-Ke-Zhong-Yao Alleviated PM2.5-Induced Lung Injury via HMGB1/NF-κB Axis. *J. Healthc. Eng.* **2022**, *2022*, 8450673. [CrossRef]
5. Tang, D.; Kang, R.; Zeh, H.J., 3rd; Lotze, M.T. High-mobility group box 1, oxidative stress, and disease. *Antioxid. Redox Signal.* **2011**, *14*, 1315–1335. [CrossRef]
6. Patel, V.S.; Sitapara, R.A.; Gore, A.; Phan, B.; Sharma, L.; Sampat, V.; Li, J.H.; Yang, H.; Chavan, S.S.; Wang, H.; et al. High Mobility Group Box–1 Mediates Hyperoxia-Induced Impairment of *Pseudomonas aeruginosa* Clearance and Inflammatory Lung Injury in Mice. *Am. J. Respir. Cell Mol. Biol.* **2013**, *48*, 280–287. [CrossRef]
7. Morrow, D.M.; Entezari-Zaher, T.; Romashko, J., 3rd; Azghani, A.O.; Javdan, M.; Ulloa, L.; Miller, E.J.; Mantell, L.L. Antioxidants preserve macrophage phagocytosis of Pseudomonas aeruginosa during hyperoxia. *Free Radic. Biol. Med.* **2007**, *42*, 1338–1349. [CrossRef]

8. Baleeiro, C.E.; Wilcoxen, S.E.; Morris, S.B.; Standiford, T.J.; Paine, R., 3rd. Sublethal hyperoxia impairs pulmonary innate immunity. *J. Immunol.* **2003**, *171*, 955–963. [CrossRef]
9. Meng, L.; Li, L.; Lu, S.; Li, K.; Su, Z.; Wang, Y.; Fan, X.; Li, X.; Zhao, G. The protective effect of dexmedetomidine on LPS-induced acute lung injury through the HMGB1-mediated TLR4/NF-κB and PI3K/Akt/mTOR pathways. *Mol. Immunol.* **2018**, *94*, 7–17. [CrossRef]
10. Bae, J.-S. Role of high mobility group box 1 in inflammatory disease: Focus on sepsis. *Arch. Pharmacal Res.* **2012**, *35*, 1511–1523. [CrossRef]
11. Han, X.; Shen, T.; Lou, H. Dietary Polyphenols and Their Biological Significance. *Int. J. Mol. Sci.* **2007**, *8*, 950–988. [CrossRef]
12. Peng-Fei, L.; Fu-Gen, H.; Bin-Bin, D.; Tian-Sheng, D.; Xiang-Lin, H.; Ming-Qin, Z. Purification and antioxidant activities of baicalin isolated from the root of huangqin (*Scutellaria baicalensis* gcorsi). *J. Food Sci. Technol.* **2012**, *50*, 615–619. [CrossRef]
13. Gao, Z.; Huang, K.; Yang, X.; Xu, H. Free radical scavenging and antioxidant activities of flavonoids extracted from the radix of *Scutellaria baicalensis* Georgi. *Biochim. Biophys. Acta (BBA) Gen. Subj.* **1999**, *1472*, 643–650. [CrossRef]
14. Liau, P.R.; Wu, M.S.; Lee, C.K. Inhibitory Effects of *Scutellaria baicalensis* Root Extract on Linoleic Acid Hydroperoxide-induced Lung Mitochondrial Lipid Peroxidation and Antioxidant Activities. *Molecules* **2019**, *24*, 2143. [CrossRef] [PubMed]
15. Chiaino, E.; Stella, R.; Peggion, C.; Micucci, M.; Budriesi, R.; Mattioli, L.B.; Marzetti, C.; Pessina, F.; Valoti, M.; Frosini, M. *Acacia catechu* Willd. Extract Protects Neuronal Cells from Oxidative Stress-Induced Damage. *Antioxidants* **2021**, *11*, 81. [CrossRef]
16. Bernatoniene, J.; Kopustinskiene, D.M. The Role of Catechins in Cellular Responses to Oxidative Stress. *Molecules* **2018**, *23*, 965. [CrossRef]
17. Musial, C.; Kuban-Jankowska, A.; Gorska-Ponikowska, M. Beneficial Properties of Green Tea Catechins. *Int. J. Mol. Sci.* **2020**, *21*, 1744. [CrossRef]
18. Burnett, B.P.; Jia, Q.; Zhao, Y.; Levy, R.M. A medicinal extract of Scutellariabaicalensis and Acacia catechu acts as a dual inhibitor of cyclooxygenase and 5-lipoxygenase to reduce inflammation. *J. Med. Food* **2007**, *10*, 442–451. [CrossRef]
19. Tseng-Crank, J.; Sung, S.; Jia, Q.; Zhao, Y.; Burnett, B.; Park, D.R.; Woo, S.S. A medicinal plant extract of Scutellaria Baicalensis and Acacia catechu reduced LPS-stimulated gene expression in immune cells: A comprehensive genomic study using QPCR, ELISA, and microarray. *J. Diet. Suppl.* **2010**, *7*, 253–272. [CrossRef] [PubMed]
20. Lewis, E.D.; Crowley, D.C.; Guthrie, N.; Evans, M. Role of Acacia catechu and Scutellaria baicalensis in Enhancing Immune Function Following Influenza Vaccination of Healthy Adults: A Randomized, Triple-Blind, Placebo-Controlled Clinical Trial. *J. Am. Nutr. Assoc.* **2023**, *42*, 678–690. [CrossRef]
21. Ge, H.; Wang, Y.-F.; Xu, J.; Gu, Q.; Liu, H.-B.; Xiao, P.-G.; Zhou, J.; Liu, Y.; Yang, Z.; Su, H. Anti-influenza agents from Traditional Chinese Medicine. *Nat. Prod. Rep.* **2010**, *27*, 1758–1780. [CrossRef] [PubMed]
22. Li, C.; Lin, G.; Zuo, Z. Pharmacological effects and pharmacokinetics properties of Radix Scutellariae and its bioactive flavones. *Biopharm. Drug Dispos.* **2011**, *32*, 427–445. [CrossRef]
23. Zhang, C.J.; Gu, L.G.; Yu, H.T. Antagonism of baicalin on cell cyclical distribution and cell apoptosis in A549 cells infected with influenza A (H1N1) virus. *Bing Du Xue Bao* **2011**, *27*, 108–116.
24. Zhi, H.-J.; Zhu, H.-Y.; Zhang, Y.-Y.; Lu, Y.; Li, H.; Chen, D.-F. In vivo effect of quantified flavonoids-enriched extract of Scutellaria baicalensis root on acute lung injury induced by influenza A virus. *Phytomedicine* **2018**, *57*, 105–116. [CrossRef] [PubMed]
25. Liu, H.; Ye, F.; Sun, Q.; Liang, H.; Li, C.; Li, S.; Lu, R.; Huang, B.; Tan, W.; Lai, L. Scutellaria baicalensis extract and baicalein inhibit rep-lication of SARS-CoV-2 and its 3C-like protease in vitro. *J. Enzyme Inhib. Med. Chem.* **2021**, *36*, 497–503. [CrossRef]
26. Sunil, M.; Sunitha, V.; Radhakrishnan, E.; Jyothis, M. Immunomodulatory activities of Acacia catechu, a traditional thirst quencher of South India. *J. Ayurveda Integr. Med.* **2019**, *10*, 185–191. [CrossRef] [PubMed]
27. Shahid, A.; Ali, R.; Ali, N.; Kazim Hasan, S.; Barnwal, P.; Mohammad Afzal, S.; Vafa, A.; Sultana, S. Methanolic bark extract of Acacia catechu ameliorates benzo(a)pyrene induced lung toxicity by abrogation of oxidative stress, inflammation, and apoptosis in mice. *Environ. Toxicol.* **2017**, *32*, 1566–1577. [CrossRef]
28. Ismail, S.; Asad, M. Immunomodulatory activity of Acacia catechu. *Indian J. Physiol. Pharmacol.* **2009**, *53*, 25–33.
29. Jia, Q. Formulation of a Mixture of Free-B-Ring Flavonoids and Flavans as a Therapeutic Agent. U.S. Patent No. 7514469, 7 April 2009.
30. Feng, T.; Zhang, M.; Xu, Q.; Song, F.; Wang, L.; Gai, S.; Tang, H.; Wang, S.; Zhou, L.; Li, H. Exploration of molecular targets and mechanisms of Chinese medicinal formula *Acacia Catechu -Scutellariae Radix* in the treatment of COVID-19 by a systems pharmacology strategy. *Phytother. Res.* **2022**, *36*, 4210–4229. [CrossRef]
31. Wang, M.; Gauthier, A.; Daley, L.; Dial, K.; Wu, J.; Woo, J.; Lin, M.; Ashby, C.; Mantell, L.L. The Role of HMGB1, a Nuclear Dam-age-Associated Molecular Pattern Molecule, in the Pathogenesis of Lung Diseases. *Antioxid. Redox Signal.* **2019**, *31*, 954–993. [CrossRef] [PubMed]
32. Wang, H.; Bloom, O.; Zhang, M.; Vishnubhakat, J.M.; Ombrellino, M.; Che, J.; Frazier, A.; Yang, H.; Ivanova, S.; Borovikova, L.; et al. HMG-1 as a Late Mediator of Endotoxin Lethality in Mice. *Science* **1999**, *285*, 248–251. [CrossRef] [PubMed]
33. Goodman, R.B.; Pugin, J.; Lee, J.S.; Matthay, M.A. Cytokine-mediated inflammation in acute lung injury. *Cytokine Growth Factor Rev.* **2003**, *14*, 523–535. [CrossRef]
34. Ware, L.B.; Matthay, M.A. The acute respiratory distress syndrome. *N. Engl. J. Med.* **2000**, *342*, 1334–1349. [CrossRef] [PubMed]
35. Windsor, A.; Mullen, P.; Fowler, A. Acute Lung Injury: What Have We Learned from Animal Models? *Am. J. Med. Sci.* **1993**, *306*, 111–116. [CrossRef] [PubMed]

36. Andersson, U.; Wang, H.; Palmblad, K.; Aveberger, A.-C.; Bloom, O.; Erlandsson-Harris, H.; Janson, A.; Kokkola, R.; Zhang, M.; Yang, H.; et al. High Mobility Group 1 Protein (Hmg-1) Stimulates Proinflammatory Cytokine Synthesis in Human Monocytes. *J. Exp. Med.* **2000**, *192*, 565–570. [CrossRef]
37. Entezari, M.; Javdan, M.; Antoine, D.J.; Morrow, D.M.; Sitapara, R.A.; Patel, V.; Wang, M.; Sharma, L.; Gorasiya, S.; Zur, M.; et al. Inhibition of extracellular HMGB1 attenuates hyperoxia-induced inflammatory acute lung injury. *Redox Biol.* **2014**, *2*, 314–322. [CrossRef]
38. Feng, T.; Zhou, L.; Gai, S.; Zhai, Y.; Gou, N.; Wang, X.; Zhang, X.; Cui, M.; Wang, L.; Wang, S. *Acacia catechu* (L.f.) Willd and *Scutellaria baicalensis* Georgi extracts suppress LPS-induced pro-inflammatory responses through NF-κB, MAPK, and PI3K-Akt signaling pathways in alveolar epithelial type II cells. *Phytother. Res.* **2019**, *33*, 3251–3260. [CrossRef]
39. Tsai, C.L.; Lin, Y.C.; Wang, H.M.; Chou, T.C. Baicalein, an active component of Scutellaria baicalensis, protects against lipopolysaccharide-induced acute lung injury in rats. *J. Ethnopharmacol.* **2014**, *153*, 197–206. [CrossRef]
40. Wang, H.; Liu, D. Baicalin Inhibits High-Mobility Group Box 1 Release and Improves Survival in Experimental Sepsis. *Shock* **2014**, *41*, 324–330. [CrossRef]

molecules

MDPI

Article

Protective Effects of Orange Sweet Pepper Juices Prepared by High-Speed Blender and Low-Speed Masticating Juicer against UVB-induced Skin Damage in SKH-1 Hairless Mice

Van-Long Truong, Razanamanana. H. G. Rarison and Woo-Sik Jeong *

Food and Bio-Industry Research Institute, School of Food Science and Biotechnology, College of Agriculture and Life Sciences, Kyungpook National University, Daegu 41566, Korea
* Correspondence: wsjeong@knu.ac.kr; Tel.: +82-53-950-5775

Abstract: Sweet pepper fruits (*Capsicum annuum* L.) contain various nutrients and phytochemicals that enhance human health and prevent the pathogenesis of certain diseases. Here, we report that oral administration of orange sweet pepper juices prepared by a high-speed blender and low-speed masticating juicer reduces UVB-induced skin damage in SKH-1 hairless mice. Sweet pepper juices reduced UVB-induced skin photoaging by the regulation of genes involved in dermal matrix production and maintenance such as collagen type I α 1 and matrix metalloproteinase-2, 3, 9. Administration of sweet pepper juices also restored total collagen levels in UVB-exposed mice. In addition, sweet pepper juices downregulated the expression of pro-inflammatory proteins such as cyclooxygenase-2, interleukin (IL)-1β, IL-17, and IL-23, which was likely via inhibiting the NF-κB pathway. Moreover, primary antioxidant enzymes in the skin were enhanced by oral supplementation of sweet pepper juices, as evidenced by increased expression of catalase, glutathione peroxidase, and superoxide dismutase-2. Immunohistochemical staining showed that sweet pepper juices reduced UVB-induced DNA damage by preventing 8-OHdG formation. These results suggest that sweet pepper juices may offer a protective effect against photoaging by inhibiting the breakdown of dermal matrix, inflammatory response, and DNA damage as well as enhancing antioxidant defense, which leads to an overall reduction in skin damage.

Keywords: high-speed blender; low-speed masticating juicer; skin health; sweet pepper juice; UVB radiation

Citation: Truong, V.-L.; Rarison, R.H.G.; Jeong, W.-S. Protective Effects of Orange Sweet Pepper Juices Prepared by High-Speed Blender and Low-Speed Masticating Juicer against UVB-induced Skin Damage in SKH-1 Hairless Mice. *Molecules* **2022**, *27*, 6394. https://doi.org/10.3390/molecules27196394

Academic Editors: José Pinela, Carla Pereira, José Ignacio Alonso-Esteban and Maria Inês Dias

Received: 10 August 2022
Accepted: 25 September 2022
Published: 27 September 2022

1. Introduction

The skin, the largest organ of the human body, is composed of the epidermis, dermis, and subcutis, which provide the first line of defense against external environmental factors, such as physical damage, temperature shock, pathogens, and ultraviolet (UV) radiation [1]. In addition, skin appearance, especially human facial skin, importantly contributes to self-confidence, social interaction, and quality of life. Various risk factors, including genetics, hormonal changes, metabolic processes, and environmental stimuli, such as UV radiation, xenobiotics, and bacteria, can affect skin structure, function, and appearance [2].

Sunlight is one of the most crucial causes of skin aging, also known as photoaging, and accounts for up to 90% of visible skin aging [3]. Skin photoaging is characterized by mottled pigmentation, roughness, dryness, visible wrinkles, and decreased elasticity. Although UVB accounts for only a small portion of total UV radiation, it can penetrate the basal cell layer of epidermal cells and causes serious damage to the epidermis and dermis of the skin and is therefore primarily responsible for sunburn and erythema [3,4]. In addition, UVB also provokes the degradation of extracellular matrix (ECM) composition in the dermis, which is composed primarily of type I collagen with lesser amounts of type II collagen, elastin, proteoglycans, and fibronectin, thereby leading to the loss of structure and integrity, laxity, and wrinkles [4,5].

Numerous epidemiological and laboratory studies suggest regular consumption of fruits and vegetables can enhance skin health and delay skin aging [6]. A wide range of phytochemicals stemming from edible plants has been reported to exert antioxidant and anti-inflammatory activities, which contribute to alleviating inflammatory response, restoring the functionality of the skin from photodamage, and preventing further progression of solar UV-induced skin disorders [7]. Sweet peppers, belonging to the genus *Capsicum* with more than 200 varieties, are considered natural sources of bioactive compounds such as carotenoids, vitamin C and E, flavonoids, phenolic compounds, and dietary fiber [8]. Previous studies alongside research in our laboratory found that orange sweet pepper fruit contains a wide range of carotenoids, such as viola, anthera, capsanthin, lutein, zeaxanthin, cryptoxanthin, carotene, and lycopene, in which lutein, β-cryptoxanthin, α-carotene, and zeaxanthin contents are higher than those in other sweet pepper fruits [9–11]. From the pharmaceutical point of view, sweet pepper exerts various health benefits, such as antioxidant [12], anti-inflammation [13], cancer chemopreventive activity [14], and prevention of cardiovascular diseases [15]. Yellow/orange/red sweet pepper fruit extracts were found to inhibit the lipopolysaccharide-induced inflammatory response by inducing heme oxygenase (HO)-1 in RAW 264.7 macrophages [16]. Using in vitro and in vivo models, fermented yellow/orange/red sweet pepper fruits with *Lactobacillus plantarum* have been indicated to exert protective effects against oxidative stress-mediated retinal damage [17]. Four colors of sweet peppers (green, yellow, orange, and red) exert antioxidant activities and inhibitory effects on Alzheimer's disease-associated key enzymes, such as acetylcholinesterase, butyrylcholinesterase, and β-secretase, and thereby may be useful for the prevention of Alzheimer's disease [18]. In addition, clinical studies have shown that the administration of red sweet pepper xanthophylls inhibited UV-induced skin damage [19] and improved moisture on facial skin in volunteers [20].

Fresh vegetable/fruit juice has been receiving great interest from consumers because of a simple household juicing preparation, retaining the major health-promoting components of whole vegetables/fruit, and the absence of additives and preservatives [21,22]. Consequently, household juicers based on several technologies such as centrifugal blending and cold-pressing were developed [22]. Although physicochemical characteristics of vegetable/fruit juice prepared by several different household juicers have been evaluated, the effects of juicing methods on the biological activity of the juice in experimental models have seldomly been investigated [23,24]. Therefore, this study aimed to evaluate and compare the photoprotective effects of fresh orange sweet pepper juices prepared by a household high-speed blender (HSB) and a low-speed masticating juicer (LSM) against UVB-induced skin damage in SKH-1 hairless mice.

2. Results and Discussion

2.1. HSB and LSM Sweet Pepper Juices Inhibit UVB-Induced Photoaging in SKH-1 Hairless Mice

The consumption of fresh vegetable/fruit juice has recently been attracting great interest in the improvement of skin health. Fresh juice contains a wide range of bioactive compounds, such as carotenoids and polyphenols that contribute to maintaining skin functions and preventing skin injury and carcinogenesis. Carotenoids have been demonstrated to exert various benefits on the skin, including protection against sunlight-induced erythema, pigmentation and collagen degradation, the inhibition of UV-induced oxidative stress and inflammation, and the prevention of photoaging and photocarcinogenesis [25]. In addition, amounting studies have indicated that carotenoid supplementation for 3–24 weeks increases serum levels of carotene, lutein, and lycopene, as well as skin levels of β-carotene and total carotenoids [26–29]. Furthermore, dietary carotenoids such as α/β-carotene, lutein, and zeaxanthin are distributed to various layers of skin and the highest levels are often found in the stratum corneum closest to the skin surface to protect skin from harmful effects of sunlight [25,30]. On the other hand, UV radiation could reduce the levels of β-carotene, lycopene, and other carotenoids in the skin, which may potentially be attributed to ROS, thereby increasing the risk of photodamage [31,32]. Taken together, regular intake of carotenoid-rich vegetables/fruits may be an effective strategy for skin health enhancement as well as complementary protection against solar UV-induced skin damage.

Protective effects of HSB and LSM sweet pepper juices against photoaging were examined using UVB-irradiated hairless mice and animal experimental design was illustrated in Figure 1a. Results showed that no significant difference in body weight and food consumption among experimental groups was observed throughout 7 weeks of treatment, suggesting that sweet pepper juices did not exhibit adverse effects (Figure 1b,c). Compared to the normal control, HSB, or LSM juice-only group, skin wrinkle was obviously observed in UVB-irradiated mice, whereas wrinkle development was suppressed by oral administration of sweet pepper juices (Figure 1d). In addition, H&E staining revealed a considerable increase in epidermal thickness in UVB-irradiated mice (Figure 1e,f). However, oral administration of sweet pepper juices significantly decreased epidermal hyperplasia and no significant difference among HSB juice, LSM juice, and positive control vitamin C groups was observed. This observation is in line with previous studies, which suggest that the hyperplasia of the epidermis in hairless mice irradiated with UV radiation is associated with the hyperproliferation of keratinocytes [33,34]. In normal human and mouse skin, the apoptotic and proliferative processes of keratinocytes are tightly regulated to construct the epidermis and chronic UV radiation may dysregulate these two events, resulting in the proliferation of keratinocytes to replace apoptotic cells and subsequently leading to the development of skin cancer [35,36].

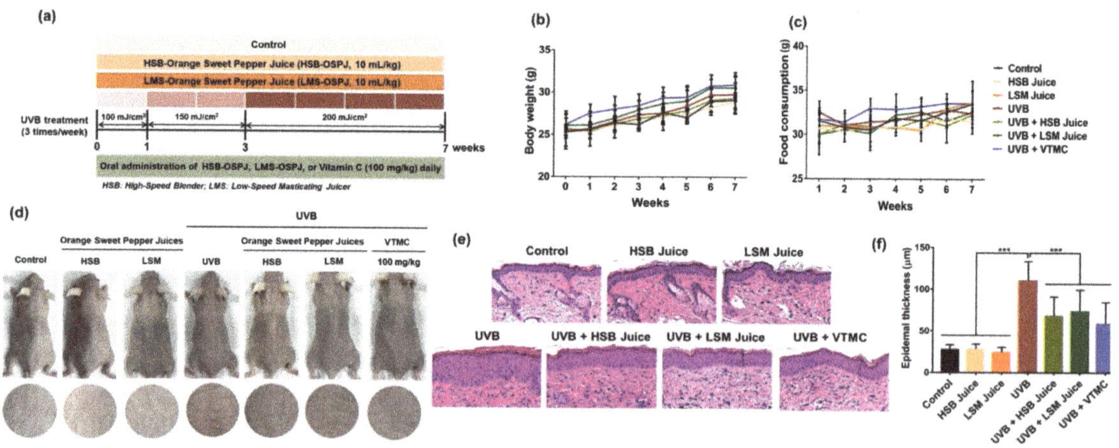

Figure 1. Protective effects of HSB and LSM sweet pepper juices against photoaging in UVB-irradiated SKH-1 hairless mice. (**a**) Animal experimental design. (**b**) Body weight. (**c**) Food consumption. (**d**) Photos of dorsal mouse skin. (**e**) Representative images of H&E staining. (**f**) Epidermal thickness. Data are expressed as the mean ±SD. *** $p < 0.001$ values are considered as statistically significant differences. HSB, high-speed blender; LSM, low-speed juicer; VTMC, vitamin C.

Skin photoaging is characterized by the atrophy of the dermal connective tissue that is associated with the destruction of extracellular matrix (ECM) components, particularly collagen fibers. UV radiation stimulates the synthesis of matrix metalloproteinase (MMP) members, such as MMP-1, -2, -3, -9, and -13 in the skin keratinocytes and fibroblasts, which in turn degrade collagen and other ECM proteins, thereby leading to wrinkle formation and skin aging [37]. Increased levels of MMPs are found in murine and human skins irradiated with UV radiation [5,38]. In addition, UV radiation also reduces collagen production, mainly through downregulation of types I and III procollagen synthesis [39]. This study confirmed that UVB considerably decreased collagen content in skin tissues, as indicated by Masson's trichrome staining and total collagen content assay (Figure 2a,b).

However, oral supplementation of HSB or LSM juice partially restored the level of collagen fiber. Consistently, UVB significantly downregulated the expression of collagen type I alpha 1 (Col1A1) but upregulated the expression of matrix collagen degradation enzymes such as MMP-2, MMP-3, and MMP-9 (Figure 2c,d). Such effects were reversed by the daily consumption of sweet pepper juices. A recent study has shown that carotenoids- and polyphenol-rich extracts reduced MMP-1 expression and simultaneously enhanced procollagen synthesis in normal human dermal fibroblasts under the condition of oxidative stress [40]. Although both HSB and LSM juices did not reduce MMP-2 expression and LSM juice was more effective in reducing MMP-9 expression, in general, the efficacy of HSB and LSM juices was similar and comparable to that of positive control vitamin C. These findings suggest that sweet pepper juice exerts anti-photoaging activity in a model of UVB-irradiated hairless mice by increasing collagen synthesis and suppressing collagen degradation.

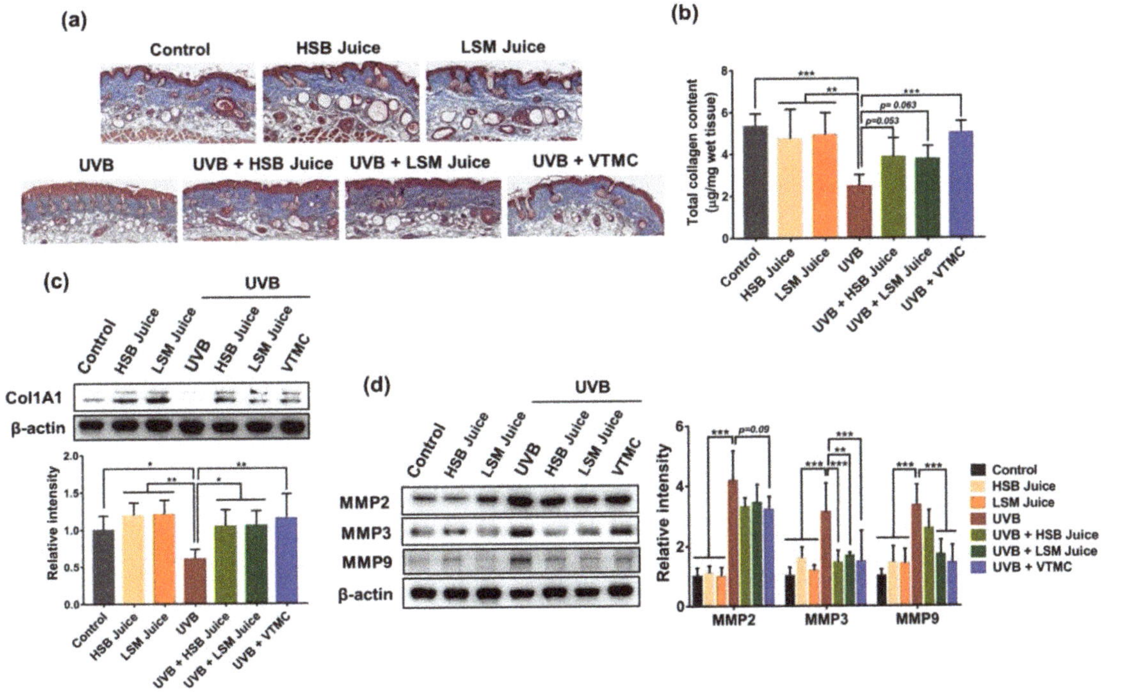

Figure 2. Effects of HSB and LSM sweet pepper juices collagen content in UVB-irradiated SKH-1 hairless mice. (**a**) Representative images of Masson's trichrome stain for collagen fibber (Blue); (**b**) Total collagen content in skin tissues from each group; (**c**) Protein expression of Col1A1 in skin tissues; (**d**) Protein expression of matrix proteinase (MMP)-2, MMP3, and MMP-9 in skin tissues. Data are expressed as the mean ±SD. * $p < 0.05$, ** $p < 0.01$, and *** $p < 0.001$ values are considered as statistically significant differences. HSB, high-speed blender; LSM, low-speed juicer; VTMC, vitamin C.

2.2. HSB and LSM Sweet Pepper Juices Attenuate Inflammatory Response in UVB-Irradiated SKH-1 Hairless Mice

Chronic UVB irradiation is reported to cause skin inflammation, as characterized by accelerating the production of pro-inflammatory mediators and cytokines. Cyclooxygenase-2 (COX2) catalyzes the rate-limiting step in the conversion of arachidonic acid into prostaglandins [41]. UV-induced COX-2 expression stimulates keratinocyte proliferation and epidermal hyperplasia, induces the production of other pro-inflammatory mediators, and triggers skin carcinogenesis [42,43]. Furthermore, COX-2 is also recognized as a molecular link between inflammation and tumorigenesis [44]. Transgenic SKH-1 mice with COX-2 overexpression promote the development of UV-induced skin carcinogenesis, whereas COX-2 knockout mice exhibit significant mitigation of UV-elicited skin tumorigenesis [43,45]. In addition, selective COX-2 inhibitors were found to considerably suppress UV-induced epidermal hyperplasia and inflammation, as well as skin tumorigenesis in terms of tumor number and tumor size [42,46,47]. Therefore, inhibition of COX-2 expression may be a beneficial strategy for preventing UV-caused skin injury and carcinogenesis.

Aberrant expression of COX-2 was observed in the models of human and murine skins acutely and chronically irradiated with UVB [48–50]. In the present study, chronic exposure of mice to UVB remarkably induced COX-2 expression in skin tissues (Figure 3a). However, UVB-induced COX-2 expression was significantly decreased by daily consumption of sweet pepper juices. In addition, under the condition of UV irradiation, keratinocytes and other skin-associated cells secrete a large number of pro-inflammatory cytokines, such as interleukin (IL)-1β, IL-6, IL-17, IL-23, and tumor necrosis factor (TNF)-α that play a crucial role in UV-induced skin inflammation due to their action of inducing edema and erythema, decreasing skin integrity, and increasing chemoattractant and other inflammatory molecules expression and ROS formation [51,52]. Moreover, these pro-inflammatory cytokines also provoke massive infiltration of immune cells, especially neutrophils that excrete matrix metalloproteinases capable of degrading extracellular matrix components and produce ROS likely to devastate vital cellular components and thereby enhance UV-induced skin damage [53,54]. Considering the suppression of pro-inflammatory cytokine production may contribute to the activity of sweet pepper juices, the levels of IL-1β, IL-23, IL-17, and TNF-α were examined. Results indicated that HSB and LSM sweet pepper juices significantly reduced protein levels of IL-1β and IL-23, but not IL-17 and TNF-α in UVB-irradiated mice (Figure 3b). Such inhibition of inflammatory factors partially explains the ability of sweet pepper juice in reducing UVB radiation-induced skin edema and inflammation, epidermal thickness, and skin aging. On the other hand, no significant difference in anti-inflammatory activity between HSB- and LSM-juices was observed, suggesting that the efficiency of fresh sweet pepper juice was independent to juicing methods.

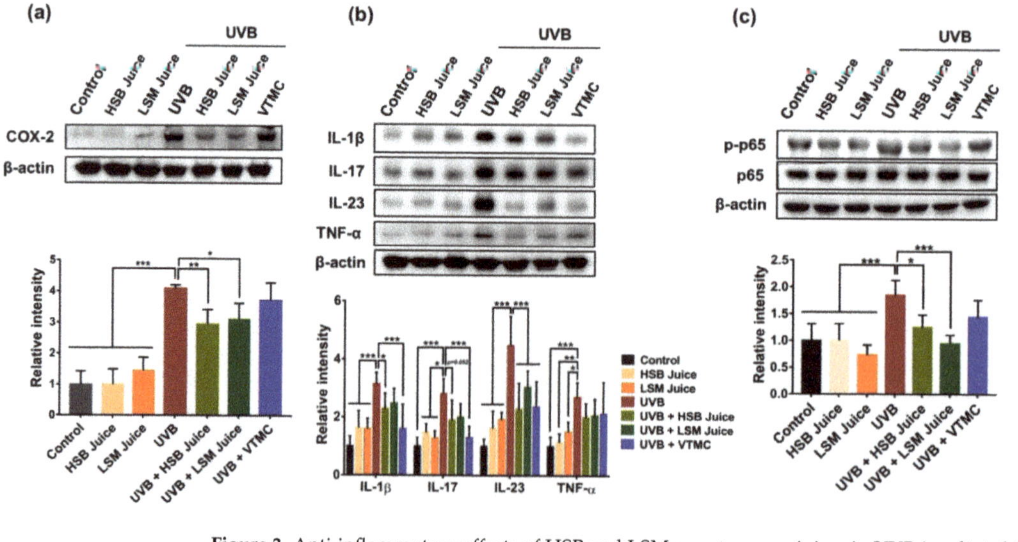

Figure 3. Anti-inflammatory effects of HSB and LSM sweet pepper juices in UVB-irradiated SKH-1 hairless mice. (**a**) Protein expression of COX-2; (**b**) Protein expression of IL-1β, IL-17, IL-23, and TNF-α in skin tissues; (**c**) Protein expression of p-p65 and p65 in skin tissues. Data are expressed as the mean ±SD. * $p < 0.05$, ** $p < 0.01$, and *** $p < 0.001$ values are considered as statistically significant differences. HSB, high-speed blender; LSM, low-speed juicer; VTMC, vitamin C.

Nuclear factor kappa (NF-κB) is a redox-sensitive transcription factor responsible for regulating the transcriptional expression of a wide range of pro-inflammatory factors, such as COX-2, cytokines, chemokines, and adhesion molecules [55]. Activation of NF- κB pathway by UV radiation is observed in many models of murine and human skins [56–59]. Thus, we further explored effects of HSB and LSM juices on NF-κB signaling pathway in UVB-exposed mice. Results showed that oral administration of sweet pepper juices significantly inhibited UVB-induced NF-κB activation, as evident from decreased p65 phosphorylation (Figure 3c). Although there was not significant difference between HSB and LSM juices, LSM juice appeared to exert more pronounced anti-inflammatory activity than HSB juice. In particular, the anti-inflammatory effect of sweet pepper juices seemed to be better than that of vitamin C. These data suggest that sweet pepper juices may inhibit UVB-induced skin inflammation by suppressing NF-κB pathway.

2.3. HSB and LSM Sweet Pepper Juices Enhance Antioxidant Defense Systems in UVB-Irradiated SKH-1 Hairless Mice

The skin is one of a few organs that come into direct contact with exogenous oxidative sources, such as sunlight. Solar UV radiation can promote the formation of reactive oxygen species (ROS) such as superoxide anion radical, hydrogen peroxides, hydroxyl radicals, and other oxidants that directly and/or indirectly damage to biomolecules, such as proteins, lipids, and nucleic acids and interrupts cellular signaling pathways, eventually leading to skin damage. Approximately 80% of ROS in the skin are thought to be produced by UVA and UVB irradiation [60].

Oxidative stress is thought to play a central role in triggering and driving the signaling events that lead to cellular responses, such as cell death, cellular senescence, and inflammation, following UVB radiation. ROS can activate redox-sensitive transcription factors, such as NF-κB and AP-1 that upregulate the production of inflammatory mediators and cytokines. In addition, UVB-induced ROS also stimulates the release of MMP enzymes, such as MMP-1, -2, -3, -9, and -13, in keratinocytes and fibroblasts, which in turn degrade collagen and other ECM proteins, and simultaneously attenuate collagen synthesis in the skin [1,61]. Therefore, oxidative stress mediates a variety of UVB radiation-induced skin damage such as edema, erythema, hyperplasia, inflammation, and skin aging through inappropriate regulation of signaling pathways, thereby increasing the risk of skin cancer.

Skin cells are equipped with an antioxidant defense system, consisting of enzymatic and non-enzymatic antioxidants, which maintain the pro-oxidant/antioxidant balance by ROS elimination. Nevertheless, flooding of ROS can overwhelm the cellular antioxidant defense capacity and further generate reactive oxidants, resulting in oxidative stress and consequently oxidative photodamaging of the skin [62]. Therefore, regular supplementation of antioxidants is a useful strategy to prevent adverse effects of solar UV radiation.

A battery of antioxidant enzymes plays a vital role in counteracting excessive ROS accumulation, thereby maintaining cellular redox homeostasis. Primary antioxidant enzymes, including catalase (CAT), glutathione peroxidase (GPx), and superoxide dismutase (SOD) are considered to be the most important antioxidant defense in the skin [62]. SOD acts as the first line of defense that catalyzes the dismutation of superoxide anion into oxygen and hydrogen peroxide, with the latter being further decomposed by CAT and GPx [63]. Earlier studies reported that chronic UV exposure reduced protein levels of activities of primary antioxidant enzymes, resulting in a wide range of skin disorders such as sunburn, erythema, photoaging, and even cancer [64–66].

To further investigate whether the protective effects of HSB and LSM sweet pepper juices were linked to the enhancement of the antioxidant defense system, we examined the expression of antioxidant enzymes in the mouse skin tissues. UVB irradiation abolished the expression of primary antioxidant enzymes, including CAT, GPx, and SOD-2, suggesting provoked oxidative stress in UVB-exposed mouse skin (Figure 4a). In contrast, both HSB and LSM juices significantly restored the levels of these antioxidant enzymes. Although there was not a significant difference between HSB and LSM juices, LSM juice appeared to be more effective in inducing the expression of antioxidant enzymes than HSB juice.

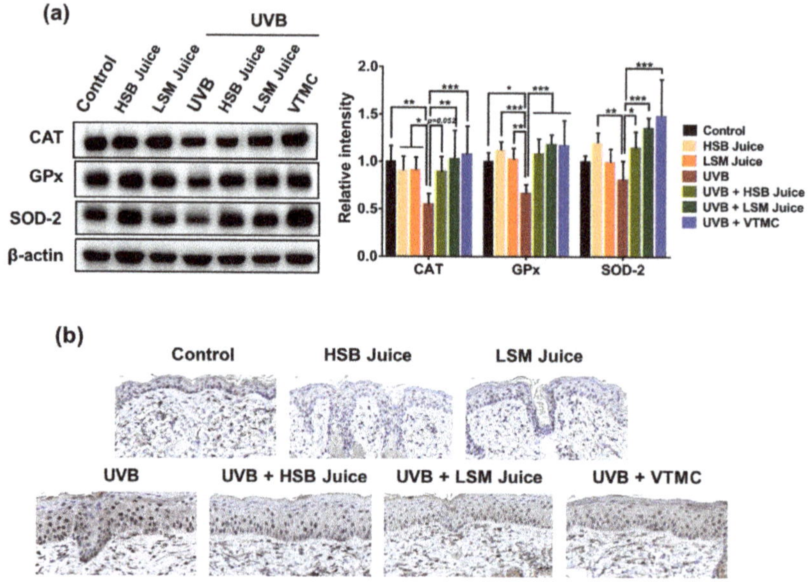

Figure 4. Antioxidant effects of HSB and LSM sweet pepper juices in UVB-irradiated SKH-1 hairless mice. (**a**) Protein expression of CAT, GPx, and SOD-2 in skin tissues; (**b**) Immunostaining for 8-OHdG in skin tissues. Data are expressed as the mean ±SD. * $p < 0.05$, ** $p < 0.01$, and *** $p < 0.001$ values are considered as statistically significant differences. HSB, high-speed blender; LSM, low-speed juicer; VTMC, vitamin C.

DNA damage induced by UVB irradiation occurs mainly via oxidative processes and 8-OHdG is the major mutagenic form of oxidative DNA damage [67]. The present study hypothesizes that sweet pepper juice acts as an antioxidant capable of protecting cells/tissues from DNA damage caused by UVB radiation. Immunohistochemical analysis indicated a high level of 8-OHdG in the mouse skin chronically irradiated with UVB radiation, compared to the normal control, HSB, or LSM juice-only group. However, both HSB and LSM sweet pepper juices exerted a protective effect against UVB-induced DNA damage, as indicated by a reduced level of 8-OHdG (Figure 4b). Taken together, these findings suggest that sweet pepper juice exerts antioxidant activity to protect against UVB-induced skin injury along with its anti-inflammatory property.

3. Conclusions

In conclusion, the regular intake of fresh vegetable/fruit juice containing various intact phytonutrients without preservatives is believed to bring about health benefits, including skin health improvement. Sweet orange pepper fruit with high levels of carotenoids, vitamin C, and other bioactive components have the potential for use in a dietary supplement for skin health. In this study, oral administration of sweet pepper juice exerted a protective effect against UVB-induced skin aging in a hairless mouse model by inhibiting MMP expression and restoring collagen synthesis, thereby preventing skin wrinkles. UVB-induced skin inflammation was suppressed by the consumption of sweet pepper juice. Simultaneously, sweet pepper juice also maintained the antioxidant defense capacity of the skin and thus prevented UVB-induced oxidative damage. In addition, juicing methods were considered in the present study and sweet pepper juice prepared by an LSM juicer may exhibit pronounced efficacies, compared to the juice by HSB. However, different experimental models with long-term consumption are required to compare the biological effects of HSB and LSM juices.

4. Materials and Methods

4.1. Materials

Formalin, hematoxylin, and eosin were purchased from Sigma-Aldrich (St. Louis, MO, USA). Anti-Col1A1, p-p65, p65, COX-2, MMP-2, MMP-3, TNF-α, and peroxidase-conjugated secondary anti-rabbit antibodies were obtained from Cell Signaling Technology (Boston, MA, USA). Anti-catalase, GPx, SOD-2, IL-1β, Il-17, Il-23, MMP-9, and peroxidase-conjugated secondary anti-mouse antibodies were purchased from Santa Cruz Biotechnology (Santa Cruz, CA, USA). Anti-8-OHdG antibody was procured from Thermo Scientific (Waltham, MA, USA). All other chemicals used in this study were of analytic grade.

4.2. Preparation of Orange Sweet Pepper Juice

Orange sweet pepper fruits (*Capsicum annuum* L.) were purchased from Hoengseong Rainbow farm (Gangwon-do, South Korea) and processed right after arrival. For purpose of this study, fresh orange sweet pepper juices were prepared daily using a high-speed blender (HSB, HC-BL 2000, HappyCall, Gimhae, South Korea) and a low-speed masticating juicer (LSM, H200-DBFA03, Hurom Co. Ltd. Gimhae, South Korea). The major bioactive constituents in orange sweet pepper juices were analyzed. Total phenolic content in LSM and HSN juices was 111.77 $\pm$ 1.88 mg/100 mL and 96.45 $\pm$ 3.56 mg/100 mL, respectively, while total carotenoid content was 3.49 $\pm$ 0.01 mg/100 mL and 4.18 $\pm$ 0.1 mg/100 mL, respectively. The level of total ascorbic acid was 119.80 $\pm$ 3.19 mg/100 mL for LSM juice and 100.63 $\pm$ 1.10 mg/100 mL for HSB juice.

4.3. Animal Experimental Design and UVB Irradiation

Six-week-old female SKH-1 hairless mice were purchased from Orient Bio (Seongnam, South Korea) and housed in plastic cages at 25 $\pm$ 2 °C and 50 $\pm$ 5% relative humidity under a 12–12 h light–dark cycle. All animals were allowed free access to standard food and water ad libitum. All experiment protocols were approved by the Institutional Animal Care and Use Committee of Kyungpook National University (No. KNU 2021-182). After a week of acclimation, mice were randomly divided into seven groups (n = 10): Normal control, HSB juice (10 mL/kg), LSM juice (10 mL/kg), UVB, UVB + HSB juice (10 mL/kg), UVB + LSM juice (10 mL/kg), and UVB + Vitamin C (positive control, 100 mg/kg). HSB juice, LSM juice, or Vitamin C were orally administered to the animals in each group daily for 7 weeks, while the mice in the normal control and UVB groups were given an equal amount of distilled water. Vitamin C is well-known to protect against UVB-induced skin damage [68,69] and vitamin C (100 mg/kg) was used as a positive control according to previous studies [70,71]. Food intake and body weight were regularly recorded throughout the study. UVB irradiation was performed three times per week for 7 weeks using a BioLink Crosslinker system (Vilber Lourmat, Paris, France) with peak emission at 312 nm. The UVB dose was started at 100 mJ/cm^2 in the first week and 150 mJ/cm^2 in the following two weeks and increased up to 200 mJ/cm^2 in the last four weeks. At the end of the experiment, the mice were sacrificed with CO_2 and dorsal skin tissues were rapidly harvested. After removing lipid and connective tissues, a part of skin tissues was cut and fixed in a 10% formalin solution, and the rest of the tissue samples were snap-frozen in liquid nitrogen and then stored at −80 °C for further analysis.

4.4. Histological and Immunohistochemical Analyses

Dorsal skin tissues of each group were fixed in 10% neutral formalin and embedded in paraffin. Sections were stained with hematoxylin and eosin (H&E) for general histopathology and Masson's trichrome to examine epidermal thickness and collagen fiber, respectively. For immunostaining, the sections were incubated with specific mouse anti-8-OHdG primary antibody in a humidified chamber overnight at 4 °C and followed by incubation with horseradish-conjugated secondary antibody and 3,3′-diaminobenzidine. Representative H&E, Masson's trichrome, and immunostained images of sections were captured using a digital microscope (Paxcam, Villa Park, IL, USA).

4.5. Measurement of Total Collagen Content

Dorsal skin tissues were homogenized in distilled water and total collagen content was measured by total collagen assay kit (Biovision, Waltham, MA, USA) according to the manufacturer's instructions.

4.6. Western Blot Analysis

Dorsal skin tissues were homogenized in RIPA buffer containing protease and phosphatase inhibitors (Thermo Fisher Scientific, Waltham, MA, USA) using Precellys 14 tissue homogenizer (Bertin Technologies, Montigny-le-Bretonneux, France). The lysates were then centrifuged at 13,000× g for 15 min at 4 °C to obtain supernatant. Protein samples were aliquoted out and stored at −80 °C. The protein concentrations were measured using a PierceTM BAC protein assay kit (Thermo Fisher Scientific). An equal amount of protein samples was resolved on SDS-PAGE gel and then transferred to polyvinylidene fluoride membrane using a semidry transfer system (Bio-rad, Hercules, CA, USA). After blocking with 5% non-fat skim milk, the membranes were incubated with specific primary antibodies overnight at 4 °C and followed by hybridization with proper secondary antibodies for 3 h at 4 °C. Eventually, protein bands were visualized using Western Blotting Luminol reagent (Santa Cruz Biotechnology).

4.7. Statistical Analysis

Data are presented as the mean ± standard deviation (SD) of at least three independent experiments. Statistical analysis was performed using analysis of variance followed by the Tukey's post hoc test. The p-value < 0.05 was considered to be statistically significant.

Author Contributions: Conceptualization, V.-L.T. and W.-S.J.; methodology, V.-L.T.; software, R.H.G.R.; formal analysis, R.H.G.R.; investigation, V.-L.T. and R.H.G.R.; data curation, R.H.G.R.; writing—original draft preparation, V.-L.T.; writing—review and editing, W.-S.J.; visualization, V.-L.T.; supervision, W.-S.J.; project administration, W.-S.J. All authors have read and agreed to the published version of the manuscript.

Funding: This research received no external funding.

Institutional Review Board Statement: The animal study protocol was approved by the Institutional Review Board of Kyungpook National University (No. KNU 2021-182; Date, 15 October 2021).

Informed Consent Statement: Not applicable.

Data Availability Statement: Not applicable.

Conflicts of Interest: The authors declare no conflict of interest.

Sample Availability: Not available.

References

1. Zhang, R.; Wei, Y.; Zhang, J.; Cai, M.; Lu, L.; Fang, L.; Qin, X.; Gu, R. Protection effects of rice protein hydrolysate on UVB-irradiated photodamage in Hartley guinea pigs skin and human skin fibroblasts. *J. Funct. Foods* **2021**, *82*, 104504. [CrossRef]
2. Rittié, L.; Fisher, G.J. UV-light-induced signal cascades and skin aging. *Ageing Res. Rev.* **2002**, *1*, 705–720. [CrossRef]
3. Choi, S.-I.; Han, H.-S.; Kim, J.-M.; Park, G.; Jang, Y.-P.; Shin, Y.-K.; Ahn, H.-S.; Lee, S.-H.; Lee, K.-T. *Eisenia bicyclis* Extract Repairs UVB-Induced Skin Photoaging In Vitro and In Vivo: Photoprotective Effects. *Mar. Drugs* **2021**, *19*, 693. [CrossRef]
4. Xiao, Z.; Yang, S.; Chen, J.; Li, C.; Zhou, C.; Hong, P.; Sun, S.; Qian, Z.-J. Trehalose against UVB-induced skin photoaging by suppressing MMP expression and enhancing procollagen I synthesis in HaCaT cells. *J. Funct. Foods* **2020**, *74*, 104198. [CrossRef]
5. Fisher, G.J.; Wang, Z.; Datta, S.C.; Varani, J.; Kang, S.; Voorhees, J.J. Pathophysiology of Premature Skin Aging Induced by Ultraviolet Light. *N. Engl. J. Med.* **1997**, *337*, 1419–1429. [CrossRef] [PubMed]
6. Fam, V.W.; Charoenwoodhipong, P.; Sivamani, R.K.; Holt, R.R.; Keen, C.L.; Hackman, R.M. Plant-Based Foods for Skin Health: A Narrative Review. *J. Acad. Nutr. Diet.* **2022**, *122*, 614–629. [CrossRef] [PubMed]

7. Działo, M.; Mierziak, J.; Korzun, U.; Preisner, M.; Szopa, J.; Kulma, A. The Potential of Plant Phenolics in Prevention and Therapy of Skin Disorders. *Int. J. Mol. Sci.* **2016**, *17*, 160. [CrossRef] [PubMed]
8. Kim, J.-S.; An, C.G.; Park, J.-S.; Lim, Y.P.; Kim, S. Carotenoid profiling from 27 types of paprika (*Capsicum annuum* L.) with different colors, shapes, and cultivation methods. *Food Chem.* **2016**, *201*, 64–71. [CrossRef] [PubMed]
9. Agarwal, R.; Hong, H.T.; Hayward, A.; Harper, S.; Mitter, N.; O'Hare, T.J. Carotenoid Profiling of Orange-Coloured Capsicums: In Search of High-Zeaxanthin Varieties for Eye Health. *Proceedings* **2021**, *70*, 84. [CrossRef]
10. Park, S.-Y.; Choi, S.R.; Lim, S.-H.; Yeo, Y.; Kweon, S.J.; Bae, Y.-S.; Kim, K.W.; Im, K.-H.; Ahn, S.K.; Ha, S.-H.; et al. Identification and quantification of carotenoids in paprika fruits and cabbage, kale, and lettuce leaves. *J. Korean Soc. Appl. Biol. Chem.* **2014**, *57*, 355–358. [CrossRef]
11. Guzman, I.; Hamby, S.; Romero, J.; Bosland, P.W.; O'Connell, M.A. Variability of carotenoid biosynthesis in orange colored *Capsicum* spp. *Plant Sci.* **2010**, *179*, 49–59. [CrossRef] [PubMed]
12. Sun, T.; Xu, Z.; Wu, C.-T.; Janes, M.; Prinyawiwatkul, W.; No, H.K. Antioxidant Activities of Different Colored Sweet Bell Peppers (*Capsicum annuum* L.). *J. Food Sci.* **2007**, *72*, S98–S102. [CrossRef] [PubMed]
13. Hernández-Ortega, M.; Ortiz-Moreno, A.; Hernández-Navarro, M.D.; Chamorro-Cevallos, G.; Dorantes-Alvarez, L.; Necoechea-Mondragón, H. Antioxidant, Antinociceptive, and Anti-Inflammatory Effects of Carotenoids Extracted from Dried Pepper (*Capsicum annuum* L.). *J. Biomed. Biotechnol.* **2012**, *2012*, 524019. [CrossRef] [PubMed]
14. Maoka, T.; Mochida, K.; Kozuka, M.; Ito, Y.; Fujiwara, Y.; Hashimoto, K.; Enjo, F.; Ogata, M.; Nobukuni, Y.; Tokuda, H.; et al. Cancer chemopreventive activity of carotenoids in the fruits of red paprika *Capsicum annuum* L. *Cancer Lett.* **2001**, *172*, 103–109. [CrossRef]
15. Aizawa, K.; Inakuma, T. Dietary capsanthin, the main carotenoid in paprika (*Capsicum annuum*), alters plasma high-density lipoprotein-cholesterol levels and hepatic gene expression in rats. *Br. J. Nutr.* **2009**, *102*, 1760–1766. [CrossRef]
16. Choi, E.-Y.; Kim, E.-S.; Lee, Y.-E. Anti-inflammatory effects of paprika fruit and leaf through heme oxygenase-1 induction in RAW264. 7 macrophages. *J. Korean Soc. Food Sci. Nutr.* **2020**, *49*, 578–585. [CrossRef]
17. Kim, H.-R.; Kim, S.; Lee, S.-W.; Sin, H.-S.; Kim, S.-Y. Protective Effects of Fermented Paprika (*Capsicum annuum* L.) on Sodium Iodate-Induced Retinal Damage. *Nutrients* **2020**, *13*, 25. [CrossRef]
18. Thuphairo, K.; Sornchan, P.; Suttisansanee, U. Bioactive Compounds, Antioxidant Activity and Inhibition of Key Enzymes Relevant to Alzheimer's Disease from Sweet Pepper (*Capsicum annuum*) Extracts. *Prev. Nutr. Food Sci.* **2019**, *24*, 327–337. [CrossRef]
19. Nishino, A.; Sugimoto, K.; Sambe, H.; Ichihara, T.; Takaha, T.; Kuriki, T. Effects of Dietary Paprika Xanthophylls on Ultraviolet Light-Induced Skin Damage: A Double-Blind Placebo-Controlled Study. *J. Oleo Sci.* **2018**, *67*, 863–869. [CrossRef]
20. Yatsuhashi, H.; Takumi, H.; Terada, Y.; Kuriki, T. Effects of Oral Supplementation with Paprika Xanthophylls on Human Skin Moisture. *J. Oleo Sci.* **2022**, *71*, 735–745. [CrossRef]
21. Agarwal, S.; Fulgoni Iii, V.L.; Welland, D. Intake of 100% Fruit Juice Is Associated with Improved Diet Quality of Adults: NHANES 2013-2016 Analysis. *Nutrients* **2019**, *11*, 2513. [CrossRef] [PubMed]
22. Khaksar, G.; Assatarakul, K.; Sirikantaramas, S. Effect of cold-pressed and normal centrifugal juicing on quality attributes of fresh juices: Do cold-pressed juices harbor a superior nutritional quality and antioxidant capacity? *Heliyon* **2019**, *5*, e01917. [CrossRef] [PubMed]
23. Kim, M.-J.; Jun, J.-G.; Park, S.-Y.; Choi, M.-J.; Park, E.; Kim, J.-I.; Kim, M.-J. Antioxidant activities of fresh grape juices prepared using various household processing methods. *Food Sci. Biotechnol.* **2017**, *26*, 861–869. [CrossRef] [PubMed]
24. Gouws, C.A.; Georgouopoulou, E.; Mellor, D.D.; Naumovski, N. The Effect of Juicing Methods on the Phytochemical and Antioxidant Characteristics of the Purple Prickly Pear (*Opuntia ficus indica*)—Preliminary Findings on Juice and Pomace. *Beverages* **2019**, *5*, 28. [CrossRef]
25. Baswan, S.M.; Klosner, A.E.; Weir, C.; Salter-Venzon, D.; Gellenbeck, K.W.; Leverett, J.; Krutmann, J. Role of ingestible carotenoids in skin protection: A review of clinical evidence. *Photodermatol. Photoimmunol. Photomed.* **2021**, *37*, 490–504. [CrossRef]
26. Lee, J.; Jiang, S.; Levine, N.; Watson, R.R. Carotenoid supplementation reduces erythema in human skin after simulated solar radiation exposure. *Proc. Soc. Exp. Biol. Med.* **2000**, *223*, 170–174. [CrossRef]
27. Heinrich, U.; Gärtner, C.; Wiebusch, M.; Eichler, O.; Sies, H.; Tronnier, H.; Stahl, W. Supplementation with β-Carotene or a Similar Amount of Mixed Carotenoids Protects Humans from UV-Induced Erythema. *J. Nutr.* **2003**, *133*, 98–101. [CrossRef]
28. Garmyn, M.; Ribaya-Mercado, J.D.; Russel, R.M.; Bhawan, J.; Gilchrest, B.A. Effect of beta-carotene supplementation on the human sunburn reaction. *Exp. Dermatol.* **1995**, *4*, 104–111. [CrossRef]
29. Baswan, S.M.; Marini, A.; Klosner, A.E.; Jaenicke, T.; Leverett, J.; Murray, M.; Gellenbeck, K.W.; Krutmann, J. Orally administered mixed carotenoids protect human skin against ultraviolet A-induced skin pigmentation: A double-blind, placebo-controlled, randomized clinical trial. *Photodermatol. Photoimmunol. Photomed.* **2020**, *36*, 219–225. [CrossRef]
30. Obana, A.; Gohto, Y.; Gellermann, W.; Ermakov, I.V.; Sasano, H.; Seto, T.; Bernstein, P.S. Skin Carotenoid Index in a large Japanese population sample. *Sci. Rep.* **2019**, *9*, 9318. [CrossRef]
31. Ribaya-Mercado, J.D.; Garmyn, M.; Gilchrest, B.A.; Russell, R.M. Skin Lycopene Is Destroyed Preferentially over β-Carotene during Ultraviolet Irradiation in Humans. *J. Nutr.* **1995**, *125*, 1854–1859. [CrossRef] [PubMed]
32. Biesalski, H.K.; Obermueller-Jevic, U.C. UV Light, Beta-carotene and Human Skin—Beneficial and Potentially Harmful Effects. *Arch. Biochem. Biophys.* **2001**, *389*, 1–6. [CrossRef] [PubMed]

33. Misawa, E.; Tanaka, M.; Saito, M.; Nabeshima, K.; Yao, R.; Yamauchi, K.; Abe, F.; Yamamoto, Y.; Furukawa, F. Protective effects of Aloe sterols against UVB-induced photoaging in hairless mice. *Photodermatol. Photoimmunol. Photomed.* **2017**, *33*, 101–111. [CrossRef] [PubMed]
34. El-Abaseri, T.B.; Putta, S.; Hansen, L.A. Ultraviolet irradiation induces keratinocyte proliferation and epidermal hyperplasia through the activation of the epidermal growth factor receptor. *Carcinogenesis* **2006**, *27*, 225–231. [CrossRef] [PubMed]
35. Lee, J.K.; Kim, J.H.; Nam, K.T.; Lee, S.H. Molecular events associated with apoptosis and proliferation induced by ultraviolet-B radiation in the skin of hairless mice. *J. Dermatol. Sci.* **2003**, *32*, 171–179. [CrossRef]
36. Ouhtit, A.; Muller, H.K.; Davis, D.W.; Ullrich, S.E.; McConkey, D.; Ananthaswamy, H.N. Temporal events in skin injury and the early adaptive responses in ultraviolet-irradiated mouse skin. *Am. J. Pathol.* **2000**, *156*, 201–207. [CrossRef]
37. Moon, J.-M.; Park, S.-H.; Jhee, K.-H.; Yang, S.-A. Protection against UVB-Induced Wrinkle Formation in SKH-1 Hairless Mice: Efficacy of Tricin Isolated from Enzyme-Treated *Zizania latifolia* Extract. *Molecules* **2018**, *23*, 2254. [CrossRef]
38. Kim, S.Y.; Lamichhane, S.; Ju, J.-H.; Yun, J. Protective Effect of Octylmethoxycinnamate against UV-Induced Photoaging in Hairless Mouse via the Regulation of Matrix Metalloproteinases. *Int. J. Mol. Sci.* **2018**, *19*, 1836. [CrossRef]
39. Zhang, J.-A.; Yin, Z.; Ma, L.-W.; Yin, Z.-Q.; Hu, Y.-Y.; Xu, Y.; Wu, D.; Permatasari, F.; Luo, D.; Zhou, B.-R. The Protective Effect of Baicalin against UVB Irradiation Induced Photoaging: An *In Vitro* and *In Vivo* Study. *PLoS ONE* **2014**, *9*, e99703. [CrossRef]
40. Darawsha, A.; Trachtenberg, A.; Levy, J.; Sharoni, Y. The Protective Effect of Carotenoids, Polyphenols, and Estradiol on Dermal Fibroblasts under Oxidative Stress. *Antioxidants* **2021**, *10*, 2023. [CrossRef]
41. Yum, H.-W.; Park, J.; Park, H.-J.; Shin, J.W.; Cho, Y.-Y.; Kim, S.-J.; Kang, J.X.; Surh, Y.-J. Endogenous ω-3 Fatty Acid Production by fat-1 Transgene and Topically Applied Docosahexaenoic Acid Protect against UVB-induced Mouse Skin Carcinogenesis. *Sci. Rep.* **2017**, *7*, 11658. [CrossRef]
42. Tripp, C.S.; Blomme, E.A.G.; Chinn, K.S.; Hardy, M.M.; LaCelle, P.; Pentland, A.P. Epidermal COX-2 Induction Following Ultraviolet Irradiation: Suggested Mechanism for the Role of COX-2 Inhibition in Photoprotection. *J. Investig. Dermatol.* **2003**, *121*, 853–861. [CrossRef]
43. Rundhaug, J.E.; Mikulec, C.; Pavone, A.; Fischer, S.M. A role for cyclooxygenase-2 in ultraviolet light-induced skin carcinogenesis. *Mol. Carcinog.* **2007**, *46*, 692–698. [CrossRef] [PubMed]
44. Kundu, J.K.; Surh, Y.-J. Breaking the relay in deregulated cellular signal transduction as a rationale for chemoprevention with anti-inflammatory phytochemicals. *Mutat. Res. Fundam. Mol. Mech. Mutagen.* **2005**, *591*, 123–146. [CrossRef] [PubMed]
45. Fischer, S.M.; Pavone, A.; Mikulec, C.; Langenbach, R.; Rundhaug, J.E. Cyclooxygenase-2 expression is critical for chronic UV-induced murine skin carcinogenesis. *Mol. Carcinog.* **2007**, *46*, 363–371. [CrossRef]
46. Wilgus, T.A.; Koki, A.T.; Zweifel, B.S.; Kusewitt, D.F.; Rubal, P.A.; Oberyszyn, T.M. Inhibition of cutaneous ultraviolet light B–mediated inflammation and tumor formation with topical celecoxib treatment. *Mol. Carcinog.* **2003**, *38*, 49–58. [CrossRef] [PubMed]
47. Rundhaug, J.E.; Fischer, S.M. Cyclo-oxygenase-2 Plays a Critical Role in UV-induced Skin Carcinogenesis. *Photochem. Photobiol.* **2008**, *84*, 322–329. [CrossRef] [PubMed]
48. Buckman, S.Y.; Gresham, A.; Hale, P.; Hruza, G.; Anast, J.; Masferrer, J.; Pentland, A.P. COX-2 expression is induced by UVB exposure in human skin: Implications for the development of skin cancer. *Carcinogenesis* **1998**, *19*, 723–729. [CrossRef]
49. Bermudez, Y.; Stratton, S.P.; Curiel-Lewandrowski, C.; Warneke, J.; Hu, C.; Bowden, G.T.; Dickinson, S.E.; Dong, Z.; Bode, A.M.; Saboda, K.; et al. Activation of the PI3K/Akt/mTOR and MAPK Signaling Pathways in Response to Acute Solar-Simulated Light Exposure of Human Skin. *Cancer Prev. Res.* **2015**, *8*, 720–728. [CrossRef]
50. Ha, S.J.; Lee, J.; Kim, H.; Song, K.-M.; Lee, N.H.; Kim, Y.E.; Lee, H.; Kim, Y.H.; Jung, S.K. Preventive effect of *Rhus javanica* extract on UVB-induced skin inflammation and photoaging. *J. Funct. Foods* **2016**, *27*, 589–599. [CrossRef]
51. Martinez, R.M.; Fattori, V.; Saito, P.; Melo, C.B.P.; Borghi, S.M.; Pinto, I.C.; Bussmann, A.J.C.; Baracat, M.M.; Georgetti, S.R.; Verri, W.A.; et al. Lipoxin A4 inhibits UV radiation-induced skin inflammation and oxidative stress in mice. *J. Dermatol. Sci.* **2018**, *91*, 164–174. [CrossRef] [PubMed]
52. Martinez, R.M.; Pinho-Ribeiro, F.A.; Steffen, V.S.; Caviglione, C.V.; Vignoli, J.A.; Barbosa, D.S.; Baracat, M.M.; Georgetti, S.R.; Verri, W.A.; Casagrande, R. Naringenin Inhibits UVB Irradiation-Induced Inflammation and Oxidative Stress in the Skin of Hairless Mice. *J. Nat. Prod.* **2015**, *78*, 1647–1655. [CrossRef] [PubMed]
53. Sharma, M.R.; Mitrani, R.; Werth, V.P. Effect of TNFα blockade on UVB-induced inflammatory cell migration and collagen loss in mice. *J. Photochem. Photobiol. B Biol.* **2020**, *213*, 112072. [CrossRef]
54. Pillai, S.; Oresajo, C.; Hayward, J. Ultraviolet radiation and skin aging: Roles of reactive oxygen species, inflammation and protease activation, and strategies for prevention of inflammation-induced matrix degradation—A review. *Int. J. Cosmet. Sci.* **2005**, *27*, 17–34. [CrossRef]
55. Truong, V.-L.; Jeong, W.-S. Cellular Defensive Mechanisms of Tea Polyphenols: Structure-Activity Relationship. *Int. J. Mol. Sci.* **2021**, *22*, 9109. [CrossRef] [PubMed]
56. Rahman, M.M.; Kundu, J.K.; Shin, J.-W.; Na, H.-K.; Surh, Y.-J. Docosahexaenoic Acid Inhibits UVB-Induced Activation of NF-κB and Expression of COX-2 and NOX-4 in HR-1 Hairless Mouse Skin by Blocking MSK1 Signaling. *PLoS ONE* **2011**, *6*, e28065. [CrossRef] [PubMed]
57. Lewis, D.A.; Spandau, D.F. UVB activation of NF-κB in normal human keratinocytes occurs via a unique mechanism. *Arch. Dermatol. Res.* **2007**, *299*, 93–101. [CrossRef]

58. Adhami, V.M.; Afaq, F.; Ahmad, N. Suppression of ultraviolet B exposure-mediated activation of NF-κB in normal human keratinocytes by resveratrol. *Neoplasia* **2003**, *5*, 74–82. [CrossRef]
59. Chang, E.-J.; Kundu, J.K.; Liu, L.; Shin, J.-W.; Surh, Y.-J. Ultraviolet B radiation activates NF-κB and induces iNOS expression in HR-1 hairless mouse skin: Role of IκB kinase-β. *Mol. Carcinog.* **2011**, *50*, 310–317. [CrossRef] [PubMed]
60. Yin, S.; Wang, Y.; Liu, N.; Yang, M.; Hu, Y.; Li, X.; Fu, Y.; Luo, M.; Sun, J.; Yang, X. Potential skin protective effects after UVB irradiation afforded by an antioxidant peptide from *Odorrana andersonii*. *Biomed. Pharmacother.* **2019**, *120*, 109535. [CrossRef] [PubMed]
61. Fu, Y.; Li, C.; Wang, Q.; Gao, R.; Cai, X.; Wang, S.; Zhang, Y. The protective effect of collagen peptides from bigeye tuna (*Thunnus obesus*) skin and bone to attenuate UVB-induced photoaging via MAPK and TGF-β signaling pathways. *J. Funct. Foods* **2022**, *93*, 105101. [CrossRef]
62. Filip, A.; Daicoviciu, D.; Clichici, S.; Bolfa, P.; Catoi, C.; Baldea, I.; Bolojan, L.; Olteanu, D.; Muresan, A.; Postescu, I.D. The effects of grape seeds polyphenols on SKH-1 mice skin irradiated with multiple doses of UV-B. *J. Photochem. Photobiol. B Biol.* **2011**, *105*, 133–142. [CrossRef] [PubMed]
63. Chen, L.; Kang, Y.-H. Anti-inflammatory and antioxidant activities of red pepper (*Capsicum annuum* L.) stalk extracts: Comparison of pericarp and placenta extracts. *J. Funct. Foods* **2013**, *5*, 1724–1731. [CrossRef]
64. Ahn, J.H.; Kim, D.W.; Park, C.W.; Kim, B.; Sim, H.; Kim, H.S.; Lee, T.-K.; Lee, J.-C.; Yang, G.E.; Her, Y.; et al. Laminarin Attenuates Ultraviolet-Induced Skin Damage by Reducing Superoxide Anion Levels and Increasing Endogenous Antioxidants in the Dorsal Skin of Mice. *Mar. Drugs* **2020**, *18*, 345. [CrossRef] [PubMed]
65. Jo, K.; Bae, G.Y.; Cho, K.; Park, S.S.; Suh, H.J.; Hong, K.-B. An Anthocyanin-Enriched Extract from *Vaccinium uliginosum* Improves Signs of Skin Aging in UVB-Induced Photodamage. *Antioxidants* **2020**, *9*, 844. [CrossRef]
66. Kwak, C.S.; Yang, J.; Shin, C.-Y.; Chung, J.H. Topical or oral treatment of peach flower extract attenuates UV-induced epidermal thickening, matrix metalloproteinase-13 expression and pro-inflammatory cytokine production in hairless mice skin. *Nutr. Res. Pract.* **2018**, *12*, 29–40. [CrossRef]
67. Yin, Y.; Li, W.; Son, Y.-O.; Sun, L.; Lu, J.; Kim, D.; Wang, X.; Yao, H.; Wang, L.; Pratheeshkumar, P.; et al. Quercitrin protects skin from UVB-induced oxidative damage. *Toxicol. Appl. Pharmacol.* **2013**, *269*, 89–99. [CrossRef] [PubMed]
68. Kawashima, S.; Funakoshi, T.; Sato, Y.; Saito, N.; Ohsawa, H.; Kurita, K.; Nagata, K.; Yoshida, M.; Ishigami, A. Protective effect of pre- and post-vitamin C treatments on UVB-irradiation-induced skin damage. *Sci. Rep.* **2018**, *8*, 16199. [CrossRef] [PubMed]
69. McArdle, F.; Rhodes, L.E.; Parslew, R.; Jack, C.I.A.; Friedmann, P.S.; Jackson, M.J. UVR-induced oxidative stress in human skin in vivo: Effects of oral vitamin C supplementation. *Free Radic. Biol. Med.* **2002**, *33*, 1355–1362. [CrossRef]
70. Han, J.-H.; Bang, J.S.; Choi, Y.J.; Choung, S.-Y. Oral administration of oyster (*Crassostrea gigas*) hydrolysates protects against wrinkle formation by regulating the MAPK pathway in UVB-irradiated hairless mice. *Photochem. Photobiol. Sci.* **2019**, *18*, 1436–1446. [CrossRef]
71. Lim, J.-Y.; Kim, O.-K.; Lee, J.; Lee, M.-J.; Kang, N.; Hwang, J.-K. Protective effect of the standardized green tea seed extract on UVB-induced skin photoaging in hairless mice. *Nutr. Res. Pract.* **2014**, *8*, 398–403. [CrossRef] [PubMed]

molecules

MDPI

Article

Salidroside Alleviates Renal Fibrosis in SAMP8 Mice by Inhibiting Ferroptosis

Sixia Yang [1,†], Tingting Pei [1,†], Linshuang Wang [2], Yi Zeng [1], Wenxu Li [1], Shihua Yan [1], Wei Xiao [1,3,*] and Weidong Cheng [1,*]

[1] School of Traditional Chinese Medicine, Southern Medical University, Guangzhou 510515, China
[2] Institute of Basic Research in Clinical Medicine, China Academy of Chinese Medical Sciences, Beijing 100700, China
[3] Key Laboratory of Glucolipid Metabolic Disorder, Ministry of Education, Guangdong Pharmaceutical University, Guangzhou 510006, China
* Correspondence: xw7688@smu.edu.cn (W.X.); chengweidong888@sina.com (W.C.)
† These authors contributed equally to this work.

Abstract: Renal fibrosis progression is closely associated with aging, which ultimately leads to renal dysfunction. Salidroside (SAL) is considered to have broad anti-aging effects. However, the roles and mechanisms of SAL in aging-related renal fibrosis remain unclear. The study aimed to evaluate the protective effects and mechanisms of SAL in SAMP8 mice. SAMP8 mice were administered with SAL and Ferrostatin-1 (Fer-1) for 12 weeks. Renal function, renal fibrosis, and ferroptosis in renal tissue were detected. The results showed that elevated blood urea nitrogen (BUN) and serum creatinine (SCr) levels significantly decreased, serum albumin (ALB) levels increased, and mesangial hyperplasia significantly reduced in the SAL group. SAL significantly reduced transforming growth factor-β (TGF-β) and α-smooth muscle actin (α-sma) levels in SAMP8 mice. SAL treatment significantly decreased lipid peroxidation in the kidneys, and regulated iron transport-related proteins and ferroptosis-related proteins. These results suggested that SAL delays renal aging and inhibits aging-related glomerular fibrosis by inhibiting ferroptosis in SAMP8 mice.

Keywords: renal fibrosis; ferroptosis; aging; salidroside; iron transport

Citation: Yang, S.; Pei, T.; Wang, L.; Zeng, Y.; Li, W.; Yan, S.; Xiao, W.; Cheng, W. Salidroside Alleviates Renal Fibrosis in SAMP8 Mice by Inhibiting Ferroptosis. *Molecules* **2022**, 27, 8039. https://doi.org/10.3390/molecules27228039

Academic Editor: Yun-Bae Kim

Received: 17 September 2022
Accepted: 15 November 2022
Published: 19 November 2022

1. Introduction

The elderly population rate is increasing, owing to improvements in living and medical standards [1]. Many organs change during aging, and the kidneys are particularly affected by aging [2]. Aging causes structural changes in the kidney which lead to conditions such as glomerular and tubular hypertrophy, glomerulosclerosis, and tubulointerstitial fibrosis [3]. Patients with chronic kidney disease (CKD) develop renal anemia and exhibit reduced erythropoietin secretion [4]. Actively combating renal anemia can improve the quality of life and survival rate of patients with CKD [5]. Therefore, research on mechanisms involved in renal aging and alleviation of circulating iron deficiency has gained attention because of their significance in the prevention of aging-related glomerular fibrosis.

Senescence-accelerated mouse (SAM) strains consist of senescence-prone inbred strains (SAMP) and senescence-resistant inbred strains (SAMR) [6]. SAMP8 mice are one of the senescence-accelerated prone mice, based on the grading score data of aging, life span, and pathological phenotype, Kyoto University conducted selective inbreeding on AKR/J mouse strain donated by Jackson Laboratory in 1968 [7], which is characterized by learning and memory deficits and impaired immune function [8]. The serum oxidative stress level of SAMP8 mice was higher than that of SAMR1 mice [9], and the expression levels of muscle atrophy, and inflammation fibrosis-related and aging gene marker genes in elderly SAMP8 mice, were significantly different from those in the young control group [10]. Elderly SAMP8 mice can show adverse cardiac remodeling [11], liver fibrosis [12], and renal fibrosis [13], which are age-related pathologies, and their incidence rate and severity increase with age [6].

Studies have shown that age-related mitochondrial dysfunction is closely related to renal fibrosis [14]. There is dysfunction in the mitochondria of elderly SAMP8 mice [15]. Renal pathological changes of 9-month-old SAMP8 mice include renal tubulointerstitial fibrosis and focal segmental glomerulosclerosis. Renal fibrosis of SAMP8 mice is related to age. The Wnt/β-catenin/RAS signaling pathway was activated in the kidney of SAMP8 mice [16]. Compared with the aging SAMR1 control group, the consumption level of GSH/GSSG ratio and MDA level in the 6-month-old-month old SAMP8 mice were increased [17]. Early CKD was observed in aged SAMP8 mice [18]. A spontaneous model such as SAM has obvious advantages over the gene-modified model, as it can better reflect the changes in age-related diseases. It is hoped that it can be used more widely for the research of biological gerontology resources [6]. There was ferroptosis and increased lipid peroxidation levels in the muscle of SAMP8 mice [19].

A recent study suggested that ferroptosis plays an important role in renal tubular injury in tubular epithelial cells (TECs), and suggested that ferroptosis plays a key role in driving kidney injury and can be used as a treatment strategy [20]. Ferroptosis is an iron-dependent cell death caused by lipid peroxidation that is controlled by integrated oxidant and antioxidant systems [21]. The iron-containing enzyme lipoxygenase is the main ferroptosis promoter by producing lipid hydroperoxides, and its function relies on the Acyl-CoA synthetase long-chain family member 4 (ACSL4)-dependent lipid biosynthesis activation [22]. In contrast, the selenium-containing enzyme Glutathione Peroxidase 4 (GPX4) is currently recognized as a central repressor of ferroptosis, and it is dependent on glutathione, which is produced by the activation of the cystine–glutamate anti-porter. Similar to transferrin (TF), it acts as a positive ferroptosis regulator by increasing iron intake [23]. Ferritin is the major intracellular iron storage protein complex and includes ferritin light polypeptide 1 (FTL1) and ferritin heavy polypeptide 1 (FTH1) [24]. Increased ferritin expression limits ferroptosis. Recent studies have indicated that FTH1 increases autophagy, which can degrade ferritin to elevate iron levels, resulting in oxidative injury through the Fenton reaction [25].

Salidroside (SAL) is a phenylpropanoid glycoside with various pharmacological benefits [26] which acts as an anti-oxidative, anti-fatigue, anti-inflammatory, anti-aging, and anti-diabetic agent. SAL has been found in sweet-scented Osmanthus, Oleifera, Striga, and Rhodiola L [27]. Current evidence suggests that SAL affects renal interstitial fibrosis in unilateral ureteral obstruction model mice [28]. However, there were few reports of SAL involved in renal interstitial fibrosis in the context of aging. In this study, we aimed to investigate the therapeutic effects of SAL in renal interstitial fibrosis in SAMP8 mice and explore its possible mechanism.

2. Results

2.1. Screened Potential Targets

After removing duplicates, 434 targets of SAL and 2586 targets of aging-related renal fibrosis were collected (Table S1). Overlap between targets of SAL and aging-related renal, 183 potential therapeutic targets of SAL were selected and presented by Venn diagrams (Figure 1A).

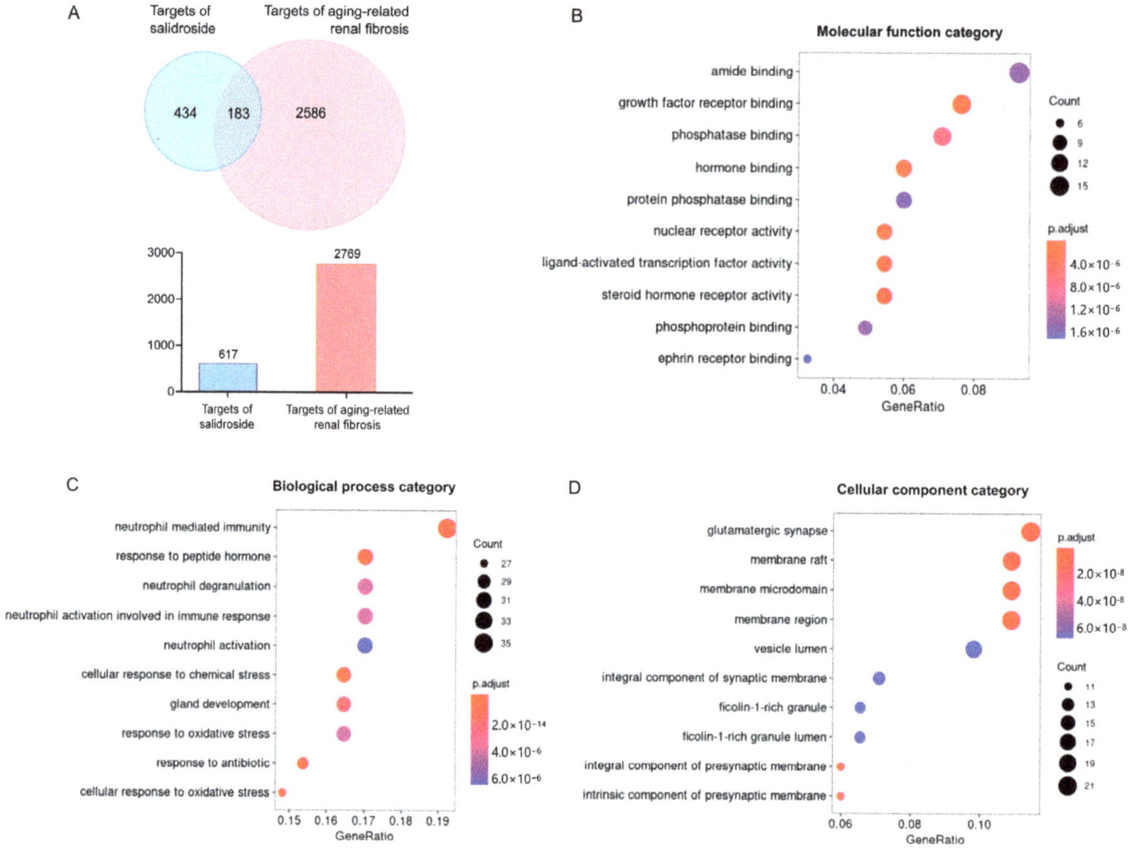

Figure 1. Potential-target screening and enrichment analysis. (**A**) The predicted targets of SAL and aging-related renal fibrosis. A total of 183 overlap targets were screened by the Venn diagram. The bubble maps of molecular function (**B**), biological process (**C**), and cellular component (**D**).

2.2. Enrichment Analysis of Potential Targets

The bubble maps provide a graphical representation of the highly enriched terms of function categories. The main molecular functions included amide binding, growth factor-receptor binding, and protein phosphatase binding (Figure 1B). The main biological processes included neutrophil degranulation, neutrophil activation, and response to oxidative stress (Figure 1C). The main cellular components included the glutamatergic synapse, the integral component of the synaptic membrane, and the vesicle lumen (Figure 1D). The main enriched pathways included ferroptosis, aging, apoptosis, dopaminegic synapse, glycolysis/gluconeogenesis, the TNF signaling pathway, MAPK signaling pathway, mTOR signaling pathway, and neurotrophin signaling pathway (Figure 2).

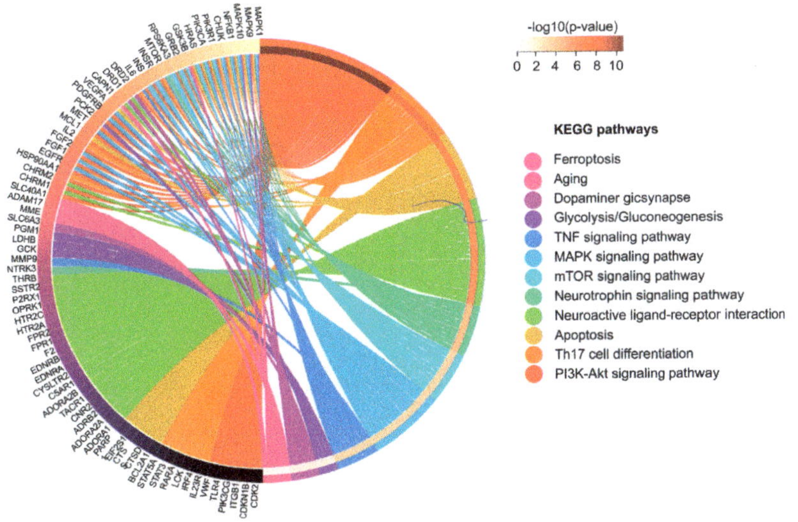

Figure 2. Circle diagram of the significant KEGG pathways.

2.3. Network Analysis of Potential Targets

The PPI network consisted of 183 nodes and 1332 edges, with an average node degree of 14.7 (Figure 3A). The PPI network contained five clusters, with a local clustering coefficient of 0.466, which was related to tau protein binding metal ion binding, response to oxygen-containing compounds, and regulation of the cell cycle, respectively (Figure 3B). The subnetworks selected by MCODE include module 1 (MCODE score = 27.758) (Figure 3C) and module 2 (MCODE score = 3.667) (Figure 3D). Clusters 1 and 2 were related to the regulation of programmed cell death and abnormal cell death, respectively. Based on the above findings, the following studies were focused on ferroptosis.

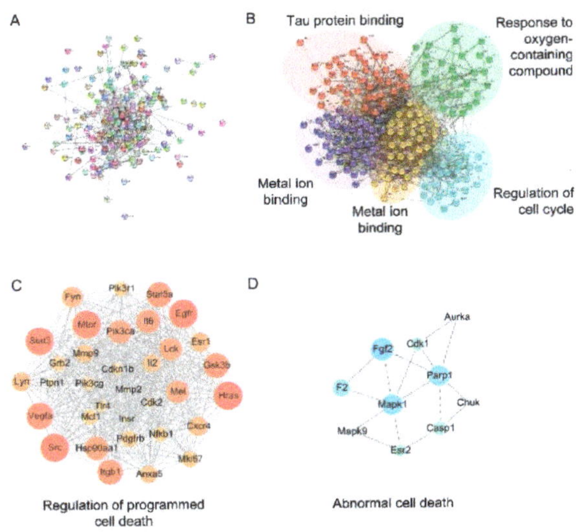

Figure 3. Network cluster analysis. (**A**) PPI network of 183 potential therapeutic targets. (**B**) Network cluster analysis. The main function of each cluster was identified by enrichment analysis. (**C**,**D**) High-scoring subnetworks and their major function.

2.4. Effects of SAL Treatment on Renal Function Parameters from the Serum of SAMP8 Mice

To investigate the protective effect of SAL against aging-related renal damage, biochemical parameters, including BUN and SCr, were evaluated. The serum BUN and SCr levels in the SAMP8 group were significantly higher than in the SAMR1 group ($p < 0.01$; Figure 4A,B). The BUN and SCr levels in the SAL-L, SAL-H, and Fer-1 groups were significantly reduced compared to the SAMP8 group, particularly in the SAL-H group ($p < 0.01$). The serum ALB levels in the SAMP8 group were significantly decreased when compared with the SAMR1 group ($p < 0.001$; Figure 4C). However, the serum ALB levels in the SAL-L, SAL-H, and Fer-1 groups significantly increased when compared with the SAMP8 group, particularly in the SAL-H group ($p < 0.01$). This indicated that senescence-related renal dysfunction had emerged in the SAMP8 group, and SAL had protective effects on the abnormal symptoms.

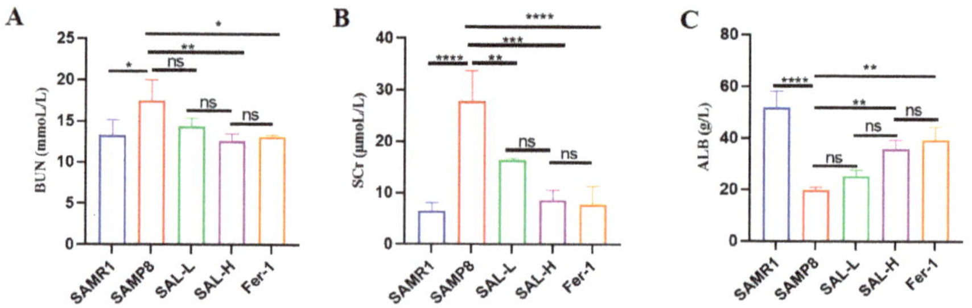

Figure 4. Effects of SAL on renal function parameters in the serum of SAMP8 mice. (**A**) The blood urea nitrogen (BUN) serum level. (**B**) The serum creatinine (SCr) level. (**C**) The albumin (ALB) serum level. Data are expressed as means ± SD, * $p < 0.05$, ** $p < 0.01$, *** $p < 0.001$, **** $p < 0.0001$.

2.5. Effects of SAL Treatment on Histopathological Changes in the Kidneys of SAMP8 Mice

The results of H&E staining revealed that the number of mesangial cells, the volume of glomeruli, and the proliferation of the mesangial matrix were further increased in the SAMP8 group. However, the number of glomerular cells, glomerular volume, and mesangial cell proliferation was significantly reduced in the SAL-L, SAL-H, and Fer-1-treated groups than in the SAMP8 group (Figure 5A). Masson staining also revealed that numerous cyanine-stained collagen fibers and very little red-stained normal kidney tissue appeared in the kidney glomeruli of the SAMP8 group when compared with the SAMR1 group ($p < 0.01$; Figure 5B). This indicated that aging significantly induces glomerular mesangial expansion and mesangial matrix accumulation. However, the SAL-L, SAL-H, and Fer-1 groups showed significantly reduced accumulation of blue-stained collagen fibers in the glomerular basal region than that in the SAMP8 group ($p < 0.05$, $p < 0.01$; Figure 5C). These results suggested that SAL treatment ameliorated renal lesions in the SAMP8 group because of aging.

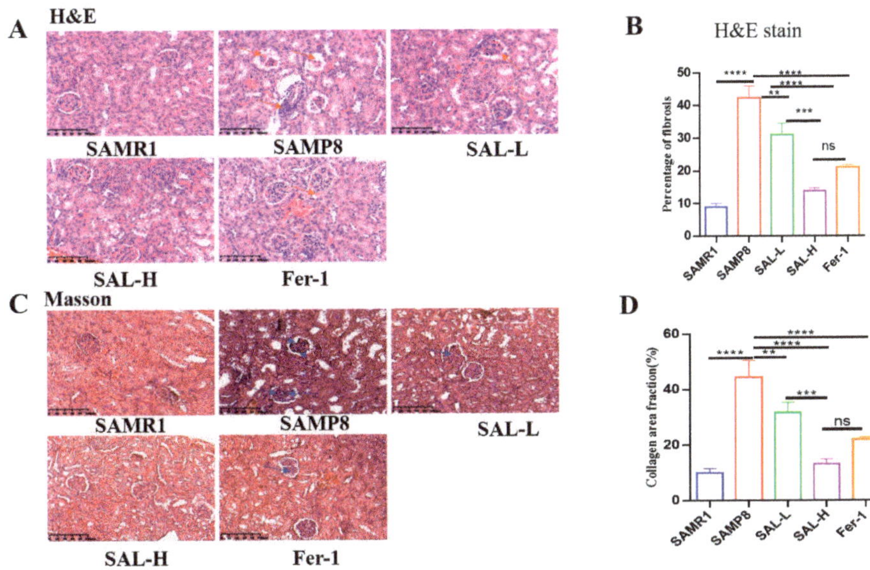

Figure 5. Effects of SAL on histopathological changes in the renal cortex in SAMP8 mice: hematoxylin–eosin (H&E) and Masson staining, 400×. (**A**,**B**) The results of HE staining in the renal cortex. The arrows indicate the glomerular mesangial cells. (**C**,**D**) The results of Masson staining in the renal cortex. The arrows indicate glycoprotein deposition. Data are expressed as means ± SD, ** $p < 0.01$, *** $p < 0.001$, **** $p < 0.0001$.

2.6. SAL Ameliorates Renal Interstitial Fibrosis in SAMP8 Mice

We assessed protein expression levels of the profibrotic factor TGF-β1. PCR and Western blot analysis revealed that TGF-β1 expression was significantly increased in the kidney tissue of mice in the SAMP8 group when compared with the SAMR1 group. TGF-β1 protein expression was significantly decreased in the SAL-L, SAL-H, and Fer-1 groups, (Figure 6A,B). Western blot analysis was used to evaluate the expression of α-SMA expression, which is the main molecular marker for myofibroblasts. The results of the Western blot indicated that α-SMA protein expression in the SAMP8 group was significantly increased compared to that in the SAMR1 group, while the level was significantly decreased after being treated with SAL (Figure 6C,D). These results suggested that SAL mediates the secretion of pro-fibrotic factors and activation of myofibroblasts in aging-induced renal tissue, thereby inhibiting aging-induced renal interstitial fibrosis.

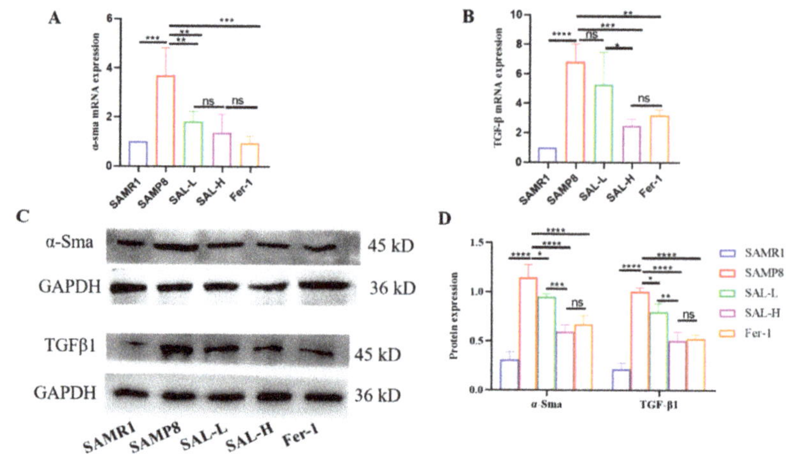

Figure 6. Effects of SAL on expressions of α-SMA and TGF-β1 in renal tissue of SAMP8 mice (Western blot and q-PCR). (**A**) The relative expression of α-SMA. (**B**) The relative expression of TGF-β1. (**C**,**D**) Western blot analysis of α-SMA and TGF-β1 expression, and quantification in the kidney. Data are expressed as means ± SD, * $p < 0.05$, ** $p < 0.01$, *** $p < 0.001$, **** $p < 0.0001$.

2.7. Effects of SAL Treatment in SAMP8 Mice on Lipid Peroxidation and GPX4

We assessed oxidative stress and lipid peroxidation in the kidney tissues of the mice. Malondialdehyde (MDA) was enhanced, while superoxide dismutase (SOD) and glutathione (GSH) were decreased in the kidneys of the SAMP8 group when compared with the SAMR1 group (Figure 7). However, the results in the SAL-L, SAL-H, and Fer-1 groups revealed a decrease in MDA, SOD, and GSH increase in the SAMP8 group (Figure 7A–C). Furthermore, we measured GPX4 expression in the mouse kidneys. Immunohistochemistry showed that GPX4 expression decreased in the kidneys of the SAMP8 group; however, the expression of GPX4 was increased in the SAL and Fer-1 groups (Figure 7D,E). These findings further confirmed that lipid peroxidation and ferroptosis are involved in aging-related kidney disease, and that SAL ameliorated aging-related kidney ferroptosis.

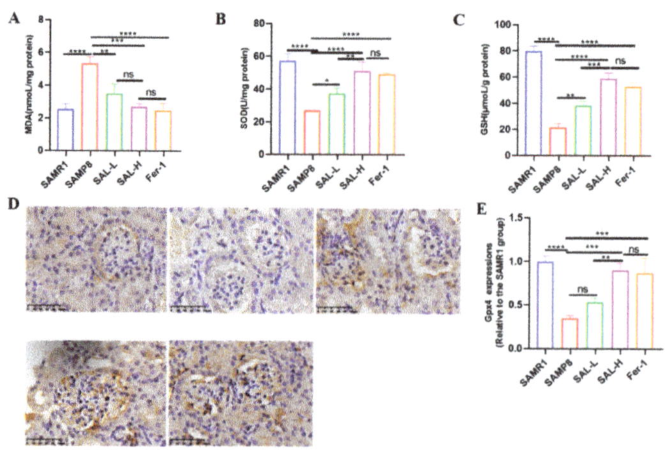

Figure 7. Effects of SAL on the lipid peroxidation and GPX4 levels of SAMP8 mice. (**A**) The MDA levels. (**B**) The SOD levels. (**C**) The GSH levels. (**D**,**E**) The GPX4 levels (IHC). Data are expressed as means ± SD, * $p < 0.05$, ** $p < 0.01$, *** $p < 0.001$, **** $p < 0.0001$.

2.8. SAL Ameliorates Renal Iron Overload in SAMP8 Mice

We measured the iron content in kidney tissue. Iron staining revealed that the iron content in the kidneys of the SAMP8 group was significantly increased (Figure 8A), and the iron content in the renal tubules was relieved after treatment with SAL and Fer-1 (Figure 8B). Western blot showed increased TFR1 expression in the kidneys of the SAMP8 group when compared with the SAMR1 group, while SAL and Fer-1 treatment decreased TFR1. FPN1 and FTH1 expression decreased in the SAMP8 group when compared with the SAMR1 group. SAL and Fer-1 treatment increased the expression of these proteins in the SAMP8 group (Figure 8B,C). These results suggested that aging causes iron overload in renal tubules; however, SAL reduces iron overload in the kidneys of the SAMP8 group.

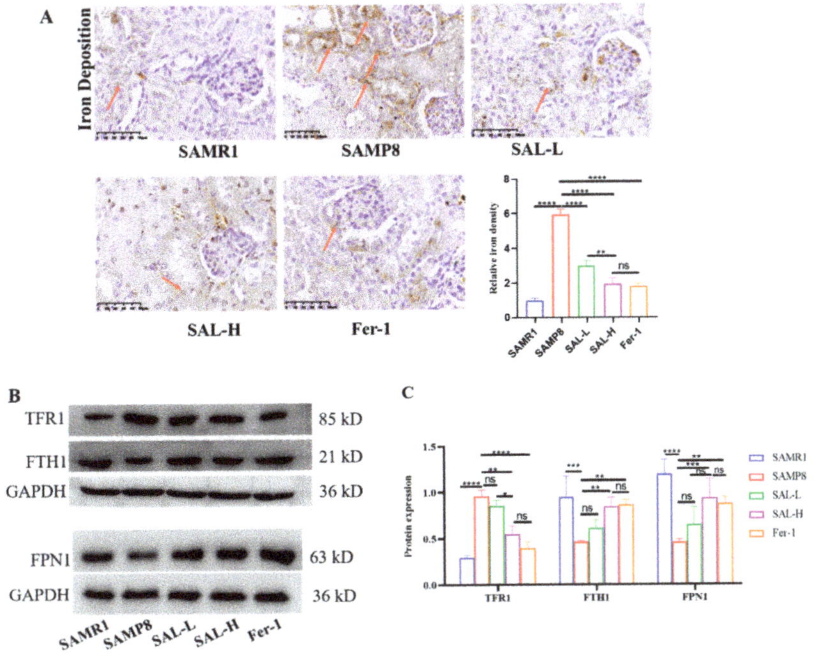

Figure 8. The effects of SAL on kidney iron deposition and the expression of iron transport-related proteins in SAMP8 mice. (**A**) Kidney iron deposition magnification × 400. (**B**,**C**) The expression of FTH1, TFR1, and FPN1 proteins in the kidney of SAMP8 mice. Data are expressed as means ± SD, * $p < 0.05$, ** $p < 0.01$, *** $p < 0.001$, **** $p < 0.0001$.

2.9. SAL Ameliorates Renal Ferroptosis in SAMP8 Mice

To determine whether ferroptosis occurs in the kidneys of SAMP8 mice, and whether SAL attenuates aging-induced kidney damage. The expressions of ACSL4, GPX4, and SLC7A11 were detected. In the SAMP8 group, ACSL4 expression was increased (Figure 9A,D,E), and SLC7A11 and GPX4 expression were decreased (Figure 9B–D,F). SLC7A11 and GPX4 expression were significantly improved, and ACSL4 expression was decreased after treatment with SAL and Fer-1 ($p < 0.05$; Figure 9A,D,E).

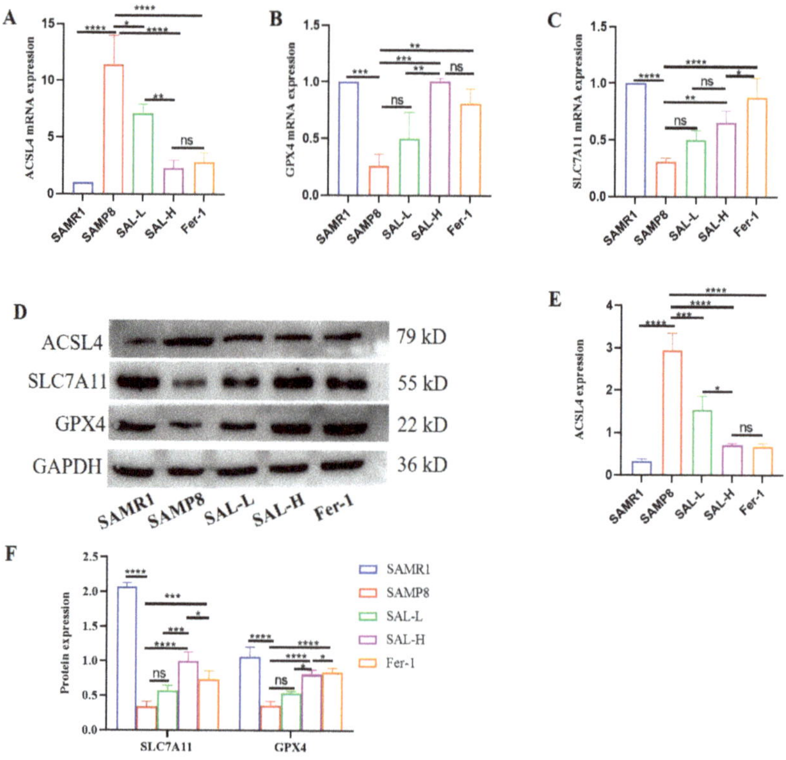

Figure 9. Effects of SAL on the expressions of ACSL4, SLC7A11, and GPX4 in renal tissue of SAMP8 mice (Western blot and q-PCR). (**A**) The relative expression of ACSL4. (**B**) The relative expression of SLC7A11. (**C**) The relative expression of GPX4. (**D–F**) Western blot analysis of ACSL4, SLC7A11, and GPX4 expression, and quantification in the kidney. Data are expressed as means ± SD, * $p < 0.05$, ** $p < 0.01$, *** $p < 0.001$, **** $p < 0.0001$.

2.10. Molecular Docking Analysis

Molecular docking can mimic the binding ability of different bound compounds and proteins. The classical targets of ferroptosis and SAL were used for molecular docking. The details of these proteins are listed in Table 1. The drug-target binding affinity and the best-scored docked position between this SAL and GPX4, SLC7A11, ACSL4, FTH1, TFR1, and FPN1 are indicated in Figure 10. To some extent, the results supported the reliability of WB and revealed the mode of action of SAL on ferroptosis.

Table 1. Potential targets ID.

Uniprot ID	PDB ID	AlphaFord ID	Gene Symbol
P36970	5l71		GPX4
D4ADU2		Q9WTR6	SLC7A11
O35547		Q9QUJ7	ACSL4
P19132	5xb1		FTH1
Q99376		Q62351	TFR1
Q923U9		Q9JHI9	FPN1

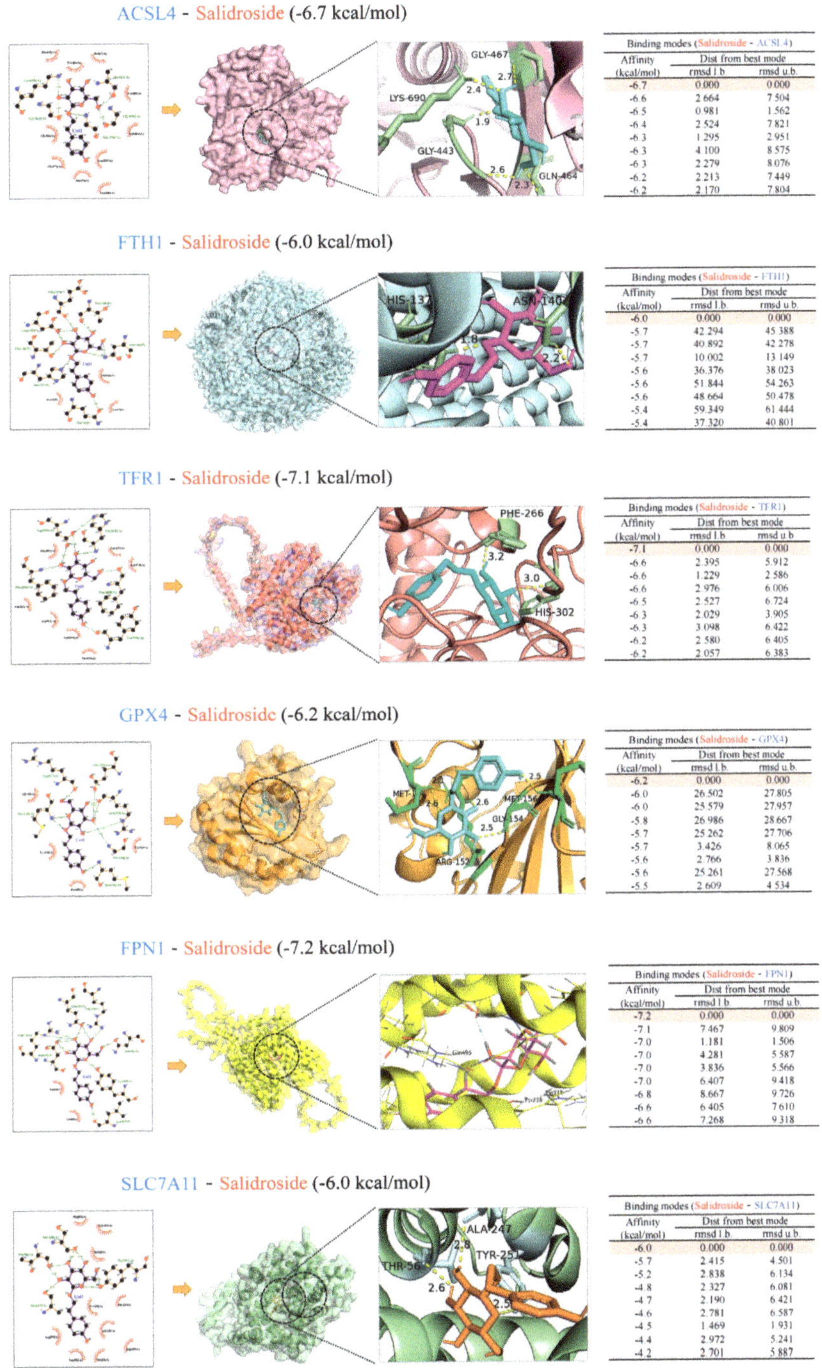

Binding modes (Salidroside - ACSL4)		
Affinity (kcal/mol)	Dist from best mode	
	rmsd l.b.	rmsd u.b.
-6.7	0.000	0.000
-6.6	2.664	7.504
-6.5	0.981	1.562
-6.4	2.524	7.821
-6.3	1.295	2.951
-6.3	4.100	8.575
-6.3	2.279	8.076
-6.2	2.213	7.449
-6.2	2.170	7.804

Binding modes (Salidroside - FTH1)		
Affinity (kcal/mol)	Dist from best mode	
	rmsd l.b.	rmsd u.b.
-6.0	0.000	0.000
-5.7	42.294	45.388
-5.7	40.892	42.278
-5.7	10.002	13.149
-5.6	36.376	38.023
-5.6	51.844	54.263
-5.6	48.664	50.478
-5.4	59.349	61.444
-5.4	37.320	40.801

Binding modes (Salidroside - TFR1)		
Affinity (kcal/mol)	Dist from best mode	
	rmsd l.b.	rmsd u.b.
-7.1	0.000	0.000
-6.6	2.395	5.912
-6.6	1.229	2.586
-6.6	2.976	6.006
-6.5	2.527	6.724
-6.3	2.029	3.905
-6.3	3.098	6.422
-6.2	2.580	6.405
-6.2	2.057	6.383

Binding modes (Salidroside - GPX4)		
Affinity (kcal/mol)	Dist from best mode	
	rmsd l.b.	rmsd u.b.
-6.2	0.000	0.000
-6.0	26.502	27.805
-6.0	25.579	27.957
-5.8	26.986	28.667
-5.7	25.262	27.706
-5.7	3.426	8.065
-5.6	2.766	3.836
-5.6	25.261	27.568
-5.5	2.609	4.534

Binding modes (Salidroside - FPN1)		
Affinity (kcal/mol)	Dist from best mode	
	rmsd l.b.	rmsd u.b.
-7.2	0.000	0.000
-7.1	7.467	9.809
-7.0	1.181	1.506
-7.0	4.281	5.587
-7.0	3.836	5.566
-7.0	6.407	9.418
-6.8	8.667	9.726
-6.6	6.405	7.610
-6.6	7.268	9.318

Binding modes (Salidroside - SLC7A11)		
Affinity (kcal/mol)	Dist from best mode	
	rmsd l.b.	rmsd u.b.
-6.0	0.000	0.000
-5.7	2.415	4.501
-5.2	2.838	6.134
-4.8	2.327	6.081
-4.7	2.190	6.421
-4.6	2.781	6.587
-4.5	1.469	1.931
-4.4	2.972	5.241
-4.2	2.701	5.887

Figure 10. Diagram of molecular docking patterns between SAL and vital ferroptosis protein targets. The docking positions of the binding mode with the best binding energy are shown from global and local perspectives, respectively. The 2D binding modes and detailed docking solutions are presented.

3. Discussion

Aging is a significant risk factor for several common human diseases [29]. CKD has become a growing problem worldwide, as it can lead to end-stage renal disease (ESRD) due to renal fibrosis. Aging is an important factor closely related to the occurrence and progression of renal fibrosis [30]. Therefore, finding suitable drugs to prevent the early onset of renal fibrosis because of aging is necessary to reduce the incidence of age-related kidney diseases. Previous studies have found that renal fibrosis occurs in SAMP8 mice [13], and this phenomenon was also found in this study.

Rhodiola Rosea is a well-known herb, and SAL is one of its main active components and is reported to have anti-aging and antioxidant activities [31]. Recent studies have found that SAL inhibits doxorubicin-induced cardiomyopathy by modulating ferroptosis-dependent pathways [32]. In CKD, the major structural kidney lesions include advanced glomerular and interstitial fibrosis, which ultimately leads to renal fibrosis, causing functional failure. It is well-known that aging is closely associated with changes in kidney structure and function [33]. A previous study showed that accumulation of senescent cells is involved in the development of renal fibrosis by upregulating the secretion of pro-inflammatory mediators and pro-fibrotic factors, ultimately preventing cell regeneration [34]. However, the protective effects of SAL against renal fibrosis because of renal aging are unclear. In this study, we found that BUN and SCr significantly decreased, ALB increased, and renal function improved in the SAMP8 group after SAL treatment. Ni et al. stated that SAL can ameliorate renal interstitial fibrosis in mice with diabetic nephropathy [35]. Rui Li et al. found that SAL treatment significantly reduced the release of inflammatory cytokines and inhibited the TLR4/NF-κB and MAPK signaling pathways. The administration of SAL may be a new therapeutic strategy for the treatment of renal fibrosis [36]. This is consistent with our findings.

With the progress of high-throughput technologies, biomedical data have greatly accumulated and continue to grow rapidly. Network pharmacology provides a feasible way to obtain an overall understanding of Traditional Chinese medicine (TCM) prescriptions from these massive clinical and experimental data [37]. Therefore, the combination of network pharmacology and experimental research is a promising approach for identifying potential targets and uncovering therapeutic mechanisms. In this study, 183 potential therapeutic targets of salidroside were screened by network pharmacology for further research. Enrichment analysis and PPI network analysis found that salidroside can regulate metal ion binding, tau protein binding, and oxygen-containing compounds via the ferroptosis pathway. Based on the network pharmacology findings, the following experimental research was focused on ferroptosis.

Mesangial expansion, massive accumulation of the mesangial matrix, glomerular hypertrophy, glomerular fibrosis, and interstitial fibrosis were significantly improved in the SAMP8 group after SAL treatment. These results suggested that intervention can delay the development of a series of structural lesions that occur in the early onset of renal aging in the SAMP8 group. A previous study showed that SAL inhibited the transformation of renal tubular epithelial cells to myofibroblasts in vitro by inhibiting the expression of α-SMA and TGF-β1 based on the dosage. This study revealed that SAL treatment for 12 weeks significantly attenuated mesangial cell proliferation and matrix deposition in the aged SAMP8 group. Furthermore, the TGF-β1 and α-sma expression levels significantly decreased in the SAMP8 group after 12 weeks of SAL administration. These results indicate that SAL treatment has a protective effect against renal fibrosis progression because of aging in the SAMP8 group. In the kidneys of patients with CKD, iron deposits lead to increased iron uptake and/or insufficient iron output. Iron accumulation triggers Fenton-mediated oxidative damage may lead to renal injury, indicating that CKD renal iron accumulation initially induces ferroptosis and that iron plays a deleterious role in CKD progression. Therefore, regulation of the expression of iron metabolism proteins plays a significance in restoring renal iron metabolism and alleviating ferroptosis.

The accumulation of intracellular iron promotes lipid peroxidation, leading to cell death [38]. The results showed that TFR1 significantly increased in the SAMP8 group, whereas FTH1 and FPN1 expression significantly decreased. After 12 weeks of SAL administration, TFR1 levels decreased, and FTH1 and FPN1 levels increased. Kidney iron staining also suggested that iron deposition in the kidneys of the SAMP8 model increased, and iron deposition reduced after SAL intervention, indicating that SAL can ameliorate renal iron overload in an aging model. Ferroptosis is a form of iron-dependent regulated cell death induced by excessive lipid peroxidation, which is morphologically and mechanistically distinct from apoptosis.

In diabetic nephropathy renal biopsies, ferroptosis-related molecules SlC7A11 and GPX4 expression were reduced compared to in non-DN patients [39]. The involvement of ferroptosis has also been demonstrated in an animal model of streptozotocin-induced DN. Significant changes in markers associated with ferroptosis included decreased GPX4 expression levels and increased ACSL4 expression, lipid peroxidation products, and ferroptosis in DN mice [40]. Iron accumulation in multiple organs, including the brain and kidneys, can lead to increased oxidative damage and decreased function in the aging process [41]. Therefore, iron levels can serve as potential ferroptosis biomarkers, and are an important causative factor of age-related diseases. GPX4 utilizes reduced glutathione to convert lipid hydroperoxides to lipid alcohols, thereby reducing lipid peroxidation and inhibiting ferroptosis [42]. Cysteine is obtained by most cells by importing extracellular cystine via the amino acid transporter SLC7A11 [43]. Interventions in the ferroptosis pathway effectively inhibit the progression of these diseases, suggesting that ferroptosis is a potential therapeutic target [44]. As a redox cycle nitrogen oxide, SAL helps reduce oxidative stress and has been reported to protect against neurodegenerative diseases in many models. A growing body of research has shown that SAL has significant beneficial antioxidant effects in many neurological diseases [28].

MDA is a major aldehyde product of lipid peroxidation. Enzymatic antioxidants include SOD, CAT, and GSH-Px [45]. Recent studies have shown increased MDA and decreased SOD, CAT, and GSH levels in kidney animals with diabetes than those in control kidneys. Our previous study showed that salidroside could inhibit the ferroptosis of neurons in AD mice [46]. In this study, results showed that MDA was elevated and SOD and GSH were decreased in SAMP8 group kidneys. However, SAL treatment inhibited MDA and increased SOD and GSH levels in the kidneys of the SAMP8 group. Furthermore, GPX4 levels were lower in the kidneys of the SAMP8 group than in SAMRI mice, whereas GPX4 levels were elevated in the kidneys of SAL-treated groups. Although GPX4 disruption is repaired by SAL, upregulation of SLC7A11 expression has been reported to cure the disease by inhibiting ferroptosis. SLC7A11 levels decreased in the kidney tissues of the SAMP8 group. However, it was restored by SAL.

This study provides evidence that SAL exerts a protective effect against aging-related kidney disease by inhibiting ferroptosis. Therefore, this study provides a strategy for the treatment of ferroptosis-related diseases that are caused by aging and lays the foundation for the development of new therapeutic drugs for ferroptosis-related diseases. However, this study explored only the relationship between SAL, ferroptosis, and renal interstitial fibrosis in the context of aging, and there is no related research conducted on the mechanisms involved, which need to be further explored in the future.

4. Materials and Methods

4.1. Screen Targets of SAL and Aging-Related Renal Fibrosis

We hypothesize that the targets of SAL that intersect with the targets of aging-related renal fibrosis were potential therapeutic targets of SAL in SAMP8 mice. The potential targets of SAL were retrieved from ChemMapper databases (http://www.lilab-ecust.cn/chemmapper/index.html) (search date: 13 July 2022), PharmMapper databases (http://lilab.ecust.edu.cn/pharmmapper/index.php) (search date: 13 July 2022), Similarity ensemble approach (SEA, http://sea.bkslab.org/) (search date: 14 July 2022), SuperPred databases (http://prediction.charite.de/) (search date: 14 July 2022), and SwissTargetPrediction (http://www.swisstargetprediction.ch/) (search date: 14 July 2022). The therapeutic targets for aging-related renal fibrosis were obtained from the GeneCards databases (https://previous.genecards.org/) (search date: 8 July 2022) and phenolyzer databases (https://phenolyzer.wglab.org/) (search date: 8 July 2022).

4.2. Enrichment Analysis and Network Construction

The enrichment analysis was performed by the SangerBox (http://sangerbox.com (accessed on 1 October 2022)) and cluster profile (http://yixuetongji.top/tools.html (accessed on 1 October 2022)), which can classify molecular function, biological process, cellular components, and the KEGG pathway. The screened common targets between SAL and aging-related renal fibrosis were analyzed to construct a PPI network. The protein subnetworks were constructed by the MCODE algorithm (http://baderlab.org/Software/MCODE (accessed on 1 October 2022)).

4.3. Experimental Design

Three-month-old male SAMP8 mice and SAMR1 mice were obtained from the Department of Medicine (Department of Experimental Animal Science) of Peking University and were fed in the SPF-level experimental animal center of Southern Medical University. The experiment was approved by the Animal Ethics Committee of Southern Medical University (Approval number: L2019185). Groups were designed when mice were 5 months, including the SAMP8+saline (i.g., n = 10), SAMP8 + 30 mg/kg/day salidroside (Macklin, lot: C10739039) (i.g., n = 8), SAMP8 + 60 mg/kg/day salidroside (i.g., n = 8), and SAMP8 + 5 mg/kg/day Ferrostatin-1 (Fer-1, Topscience, lot: 347174-05-4) (i.p., n = 8) groups. Fer-1 is a potent inhibitor of ferroptosis and was used as a positive control [47]. SAMR1 mice were used as the control group (saline, i.g., n = 10). All the mice were treated for 3 months. After the intraperitoneal injection of pentobarbital sodium, blood was collected from the heart. The abdominal cavity was opened and the kidney removed. One kidney tissue was fixed in formalin and embedded in paraffin for pathological staining, while the rest was quickly frozen in liquid nitrogen and preserved at −80 °C for further study.

4.4. Biochemical Analysis

Renal function was assessed by measuring albumin (ALB), blood urea nitrogen (BUN), and serum creatinine (SCr) in mice. Aortic blood obtained from anesthetized mice was used to measure SCr (C011-2-1JianChen Nanjing), BUN (C013-1-1JianChen Nanjing), and ALB (A028-2-1JianChen Nanjing) levels.

4.5. Histology and Morphometry

Kidneys were removed and fixed with 4% paraformaldehyde for 24 h at 4 °C. Sections (5 μm) were cut from paraffin-embedded kidney tissues. Sections were stained with hematoxylin–eosin (HE) and Masson's trichrome staining for h analysis. Masson staining was used to quantify fibrosis. Sections were made to detect collagen fibers. The area of interstitial fibrosis was identified, excluding the vascular area of the area of interest, such as interstitial fibrosis or interstitial fibrosis. Collagen was deposited to the total tissue area. Tubular. ImageJ software (http://rsb.info.nih.gov/ij, access date: 17 April 2022) was used. In addition, a qualitative assessment was included, which was performed blindly by a renal pathologist. The sections for immunohistochemical staining were incubated with a primary antibody against GPX4 overnight at 4 °C, and secondary anti-goat IgG (Proteintech, SA00001-2, Chicago, IL, USA) for 30 min at room temperature. DAB color development, hematoxylin counter-staining, and dehydrated transparent mount section were obtained. The sections were stored in the dark and eventually photographed with digital slice scanner (KF-PRO-005, KFBIO, China).

4.6. Kidney Iron Staining

The paraffin sections (5 μm) were dewaxed, hydrated, and then immersed in TBST containing 3% H_2O_2 for 10 min. The sections were then treated with an equal-ratio mixture of 4% aqueous potassium ferrocyanide and 4% hydrochloric acid for 30 min. Iron staining was amplified with TBS containing 0.025% DAB and 0.0033% H_2O_2 for 10 min. All sections were handled simultaneously to maintain consistency in the dyeing conditions. The sections were stained with hematoxylin, differentiated, sealed, and observed.

4.7. Measurement of Malondialdehyde (MDA), Superoxide Dismutase (SOD), and Glutathione (GSH)

The MDA, SOD, and GSH levels in tissues were detected using a lipid peroxidation MDA assay kit, GSH assay kit, and SOD assay kit (Beyotime, Shanghai, China), respectively, following the instructions of the manufacturer.

4.8. Polymerase Chain Reaction

Total RNA was extracted from the kidney tissues using TRIzol reagent (Invitrogen) and converted to cDNA using the HiScript ® III-RT SuperMix for qPCR (+gDNA wiper) kit (Vazyme Biotech Co., Ltd.); qPCR was performed using ChamQ Universal SYBR qPCR Master Mix (Vazyme Biotech Co., Ltd.) in Roche LightCycler96. The relative mRNA levels were calculated by normalization to the GAPDH levels (B661304, Sangon Biotech, Shanghai, China). Relative gene expression was analyzed based on fold-change ($2^{-\Delta\Delta Ct}$ method). Primer sequences were shown in Table 2.

Table 2. Gene primer sequence used for qRT-PCR.

GPX4	Forward primer	CCGCCGAGATGAGCTGG
	Reverse primer	GTCGATGTCCTTGGCTGAG
SLC7A11	Forward primer	ATGGTCAGAAAGCCAGTTGTG
	Reverse primer	CAGGGCGTATTACGAGCAGT
ACSL4	Forward primer	TCCCTGGACTAGGACCGAAG
	Reverse primer	GGGGCGTCATAGCCTTTCTT
TGF-β	Forward primer	CCAGATCCTGTCCAAACTAAGG
	Reverse primer	CTCTTTAGCATAGTAGTCCGCT
α-sma	Forward primer	CTATGAGGGCTATGCCTTGCC
	Reverse primer	GCTCAGCAGTAGTAACGAAGGA
β-actin	Forward primer	CCACCATGTACCCAGGCATT
	Reverse primer	CGGACTCATCGTACTCCTGC

4.9. Western Blotting

The kidney tissues of mice in each group were added with protein lysate, and the total protein was extracted after homogenization. The protein concentration of each group was calculated by the BCA method and denatured at 5 × loading buffer at 95 °C for 10 min. The total protein loading of each group was 20 μg. Protein samples were electrophoresed in SDS–polyacrylamide gel electrophoresis (SDS-PAGE) and transferred onto PVDF membranes (Millipore, Billerica, MA, USA). The blots were blocked with 5% skimmed milk. Then, they were probed with primary antibodies: GPX4 (ab125066, 1:2000, Abcam, Cambridge Science Park, Cambridge, UK), SLC7A11 (26864-1-AP, 1:1000, Proteintech, Chicago, IL, USA), ACSL4 (ab155282, 1:10,000, Abcam, Cambridge Science Park, Cambridge, UK), TFR1 (bs-21319R, 1:1000, Bioss, Bejing, China), FTH1 (#3998, 1:1000, CST, Boston, MA, USA), FPN1 (2660, 1-1-AP, Proteintech, Chicago, IL, USA), α-sma (# 19245, 1:2000, CST, Boston, MA, USA), and TGF-β1 (#84912, 1:1000, CST, Boston, MA, USA) overnight at 4 °C. The blots were washed and incubated for 1 h at room temperature with the HRP-conjugated secondary antibody and then developed with enhanced chemiluminescence reagents (E412-01, Vazyme, Nangjing, Jiangsu, China). The densitometry of the protein bands was quantified using ImageJ software.

4.10. Molecular Docking

The main compounds of SAL and aging-related renal fibrosis protein targets were analyzed by molecular docking using the AutoDock Vina and AutoDock. The 3D structures of key protein targets were obtained from the RCSB Protein Data Bank (PDB) and AlphFold databases. The figures of the active binding site were generated with the PyMOL software.

4.11. Statistical Analysis

Quantitative data are represented as mean ± Standard Deviation (SD). Statistical analyses were completed using SPSS version 25. Comparison among groups was analyzed using a one-way analysis of variance (ANOVA) test with Tukey post hoc multiple comparisons. Results were considered statistically significant when $p < 0.05$. Data were tabulated and plotted using GraphPad Prism, version 8.

5. Conclusions

After the intervention of SAL, renal fibrosis and ferroptosis in 8-month-old SAMP8 mice can be alleviated, which may be related to regulating renal iron metabolism, reducing iron deposition, regulating SLC7A11 and GPX4 protein expression, and finally alleviating renal ferroptosis. SAL delays renal aging and inhibits aging-related glomerular fibrosis by inhibiting ferroptosis in SAMP8 mice.

Supplementary Materials: The following supporting information can be downloaded at: https://www.mdpi.com/article/10.3390/molecules27228039/s1, Table S1: The targets of SAL and 2586 aging-related renal fibrosis.

Author Contributions: Conceptualization, supervision, project administration, funding acquisition, W.C. and W.X.; methodology, software, validation, formal analysis, writing, original draft preparation, S.Y. (Sixia Yang), T.P. and L.W.; investigation, review and editing, Y.Z., W.L. and S.Y. (Shihua Yan). All authors have read and agreed to the published version of the manuscript.

Funding: This research was funded by the National Natural Science Foundation of China, (82174149, 81973641, 81973804), the Administration of Traditional Chinese Medicine Project (20200505160744).

Institutional Review Board Statement: The animal study protocol was approved by Experimental Animal Ethics Committee of Southern Medical University (protocol code L2019185, approval date: 14 October 2019).

Informed Consent Statement: Not applicable.

Data Availability Statement: Data are contained within the article and Supplementary Materials.

Conflicts of Interest: The authors declare no conflict of interest.

Sample Availability: Samples of the compounds are available from the authors.

References

1. Partridge, L.; Deelen, J.; Slagboom, P.E. Facing up to the global challenges of ageing. *Nature* **2018**, *561*, 45–56. [CrossRef] [PubMed]
2. Xie, Y.; Bowe, B.; Mokdad, A.H.; Xian, H.; Yan, Y.; Li, T.; Maddukuri, G.; Tsai, C.Y.; Floyd, T.; Al-Aly, Z. Analysis of the Global Burden of Disease study highlights the global, regional, and national trends of chronic kidney disease epidemiology from 1990 to 2016. *Kidney Int.* **2018**, *94*, 567–581. [CrossRef] [PubMed]
3. O'Sullivan, E.-D.; Hughes, J.; Ferenbach, D.-A. Renal Aging: Causes and Consequences. *J. Am. Soc. Nephrol.* **2017**, *28*, 407–420. [CrossRef]
4. Glassock, R.J.; Warnock, D.G.; Delanaye, P. The global burden of chronic kidney disease: Estimates, variability and pit-falls. *Nat. Rev. Nephrol.* **2017**, *13*, 104–114. [CrossRef]
5. Webster, A.C.; Nagler, E.V.; Morton, R.L.; Masson, P. Chronic Kidney Disease. *Lancet* **2017**, *389*, 1238–1252. [CrossRef]
6. Takeda, T. Senescence-accelerated mouse (SAM) with special references to neurodegeneration models, SAMP8 and SAMP10 mice. *Neurochem. Res.* **2009**, *34*, 639–659. [CrossRef]
7. Takeda, T. Senescence-accelerated mouse (SAM): With special reference to age-associated pathologies and their modulation. *Nihon Eiseigaku Zasshi* **1996**, *51*, 569–578. [CrossRef] [PubMed]

8. Takeda, T.; Matsushita, T.; Kurozumi, M.; Takemura, K.; Higuchi, K.; Hosokawa, M. Pathobiology of the senes-cence-accelerated mouse (SAM). *Exp. Gerontol.* **1997**, *32*, 117–127. [CrossRef]
9. Taniguchi, S.; Hanafusa, M.; Tsubone, H.; Takimoto, H.; Yamanaka, D.; Kuwahara, M.; Ito, K. Age-dependency of the serum oxidative level in the senescence-accelerated mouse prone 8. *J. Vet. Med. Sci.* **2016**, *78*, 1369–1371. [CrossRef]
10. Yoshida, N.; Endo, J.; Kinouchi, K.; Kitakata, H.; Moriyama, H.; Kataoka, M.; Yamamoto, T.; Shirakawa, K.; Morimoto, S.; Nishiyama, A.; et al. (Pro)renin receptor accelerates develop-ment of sarcopenia via activation of Wnt/YAP signaling axis. *Aging Cell* **2019**, *18*, e12991. [CrossRef]
11. Karuppagounder, V.; Arumugam, S.; Babu, S.S.; Palaniyandi, S.S.; Watanabe, K.; Cooke, J.P.; Thandavarayan, R. The senescence accelerated mouse prone 8 (SAMP8): A novel murine model for cardiac aging. *Ageing Res. Rev.* **2017**, *35*, 291–296. [CrossRef]
12. Li, Y.; Zhang, D.; Li, L.; Han, Y.; Dong, X.; Yang, L.; Li, X.; Li, W.; Li, W. Ginsenoside Rg1 ameliorates aging-induced liver fibrosis by in-hibiting the NOX4/NLRP3 inflammasome in SAMP8 mice. *Mol. Med. Rep.* **2021**, *24*, 801. [CrossRef]
13. Shen, X.; Dong, X.; Han, Y.; Li, Y.; Ding, S.; Zhang, H.; Sun, Z.; Yin, Y.; Li, W.; Li, W. Ginsenoside Rg1 ameliorates glomerular fibrosis during kidney aging by inhibiting NOX4 and NLRP3 inflammasome activation in SAMP8 mice. *Int. Immunopharmacol.* **2020**, *82*, 106339. [CrossRef] [PubMed]
14. Miao, J.; Liu, J.; Niu, J.; Zhang, Y.; Shen, W.; Luo, C.; Liu, Y.; Li, C.; Li, H.; Yang, P.; et al. Wnt/β-catenin/RAS signaling mediates age-related renal fibrosis and is associated with mitochondrial dysfunction. *Aging Cell* **2019**, *18*, e13004. [CrossRef] [PubMed]
15. Nakahara, H.; Kanno, T.; Inai, Y.; Utsumi, K.; Hiramatsu, M.; Mori, A.; Packer, L. Mitochondrial dysfunction in the senescence accelerated mouse (SAM). *Free. Radic. Biol. Med.* **1998**, *24*, 85–92. [CrossRef]
16. Cao, B.; Zeng, M.; Si, Y.; Zhang, B.; Wang, Y.; Xu, R.; Huang, Y.; Feng, W.; Zheng, X. Extract of Corallodiscus flabellata attenuates renal fibrosis in SAMP8 mice via the Wnt/β-catenin/RAS signaling pathway. *BMC Complement. Med. Ther.* **2022**, *22*, 52. [CrossRef]
17. Baltanás, A.; Solesio, M.E.; Zalba, G.; Galindo, M.F.; Fortuño, A.; Jordán, J. The senescence-accelerated mouse prone-8 (SAM-P8) oxidative stress is associated with upregulation of renal NADPH oxidase system. *J. Physiol. Biochem.* **2013**, *69*, 927–935. [CrossRef]
18. Zeng, Y.; Wang, P.H.; Zhang, M.; Du, J.R. Aging-related renal injury and inflammation are associated with downregulation of Klotho and induction of RIG-I/NF-κB signaling pathway in senescence-accelerated mice. *Aging Clin. Exp. Res.* **2016**, *28*, 69–76. [CrossRef]
19. Huang, Y.; Wu, B.; Shen, D.; Chen, J.; Yu, Z.; Chen, C. Ferroptosis in a sarcopenia model of senescence accelerated mouse prone 8 (SAMP8). *Int. J. Biol. Sci.* **2021**, *17*, 151–162. [CrossRef]
20. Belavgeni, A.; Meyer, C.; Stumpf, J.; Hugo, C.; Linkermann, A. Ferroptosis and Necroptosis in the Kidney. *Cell Chem. Biol.* **2020**, *27*, 448–462. [CrossRef]
21. Yang, W.S.; Stockwell, B.R. Ferroptosis: Death by Lipid Peroxidation. *Trends Cell Biol.* **2016**, *26*, 165–176. [CrossRef] [PubMed]
22. Wei, X.; Yi, X.; Zhu, X.H.; Jiang, D.S. Posttranslational Modifications in Ferroptosis. *Oxid. Med. Cell. Longev.* **2020**, *2020*, 8832043. [CrossRef] [PubMed]
23. Tang, D.; Kroemer, G. Ferroptosis. *Curr. Biol.* **2020**, *30*, R1292–R1297. [CrossRef]
24. Ajoolabady, A.; Aslkhodapasandhokmabad, H.; Libby, P.; Tuomilehto, J.; Lip, G.Y.; Penninger, J.M.; Richardson, D.R.; Tang, D.; Zhou, H.; Wang, S.; et al. Ferritinophagy and ferroptosis in the management of metabolic diseases. *Trends Endocrinol. Metab.* **2021**, *32*, 444–462. [CrossRef] [PubMed]
25. Bogdan, A.R.; Miyazawa, M.; Hashimoto, K.; Tsuji, Y. Regulators of Iron Homeostasis: New Players in Metabolism, Cell Death, and Disease. *Trends Biochem. Sci.* **2016**, *41*, 274–286. [CrossRef] [PubMed]
26. Fan, F.; Yang, L.; Li, R.; Zou, X.; Li, N.; Meng, X.; Zhang, Y.; Wang, X. Salidroside as a potential neuroprotective agent for ischemic stroke: A review of sources, pharmacokinetics, mechanism and safety. *Biomed. Pharmacother.* **2020**, *129*, 110458. [CrossRef] [PubMed]
27. Amsterdam, J.D.; Panossian, A.G. Rhodiola rosea L. as a putative botanical antidepressant. *Phytomedicine* **2016**, *23*, 770–783. [CrossRef]
28. Li, R.; Chen, J. Salidroside Protects Dopaminergic Neurons by Enhancing PINK1/Parkin-Mediated Mitophagy. *Oxid. Med. Cell. Longev.* **2019**, *2019*, 9341018. [CrossRef]
29. Davalli, P.; Mitic, T.; Caporali, A.; Lauriola, A.; D'Arca, D. ROS, Cell Senescence, and Novel Molecular Mechanisms in Aging and Age-Related Diseases. *Oxid. Med. Cell. Longev.* **2016**, *2016*, 3565127. [CrossRef]
30. Humphreys, B.D. Mechanisms of Renal Fibrosis. *Annu. Rev. Physiol.* **2018**, *80*, 309–326. [CrossRef]
31. Jin, H.; Pei, L.; Shu, X.; Yang, X.; Yan, T.; Wu, Y.; Wei, N.; Yan, H.; Wang, S.; Yao, C.; et al. Therapeutic Intervention of Learning and Memory Decays by Salidroside Stimulation of Neurogenesis in Aging. *Mol. Neurobiol.* **2016**, *53*, 851–866. [CrossRef]
32. Chen, H.; Zhu, J.; Le, Y.; Pan, J.; Liu, Y.; Liu, Z.; Wang, C.; Dou, X.; Lu, D. Salidroside inhibits doxorubicin-induced cardiomyopathy by modulating a ferroptosis-dependent pathway. *Phytomedicine* **2022**, *99*, 153964. [CrossRef] [PubMed]
33. Anders, H.J.; Huber, T.B.; Isermann, B.; Schiffer, M. CKD in diabetes: Diabetic kidney disease versus nondiabetic kidney disease. *Nat. Rev. Nephrol.* **2018**, *14*, 361–377. [CrossRef] [PubMed]
34. Gong, W.; Luo, C.; Peng, F.; Xiao, J.; Zeng, Y.; Yin, B.; Chen, X.; Li, S.; He, X.; Liu, Y.; et al. Brahma-related gene-1 promotes tubular senescence and renal fibrosis through Wnt/beta-catenin/autophagy axis. *Clin. Sci.* **2021**, *135*, 1873–1895. [CrossRef] [PubMed]
35. Ni, J.; Li, Y.; Xu, Y.; Guo, R. Salidroside protects against cardiomyocyte apoptosis and ventricular remodeling by AKT/HO-1 signaling pathways in a diabetic cardiomyopathy mouse model. *Phytomedicine* **2021**, *82*, 153406. [CrossRef] [PubMed]

36. Li, R.; Guo, Y.; Zhang, Y.; Zhang, X.; Zhu, L.; Yan, T. Salidroside Ameliorates Renal Interstitial Fibrosis by Inhibiting the TLR4/NF-kappa B and MAPK Signaling Pathways. *Int. J. Mol. Sci.* **2019**, *20*, 1103. [CrossRef]
37. Wang, X.; Wang, Z.Y.; Zheng, J.H.; Li, S. TCM network pharmacology: A new trend towards combining computational, experimental and clinical approaches. *Chin. J. Nat. Med.* **2021**, *19*, 1–11. [CrossRef]
38. Bai, T.; Lei, P.; Zhou, H.; Liang, R.; Zhu, R.; Wang, W.; Zhou, L.; Sun, Y. Sigma-1 receptor protects against ferroptosis in hepatocellular carcinoma cells. *J. Cell Mol. Med.* **2019**, *23*, 7349–7359. [CrossRef]
39. Kim, S.; Kang, S.W.; Joo, J.; Han, S.H.; Shin, H.; Nam, B.Y.; Park, J.; Yoo, T.H.; Kim, G.; Lee, P.; et al. Characterization of ferroptosis in kidney tubular cell death under diabetic conditions. *Cell Death Dis.* **2021**, *12*, 160. [CrossRef]
40. Tan, H.; Chen, J.; Li, Y.; Li, Y.; Zhong, Y.; Li, G.; Liu, L.; Li, Y. Glabridin, a bioactive component of licorice, ameliorates diabetic nephropathy by regulating ferroptosis and the VEGF/Akt/ERK pathways. *Mol. Med.* **2022**, *28*, 58. [CrossRef]
41. Fang, X.; Cai, Z.; Wang, H.; Han, D.; Cheng, Q.; Zhang, P.; Gao, F.; Yu, Y.; Song, Z.; Wu, Q.; et al. Loss of Cardiac Ferritin H Facilitates Cardiomyopathy via Slc7a11-Mediated Ferroptosis. *Circ. Res.* **2020**, *127*, 486–501. [CrossRef] [PubMed]
42. Sha, R.; Xu, Y.; Yuan, C.; Sheng, X.; Wu, Z.; Peng, J.; Wang, Y.; Lin, Y.; Zhou, L.; Xu, S.; et al. Predictive and prognostic impact of ferroptosis-related genes ACSL4 and GPX4 on breast cancer treated with neoadjuvant chemotherapy. *Ebiomedicine* **2021**, *71*, 103560. [CrossRef] [PubMed]
43. Weigand, I.; Schreiner, J.; Rohrig, F.; Sun, N.; Landwehr, L.S.; Urlaub, H.; Kendl, S.; Kiseljak-Vassiliades, K.; Wierman, M.E.; Angeli, J.P.F.; et al. Active steroid hormone synthesis renders adrenocortical cells highly susceptible to type II ferroptosis induction. *Cell Death Dis.* **2020**, *11*, 192. [CrossRef]
44. Lei, G.; Zhang, Y.; Koppula, P.; Liu, X.; Zhang, J.; Lin, S.H.; Ajani, J.A.; Xiao, Q.; Liao, Z.; Wang, H.; et al. The role of ferroptosis in ionizing radiation-induced cell death and tumor suppression. *Cell Res.* **2020**, *30*, 146–162. [CrossRef] [PubMed]
45. Su, L.J.; Zhang, J.H.; Gomez, H.; Murugan, R.; Hong, X.; Xu, D.; Jiang, F.; Peng, Z.Y. Reactive Oxygen Species-Induced Lipid Peroxidation in Apoptosis, Autophagy, and Ferroptosis. *Oxid. Med. Cell. Longev.* **2019**, *2019*, 5080843. [CrossRef]
46. Yang, S.; Xie, Z.; Pei, T.; Zeng, Y.; Xiong, Q.; Wei, H.; Wang, Y.; Cheng, W. Salidroside attenuates neuronal ferroptosis by activating the Nrf2/HO1 signaling pathway in $A\beta_{1\text{-}42}$-induced Alzheimer's disease mice and glutamate-injured HT22 cells. *Chin. Med.* **2022**, *17*, 82. [CrossRef]
47. Jiang, X.; Stockwell, B.R.; Conrad, M. Ferroptosis: Mechanisms, biology and role in disease. *Nat. Rev. Mol. Cell Biol.* **2021**, *22*, 266–282. [CrossRef]

molecules

MDPI

Article

Can Dietary n-3 Polyunsaturated Fatty Acids Affect Apelin and Resolvin in Testis and Sperm of Male Rabbits?

Simona Mattioli [1], Elena Moretti [2,*], Cesare Castellini [1], Cinzia Signorini [2], Roberta Corsaro [2], Elisa Angelucci [1] and Giulia Collodel [2]

1 Department of Agricultural, Environmental, and Food Science, University of Perugia, Borgo XX Giugno 74, 06123 Perugia, Italy; simona.mattioli@unipg.it (S.M.); cesare.castellini@unipg.it (C.C.); elisa.angelucci@unipg.it (E.A.)

2 Department of Molecular and Developmental Medicine, University of Siena, Policlinico Santa Maria alle Scotte, Viale Bracci 14, 53100 Siena, Italy; cinzia.signorini@unisi.it (C.S.); roberta.corsaro@student.unisi.it (R.C.); giulia.collodel@unisi.it (G.C.)

* Correspondence: elena.moretti@unisi.it; Tel.: +39-05-7723-2452; Fax: +39-05-7723-3527

Abstract: Apelin and other novel adipokines have been associated with normal and pathological reproductive conditions in humans and animals. In this paper, we used a rabbit model to investigate if apelin and resolvin (RvD1) in testis and sperm are associated with the oxidative status of semen and serum testosterone of rabbits fed different diets enriched with flaxseed (alpha-linolenic acid, ALA) or with fish oil (eicosapentaenoic acid, EPA, docosapentaenoic acid, DPAn-3, and docosahexaenoic acid, DHA). Apelin and RvD1 were detected by ELISA and apelin and the apelin receptor by immunofluorescence. Increased levels of apelin in testes from both enriched diets were shown, particularly in the interstitial tissue of the FLAX group. The FLAX diet enhanced serum testosterone, and both enriched diets showed higher levels of malondialdehyde and RvD1 in the testis. In ejaculated sperm, apelin and its receptor were localized in the entire tail of the control and both treated groups. The ryanodine receptor was investigated in rabbit testis; the fluorescent signal was increased in mature elongated spermatids of the FLAX group. In conclusion, this data seems to indicate that FLAX increases the amount of apelin in testis, suggesting an involvement of this adipokine in male reproduction and probably a role in the resolution of the inflammatory status.

Keywords: angiotensin-like-receptor 1; apelin; diet; ejaculated sperm; oxidative stress; PUFA; resolvins; ryanodine receptor; testes

Citation: Mattioli, S.; Moretti, E.; Castellini, C.; Signorini, C.; Corsaro, R.; Angelucci, E.; Collodel, G. Can Dietary n-3 Polyunsaturated Fatty Acids Affect Apelin and Resolvin in Testis and Sperm of Male Rabbits? *Molecules* **2023**, *28*, 6188. https://doi.org/10.3390/molecules28176188

Academic Editors: José Virgílio Santulhão Pinela, Carla Susana Correia Pereira, Maria Inês Moreira Figueiredo Dias, Zhaojun Wei and José Ignacio Alonso-Esteban

Received: 17 July 2023
Revised: 12 August 2023
Accepted: 17 August 2023
Published: 22 August 2023

1. Introduction

The lipid profile of a diet deeply affects the reproductive functions of humans and other animal species [1]. The main effects are related to several changes in the lipid profile of germ cells and reproductive tissues, and to modification of hormonal assets and the cascade of some lipid-derived molecules (isoprostanes, sterols, etc.) [2]. Oxidation and inflammation are relevant endpoints of this complex interaction among dietary lipids and derived molecules since dietary lipids modulate the fatty acids profile of tissues and, thus, their oxidative onset [3].

Supplemental dietary n-3 PUFA improved sperm motion traits and resulted in an enrichment of membrane fatty acid in the sperm and testes of the rabbits [2]. Testes showed high gene expression of enzymes involved in PUFA synthesis, as Δ6 desaturase (FADS2), elongase (ELOVL)2, and ELOVL5, and the low expression of Δ5 desaturase (FADS1). Intermediate metabolites, enzymes, and final products were found differently in Leydig, Sertoli, and germinal cells [4].

In recent studies, adipokines, such as adiponectin, chemerin, visfatin, resistin, and omentin, have been associated with normal and pathological reproductive conditions in humans and animals [5–7].

Apelin is an endogenous peptide that is expressed in the brain, placenta, heart, lungs, kidneys, pancreas, testis, prostate, and adipose tissues. It has been suggested that apelin affects energy metabolism [8] and the release of inflammatory mediators [9]. Apelin also inhibits the release of reactive oxygen species (ROS) in adipocytes, promotes the expression of antioxidant enzymes [10], and seems to play a protective role in the progression of lymphatic tumors. Moreover, apelin is locally synthesized in the hypothalamus, pituitary gland, ovaries, and testes of many species, showing autocrine and/or paracrine effects [11]. The receptor of apelin, called angiotensin-like-receptor 1 (APJ), and the apelinergic system are expressed in several tissues of the brain, spleen, placenta, heart, liver, intestine, prostate, thymus, ovary, lung, kidney, stomach, adipose tissue, and testis, where many pleiotropic effects are observed.

Moustafa [12] observed that in rats, the administration of n-3 or n-6 fatty acids impaired steroidogenesis but improved the antioxidant and anti-inflammatory status of the reproductive system via modulation of adipokines in the testicles. Das et al. [13] indicated that apelin may be a signal during the early postnatal stage for the regulation of germ cell proliferation, apoptosis, and expression of androgen receptors. Accordingly, it seems that the diet and the availability of some nutrients (e.g., fatty acids) have a relevant role in sperm physiology and in the homeostasis of the apelinergic system [14].

Recently apelin/APJ-R system has been localized in human spermatozoa and testicular tissue and is probably involved in human fertility [15].

At the same time, it appears that other molecules derived from omega-3 fatty acids (eicosapentaenoic acid, EPA, docosapentaenoic acid, DPAn-3, and docosahexaenoic acid, DHA), like resolvin (Rv)D1, play a major role in promoting the restoration of normal cellular function following an inflammation occurring after tissue injury [16,17].

RvD1 amount was positively correlated with F_2-IsoPs and reduced sperm quality, and it increased along with other markers of oxidative stress and inflammation as fatty acids content and clinical biomarkers [18].

In this paper, we used the rabbit as a model to investigate if the apelinergic system (apelin and APJ) in sperm and testis and RvD1 may be associated with the fatty acids profile and the oxidative status semen and serum testosterone of rabbits fed different diets enriched with flaxseed, which has a very high alpha-linolenic acid (ALA) level, or with fish oil that directly supplies ALA derivatives (EPA, DHA, and DPA). The possible involvement of the ryanodine receptor (RyR), a lipid mediator of resolution of inflammation, the major cellular mediator of induced calcium release, was also detected.

2. Results

Testosterone level, sperm parameters and fatty acid profile of sperm, and oxidative status are shown in Table 1.

Table 1. Serum testosterone (pg/mL) and sperm characteristics (progressive motility %, curvilinear velocity, VCL µm/sec, malondialdehyde, MDA nmol MDA/mL, fatty acids, and FA % of total FA) of rabbits fed experimental diets. Polyunsaturated fatty acids, PUFA; very long chain PUFA, VLCP.

Groups	Serum Testosterone (pg/mL)	Sperm Motility (%)	VCL (µm/sec)	MDA (nmol MDA/mL)	n-3 PUFA (% of Total FA)	n-6 PUFA (% of Total FA)	n-3 VLCP (% of Total FA)	n-6 VLCP (% of Total FA)
Control	3.63 b	62.13 a	184.7 ab	2.82 a	0.63 a	38.56 c	0.39 a	26.12 c
FLAX	4.60 c	76.31 b	236.5 b	14.79 b	4.25 b	22.45 b	2.83 b	18.31 a
FISH	2.82 a	77.27 b	226.9 b	20.64 c	13.12 c	19.06 a	12.62 c	23.62 b
SE	0.15	1.29	7.05	0.31	0.15	0.21	0.11	0.29
p-value	<0.001	<0.001	<0.001	<0.001	<0.001	<0.001	<0.001	<0.001

On the same column with different letters (a–c) means $p < 0.001$.

The FLAX group showed the highest testosterone concentration, whereas the FISH group showed the lowest level (3.63 vs. 4.60 and 2.82 pg/mL, respectively, in the control, FLAX, and FISH groups).

The sperm motility and curvilinear velocity (VCL) increased in the rabbits fed FLAX and FISH diets.

Sperm MDA, as well as n-3 PUFA and n-3 VLCP, were higher in enriched dietary groups compared to the control. In the latter groups, n-6 PUFA and n-6 VLCP resulted in significantly lower values with respect to the control (Table 1).

Apelin and RvD1, as well as MDA level, n-3 and n-6 PUFA, and n-3 and n-6 VLCP in testis, are shown in Table 2. Apelin resulted higher in the FLAX group compared to the control and FISH groups. RvD1 showed higher values in both enriched dietary groups with respect to that found in the control. MDA, n-3 PUFA, and n-3 VLCP were higher in the FLAX and FISH groups compared to the control, while n-6 PUFA and n-6 LCP resulted in lower values (Table 2).

Table 2. Apelin (ng/g), resolvin (RvD1) (pg/g), malondialdehyde MDA (nmol/g), polyunsaturated fatty acids (PUFA), and vary long chain PUFA (VLCP %) in testes of rabbits fed experimental diets.

Groups	Apelin (ng/g)	RvD1 (pg/g)	MDA (nmol/g)	n-3 PUFA (%)	n-6 PUFA (%)	n-3 VLCP (%)	n-6 VLCP (%)
Control	1.187 a	111.578 a	34.755 a	2.305 a	38.695 b	0.078 a	26.525 b
FLAX	2.400 b	382.058 b	42.295 c	7.747 c	32.428 a	1.602 c	18.700 a
FISH	1.166 a	394.157 b	37.370 b	4.470 b	32.262 a	0.950 bc	17.930 a
SE	0.197	20.971	0.270	0.192	0.312	0.036	0.325
p value	<0.001	<0.001	<0.001	<0.001	<0.001	<0.001	<0.001

On the same column with different letters (a–c) means $p < 0.001$.

The correlation among the previous traits is shown in Table 3 and Supplementary Table S1. A high positive ($p > 0.01$) correlation was recorded between apelin, RvD1, and lipid oxidation in the testis (MDA), whereas only the RvD1 was positively correlated with the sperm lipid oxidation (MDA). RvD1 also showed a positive correlation with the sperm kinetic parameters (VCL and motility), whereas the apelin was positively correlated with serum testosterone (T, $p < 0.01$) and VCL ($p < 0.05$).

n-3 PUFA in the testis showed a higher positive correlation with these latter proteins, whereas the sperm n-3 PUFA was only correlated with RvD1 ($p < 0.01$). On the contrary, n-6 PUFA was negatively correlated only with the RvD1 in both testis and sperm ($p < 0.01$). Testis and sperm lipid oxidation exhibited a strong correlation with the kinetic traits of sperm. As expected, n-6 and n-3 PUFA were inversely correlated both in the testis and sperm.

The n-3 intake (estimated considering that the feed intake was no different between groups, ~150 g/d) was correlated with almost all the parameters (positive with apelin, RvD1, MDA, T, VCL, Motility, and n-3 testis profile; negative with n-6 PUFA of sperm and testis) except n-3 PUFA of sperm, whereas n-6 intake was significantly correlated with apelin, testis MDA, and T.

Immunofluorescence performed on ejaculated sperm showed that apelin was located on the entire tail (Figure 1a,b) of the sample from the control (Figure 1a) and enriched dietary groups (e.g., FLAX; Figure 1b). The APJ receptor was also detected in the tail in all the samples (control and both enriched diets) (Figure 1c).

Table 3. Correlations (Pearson's coefficient) between all considered variables. Apelin, resolvin (RvD1), malondialdehyde (MDA), polyunsaturated acids (PUFA) n-3, n-6, final testosterone (T), and curvilinear velocity (VCL).

	RvD1 Testis	MDA Testis	n-3 PUFA Testis	n-6 PUFA Testis	T	MDA Sperm	n-3 PUFA Sperm	n-6 PUFA Sperm	VCL	Sperm Motility	n-3 PUFA Intake	n-6 PUFA Intake
Apelin testis	0.360	0.838 **	0.798 **	−0.408	0.692 **	0.145	−00.210	−0.292	0.567 *	0.376	0.819 **	0.859 **
RvD1 testis		0.608 **	0.736 **	−0.947 **	−0.002	0.930 **	0.721 **	−0.960 **	0.776 **	0.949 **	0.724 **	0.248
MDA testis			0.964 **	−0.639 **	0.742 **	0.387	−0.033	−0.531 *	0.708 **	0.607 **	0.980 **	0.900 **
n-3 PUFA testis				−0.780 **	0.593 **	0.560 *	0.168	−0.684 **	0.828 **	0.718 **	0.990 **	0.799 **
n-6 PUFA testis					−0.029	−0.936 **	−0.730 **	0.975 **	−0.888 **	−0.945 **	−0.747 **	−0.267
T						−0.262	−0.621 **	0.105	0.192	−0.023	0.643 **	0.925 **
MDA sperm							0.904 **	−0.984 **	0.777 **	0.930 **	0.524 *	−0.031
n-3 PUFA sperm								−0.825 **	0.540 *	0.724 **	0.119	−0.448
n-6 PUFA sperm									−0.840 **	−0.956 **	−0.657 **	−0.133
VCL										0.815 **	0.779 **	0.406
Sperm motility											0.704 **	0.228
n-3 PUFA intake												0.406

* Correlation is significant at the 0.05 level (2-tailed). ** Correlation is significant at the 0.01 level (2-tailed).

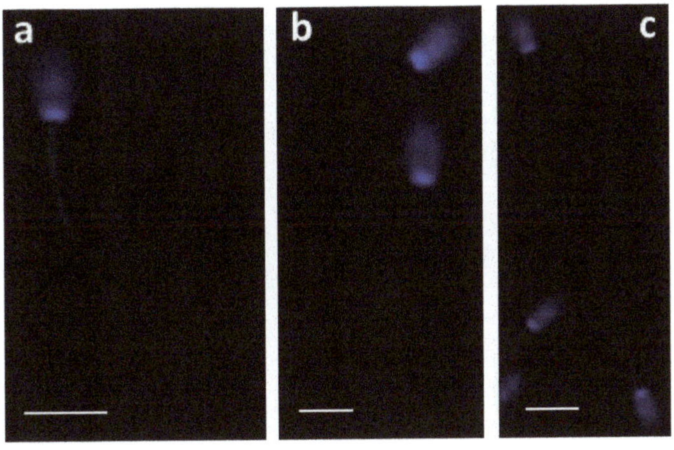

Figure 1. Ejaculated spermatozoa of the control (**a**) and FLAX (**b**) rabbits incubated with anti-apelin antibody. A fluorescent label was detected in all samples along the sperm tail (**a**,**b**). The APJ receptor was localized in the entire tail (**c**) in all the samples. Bars: 5 µm.

As shown in Figure 2, apelin was strongly expressed in the interstitial tissue of the testis from the FLAX (Figure 2b) compared with the control (Figure 2a) and FISH groups (Figure 2c). Spermatids were labeled in both enriched dietary groups. As in sperm, the localization of the APJ was similar to what showed for apelin.

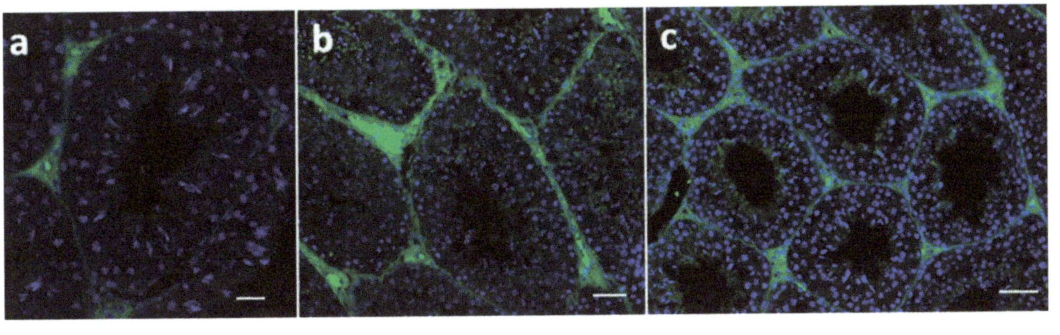

Figure 2. UV micrographs of rabbit testicular tissue treated with anti-apelin antibody from control (**a**) and enriched diets ((**b**) FLAX; (**c**) FISH). A weak fluorescent signal is present in the interstitial tissue of controls and FISH diets (**a**,**c**), respectively). In (**b**), high labeling intensity in the interstitial tissue is evident. In (**b**,**c**), spermatids are labeled. Bars: 50 µm.

Finally, the amount and distribution of ryanodine receptors were investigated in the testes. In the testes of the control rabbits, a faint label was observed in interstitial cells, and fluorescent spots were detected in a few numbers of spermatids (Figure 3a). A weak presence of the signal in samples from rabbits fed both enriched diets was observed in the interstitial tissue, and a high-dotted localization was detected in seminiferous tubules (Figure 3b,c). In particular, the fluorescence was increased in mature elongated spermatids of the FLAX group (Figure 3b).

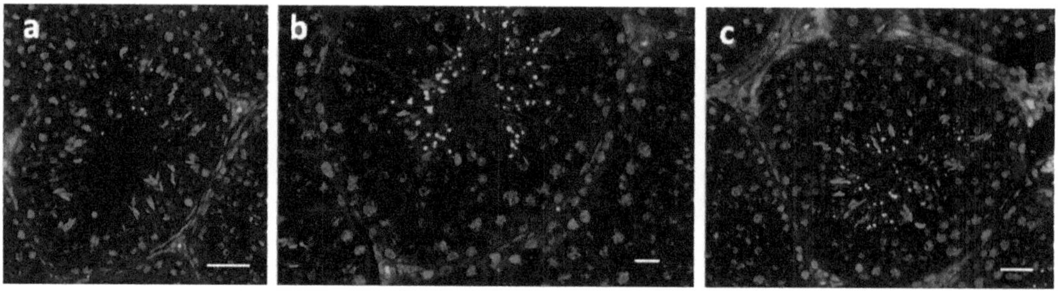

Figure 3. Immunolocalization of ryanodine receptors in rabbit testicular tissue from the control (**a**,**b**) FLAX, and (**c**) FISH groups. A dotted fluorescent signal is present in the seminiferous tubules in a reduced number of spermatids (**a**); the number of spots appears increased in (**b**,**c**), and strongly present in b, in particular. A faint localization in the interstitial tissue is shown (**b**,**c**). Bars: 50 μm.

3. Discussion

In this study, the possible role of apelin and its receptors, RvD1, and ryanodine receptors were investigated in the reproductive system and mature sperm of rabbits fed different diets enriched with ALA (FLAX group), or VLCP (FISH group). Rabbit represents a good animal model for studying sperm modifications due to the diet of other stressors (infection, inflammation, etc.) [19–21].

It is widely known that some adipokines, such as apelin, can regulate male and female reproduction according to the energy balance of the body. Chemerin, apelin, resistin, and visfatin are expressed in the ovaries of various animal species; however, the role of apelin in male reproduction has not yet been clarified [22]. Moretti et al. [15] detected increased levels of apelin and IL-1β concentrations in patients' samples with varicocele and infections.

It has been reported that diets rich in n-3 fatty acids and low in saturated fatty acids and trans-fatty acids, if adequately protected with antioxidants (vitamins E, C, D, β-carotene, selenium, zinc, folate), are positively associated with sperm quality [2,23]. Different dietary sources of n-3 polyunsaturated fatty acids modify the lipid plasma membrane of the somatic and germ cells of the rabbit testis, even if they do not change the spermatogenesis and/or the ultrastructure [2]. This fact was confirmed in this study, where spermatozoa from rabbits of FLAX and FISH groups displayed increased progressive motility.

In this paper, ELISA and immunofluorescence analysis in the testes from both enriched diets showed that apelin was widely expressed, particularly in the interstitial testis of the FLAX group, in agreement with other authors that have reported high intakes of PUFA and VLCP [24]. Moreover, spermatids appeared labeled in both enriched dietary groups.

Apelin and APJ were detected in the entire tail of ejaculated sperm in the control and both treated groups. In other cell models, apelin can reduce mitochondrial ROS-triggered oxidative damage [25], mitochondria apoptosis, and inflammatory responses stimulated by NF-κB and NLRP3 inflammasome [26]. Overproduction of ROS has been linked to sperm damage and male infertility. There are many pathological conditions, including varicocele, tobacco usage, alcohol, obesity/metabolic syndrome, leukocytospermia, and sexually transmitted infections [27], which enhance the production of radicals.

Recently, in mouse, apelin and its receptor were localized in the Leydig and germ cells, where they can influence testis steroidogenesis [13]. It has been reported that, in rats, intraperitoneal administration of apelin resulted in a decrease of serum testosterone, luteinizing, and follicle-stimulating hormone [28]. In the canine testis, apelin was detected in spermatids and mature sperm, and the APJ receptor was also detected in the cytoplasm of Leydig cells [29].

Exogenous treatment with other adipokines, as well as adiponectin, positively influenced testicular mass, insulin receptor expression, and testosterone synthesis in the testis of aged mice [30].

FLAX diet increased serum testosterone, and both the enriched diets enlarged MDA and RvD1 in the testis. Previously, Castellini et al. [2] showed persistent higher blood testosterone in rabbits fed a three-month FLAX diet.

Different authors [31–33] suggested that dietary flaxseed, being one of the richest sources of phytoestrogens (lignans), may increase testosterone hormone secretion by modulating blood cholesterol; indeed, lignans are able to reduce blood cholesterol (mainly on low-density lipoprotein, LDL) competing with the cholesterol anabolism of liver and favoring the sterol-hormone one [34]. Furthermore, the flaxseed phytoestrogens possess, to a greater or lesser extent, functional similarity to the 17β-estradiol, thus, binding the same receptors of estrogen hormones (α and β-estrogen receptors) [35].

Serum testosterone, oxidation, and the amount of n-3 PUFA in the testis were positively correlated with apelin and RvD1. RvD1 is a member of the specialized pro-resolving lipid mediator family derived from LC, and consequently, it increased the lipid oxidation (testis and sperm MDA) induced by the higher PUFA concentration of enriched diets. However, the reproductive apparatus was able to counteract this higher oxidative thrust since RvD1 was correlated with sperm motility and VCL, suggesting that its activation improves sperm kinetic, probably operating an anti-inflammatory/antioxidant action on reproductive tissues, as hypothesized hereinafter (Figure 4).

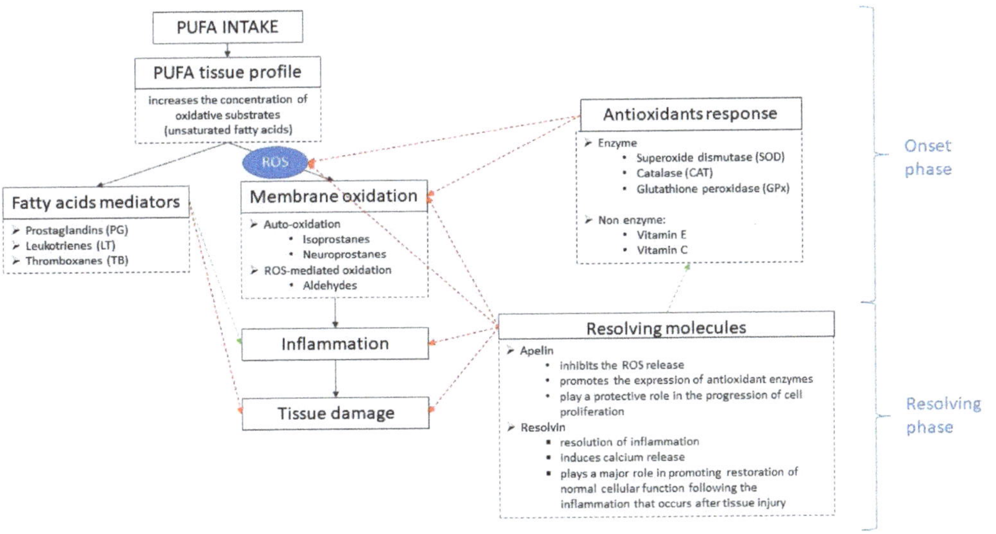

Figure 4. Tentative scheme of action of antioxidants and resolving molecules following high dietary PUFA intake. Solid line = mechanism flow; dotted line (arrows and squares) = molecule activity; red line = reducing activity; green line = enhancing activity. PUFA: polyunsaturated fatty acids; ROS: reactive oxygen species.

In humans, the RvD1 amount was increased in patients with leukocytospermia, varicocele, and idiopathic infertility compared to fertile men and with other markers of oxidative stress and inflammation, such as fatty acids content and clinical biomarkers suggesting a possible role for a diagnosis of inflammatory status and a subsequent appropriate therapeutic approach [18].

Sperm motility is influenced by the lipid composition of the plasma membrane, largely determined by VLCP that ameliorates the flexibility of cells. The lipid bilayer of the rabbit sperm membrane consists mainly of cholesterol and phospholipids [36], which contain about 43% VLCP with more than 20 carbon atoms, with DPA n-6 being the most representative [4]. The VLCP in the sperm membrane is derived from linoleic acid (C18:2n-

6; LA) and ALA resulting, in turn, from the dietary intake and are converted into their derivatives by the liver, or to a lesser extent, by testicular cells [4]. Furthermore, the lipid bilayers are constituted by different sterols (cholesterol and desmosterol, a cholesterol precursor), which, together with the phospholipids, modulate the membrane fluidity, sperm capacitation, and acrosome reaction [23]. Nordgren et al. [37] suggested that higher n-3 PUFA intake provides more substrate to produce pro-resolving lipid mediators during high-risk pregnancy/delivery conditions. Zirpoli et al. [38] observed that acute or chronic administration of n-3 PUFA enhanced RvD1, eliciting cardio- and neuroprotection through the activation of several interrelated anti-inflammatory pathways.

Probably, increasing PUFA intake activates a complex mechanism consisting in triggering the cascade of several molecules, as follows: PUFA intake > PUFA tissues > oxidative thrust > oxidative markers (e.g., MDA, Isoprostanes) > antioxidants response (enzymatic and non-enzymatic) and resolving molecules (i.e., apelin and resolvin) = inflammation controlling or solving (Figure 4).

Indeed, it is reported that resolvins were enzymatically produced by oxidized VLCP metabolites; in particular, the RvD1 comes from DHA oxidation [39,40] in response to inflammation. There are two phases in acute inflammation: the onset phase and the resolving phase [39]. During the onset phase, fatty acids-mediators such as prostaglandins (PG) and leukotrienes (LT) operate as pro-inflammatory mediators, whereas after tissue injury or trauma, some specific mediators (i.e., PG-E2 and LT-B4) drives the neutrophils infiltration into the injured tissue, and remove dead cells. After that, it initiates the resolving phase, where pro-resolving lipid mediators act against acute inflammation with a mechanism that is not yet well defined.

PUFA intake increases the PUFA content of tissues, inducing, on the one hand, the oxidative thrust ROS-mediated (malondialdehyde MDA and isoprostanes) and from the other, the fatty acid-mediators release (prostaglandins PG, leukotrienes LT, and thromboxanes TB). The increase of fatty acid mediators modulates tissue inflammation (onset phase). Contemporarily, the antioxidant's defense (enzymatic and non-enzymatic) acts against oxidation in the onset phase, followed by resolving molecules activation (resolving phase), which reduces ROS generation and tissue inflammation.

As previously reported, in physiological conditions, a higher intra-testicular expression of the apelinergic system in the mouse testes was associated with a reduction in testosterone secretion [13]; conversely, in this research, apelin was negatively correlated with the serum testosterone in the FLAX diet (Supplementary Table S1), probably due to phytoestrogens competition, as previously stated.

Moreover, higher levels of apelin in the testis of the FLAX group are associated with increased oxidation in the testis and sperm. These results underline that the number of antioxidants in the diet (i.e., α-tocopherol), even if considered supranutritional (50 vs. 200 mg/kg biblio), is not sufficient to balance the oxidation thrust induced by lipoperoxidation of VLCP. Compensatory mechanisms have probably been activated with possible effects on the reproductive system [3]. For example, the effect of vitamin E is partially compensated by the activation of enzymes (i.e., cytochrome p450) and by some other mechanism studied in this experiment (Figure 4), namely the increase of apelin and RvD1 for reducing the oxidation and/or its deleterious effect on cells of reproductive apparatus [41].

It is reported that in human neuroblastoma SH-SY5Y cells, apelin reduced calcium release, caspase-3, and cytochrome c, preventing apoptosis, oxidative stress, and mitochondrial toxicity [42].

The resolution of inflammatory processes may be also modulated by calcium channels. Therefore, finally, the presence of ryanodine receptors (RyRs) was investigated in the testes of rabbits. RyRs are intracellular calcium release channels that are highly expressed in striated muscles and neurons but are also detected in several non-excitable cells [43]. Spermatogenic cells express transcripts for all three RyR isoforms. However, there is no consensus regarding the presence and exact localization of RyRs in mature sperm [44]. This receptor was localized in the mitochondrial helix or in the entire sperm tail and may be involved in the capacitation process [45]. It is known that the atrial activity of the Ca^{2+}/calmodulin-dependent protein kinase II (CaMKII) is higher during sepsis and causes hyperphosphorylation of cardiac ryanodine receptor 2 (RYR2) channels preventing neuronal excitotoxicity, smooth muscle relaxation, vasodilation, and immunomodulation [46]. RvD1 can reduce the pro-inflammatory phenotype of microglia and enhance phagocytosis of Aβ by microglia of AD patients and ameliorates the decline of phagocytosis of FAM-Aβ through binding to different receptors. The in vitro effects were concentration-dependent on MAPK, PI3K, and calcium signaling pathways [47].

On the other side, it has been recently reported that apelin influences the expression of mitochondrial calcium uniporter, which increases mitochondrial calcium uptake [48], suggesting a relationship between apelin levels and calcium that may have a fundamental role not only in inflammation resolution but in sperm physiology and fertilization ability (i.e., spermatozoa capacitation).

4. Materials and Methods

4.1. Animals and Experimental Design

Fifteen New Zealand White male rabbits were trained for semen collection with an artificial vagina (50 days), then, they were divided into three homogeneous groups (5/group): the control group was fed ad libitum with the standard diet, the FLAX group was fed a standard diet, which was supplemented with 10% of extruded flaxseed, the FISH group was fed a standard diet, which contained 3.5% of fish oil (Nordic Naturals Omega-3®, Table 4). The dietary protocol involved 60 days. Animals were handled as reported in Boiti et al. [49]. The feed intake was registered weekly to calculate the approximate intake of n-3 and n-3 VLCP. Semen samples were evaluated at the beginning and at the end of the protocol. This study was conducted in accordance with the Guiding Principles in the Use of Animals and approved by the Animal Ethics Monitoring Committee of the University of Siena (CEL AOUS; authorization no. 265/2018-PR, ISOPRO 7DF19.23).

Table 4. Formulation (g/kg of diet), chemical composition (g/kg of diet), and main fatty acids (% of total fatty acids) of diets. Linoleic acid, LA; alpha-linolenic acid, ALA; polyunsaturated acids, PUFA; long-chain PUFA, VLCP.

	Control	FLAX	FISH
Dehydrated alfalfa meal	300	380	380
Soybean meal 44%	150	100	150
Barley meal	410	310	335
Wheat bran	52	52	52
Soybean oil	30	-	-
Extruded flaxseed	-	100	-
Fish oil	-	-	35
Beet molasses	20	10	10
Calcium carbonate	7	7	7
Calcium diphosphate	13.5	13.5	13.5
Salt	7	7	7
DL-methionine	0.5	0.5	0.5
Vitamin-mineral premix †	10	10	10
Crude protein	175	174	175
Ether extract	480	472	425
Crude fiber	124	137	130
Ash	89	84	90
LA	50.45	22.30	20.50
ALA	11.15	45.80	18.50
n-6 PUFA	51.45	22.80	21.00
n-3 PUFA	11.35	46	26.40
n-3 VLCP	-	-	10.50

Nordic Naturals Omega-3® = purified deep sea fish oil (from anchovies and sardines) containing EPA—330 mg/100 g, DHA—220 mg/100 g, and other VLCP—140 mg/100 g + α-tocopherol for preservation. † Per kg diet: vitamin A—11.000 IU; vitamin D_3—2000 IU; vitamin B_1—2.5 mg; vitamin B_2—4 mg; vitamin B_6—1.25 mg; vitamin B_{12}—0.01 mg; alpha-tocopheryl acetate—200 mg; biotine—0.06 mg; vitamin K—2.5 mg; niacin—15 mg; folic acid—0.30 mg; D-pantothenic acid—10 mg; choline—600 mg; Mn—60 mg; Fe—50 mg; Zn—15 mg; I—0.5 mg; Co—0.5 mg.

Blood samples drawn from the marginal ear vein were collected in tubes containing Na_2-EDTA and centrifuged at 5000× *g* for 15 min at 4 °C. Serum was obtained from blood samples coagulated at room temperature for 2 h, and then the collection tubes were rimmed and refrigerated at 4 °C for 24 h before analysis.

At the end of the trial (110 days), the rabbits were killed, and their testes were accurately removed; a part was fixed for immunofluorescence, and another part was stored at −80 °C for the evaluation of malondialdehyde (MDA), fatty acid content, apelin, and RvD1.

4.2. Testosterone Evaluation in Blood Serum

The testosterone concentration in rabbit serum was performed by radioimmunoassay (RIA) using the Testosterone RIA KIT (Ref: RK-61M Institute of Isotopes Co. Ltd., Bucharest, Romania) as reported in Castellini et al. [2].

4.3. Semen Quality Assessment

Semen samples were collected by means of an artificial vagina heated to 38 °C with water and immediately transferred to the laboratory. Sperm quality was immediately evaluated on raw samples, as reported by Castellini et al. [2]. The concentration of sperm was evaluated with a Thoma-Zeiss chamber, and record sperm motility (%) and curvilinear velocity (VCL, μm/s) was evaluated by means of a computer assisted sperm analysis (CASA) system.

After, semen samples were centrifuged for 15 min, and the sperm cells were divided into three aliquots. One aliquot was processed for immunocytochemistry, and the other two aliquots of 10^8 spermatozoa/mL were stored at −80 °C for the evaluation of the oxidative status and fatty acid profile.

4.4. Oxidative Status of Testis and Sperm

Lipid peroxidation in the testis was assessed by the MDA level. Rabbit tissue samples were homogenized in a 0.04 M K^+-phosphate buffer (pH 7.4) containing 0.01% BHT (1:5 *w*/*v*). The homogenate was deproteinized with acetonitrile (1:1) and then centrifuged at 3000× *g* for 1 min. The supernatants were used for MDA analysis after pre-column derivatization with 2,4-dinitrophenylhydrazine according to the method published by Shara et al. [50] with minor modifications. The samples were immediately stirred, extracted with 5 mL of pentane, and dried using nitrogen. MDA hydrazone was quantified by isocratic HPLC using a Waters 600 E system controller HPLC instrument equipped with a Waters Dual 2487 UV detector set at 307 nm. A 5 μm Ultrasphere ODS C18 column was used with a mobile phase composed of acetonitrile (45%) and HCl 0.01 N (55%) at a flow rate of 0.8 mL/min. A calibration curve with MDA concentrations ranging from 0.2 to 10 nmol/mL was used for quantification. The MDA concentration was calculated by peak areas using an Agilent 3395 integrator. The results are expressed as nmol/g tissue.

The extent of sperm lipid peroxidation (thiobarbituric reactive substances, TBARs) was assessed by measuring malondialdehyde (MDA) along with other substances that are reactive to 2-thiobarbituric acid (TBA), as reported by Mourvaki et al. [51]. The results were expressed as nmol MDA/mL.

4.5. Fatty Acid Profiles of Diets, Testis and Sperm

Lipids were extracted from the sperm and testis according to Folch et al. [52]; the esterification was performed according to Christie [53]. The trans-methylation procedure was performed using eicosenoic acid methyl esters (Sigma-Aldrich, Bellefonte, PA, USA) as an internal standard.

The fatty acid composition was determined using a Varian gas chromatograph (CP-3800) equipped with a flame ionization detector and a capillary column 100 m long × 0.25 mm × 0.2 μm film (Supelco, Bellefonte, PA, USA). Helium was used as the carrier gas with a flow of 0.6 mL/min. The split ratio was 1:20. The oven temperature was programmed, as reported by Mattioli et al. [54]. Individual fatty acid methyl esters (FAMEs) were detected by comparing the relative retention times of peaks in the sample with those of a standard mixture (FAME Mix Supelco; 4: 0 to 24: 0) plus cis- 395 9 cis-12 C18: 2; cis-9 cis-12 cis-15 C18: 3; and cis-9 cis-12 cis-15 C18: 3 (all from Sigma-Al- 396 drich). The fatty acids were expressed as % of total fatty acids. The average amount of each 397 fatty acid was used to calculate the sum of the total saturated fatty acid, monounsaturated 398 fatty acid, and polyunsaturated fatty acids (PUFA) n-3 and n-6. The very long-chain PUFA 399 (VLCP) included: EPAn-3, DPAn-3, and DHAn-3 acids, and arachidonic (ARAn-6) and 400 DPAn-6 were determined. The fatty acids profile of diets and registered feed intake was 401 used to estimate the daily intake of -3 PUFA and n-6 PUFA (g/day).

4.6. *Immunofluorescence in Testis and Sperm*

Small pieces of rabbit testes were fixed with 10% buffered formalin for 24 h at 4 °C, and then washed in water for 1 h. The samples were dehydrated in an increasing series of ethanol and cleared with xylene. Then, the specimens were incubated with three infiltrations of molten paraffin at 60 °C for 1 h, and solidified at room temperature. Paraffin sections were obtained by a Leica RM2125 RTS microtome (Leica Biosystem, Wetzlar, Germany); sections were deparaffinized with xylene, dehydrated in a series of ethanol concentrations for 5 min, and finally, in water. For antigen retrieval, the sections were washed and treated with heat-induced epitope retrieval 1 (HIER 1) buffer (10 mM sodium citrate) at pH 6 for 20 min at 95 °C.

Sperm were smeared on slides, fixed in 4% paraformaldehyde for 15 min and treated with a blocking solution (phosphate-buffered saline (PBS)–bovine serum albumin (BSA) 1% Normal Goat Serum (NGS) 5%) for 20 min.

Specimens were treated overnight at 4 °C with the rbbit anti-apelin polyclonal antibody (Abcam, Cambridge, UK) diluted 1:500 or rabbit anti-APJ-R polyclonal antibody (Thermo Fisher Scientific, Waltham, MA, USA) diluted 1:100 or anti-ryanodine receptor polyclonal antibody (Invitrogen, Thermo Fisher Scientific, Carlsbad, CA, USA) diluted 1:500. After three washes for 10 min in PBS, reactions were revealed by an anti-rabbit antibody raised in goat Alexa Fluor® 488 conjugate (Invitrogen, Thermo Fisher Scientific, Carlsbad, CA, USA), diluted at 1:500 for 1 h at room temperature. Primary antibodies were omitted in the control slides. Nuclei were stained with 4,6-diamidino-2-phenylindole (DAPI) solution (Vysis, Downers Grove, IL, USA) for 10 min; the slides were washed in PBS and mounted with 1,4-diazabicyclo[2.2.2]octane (DABCO, Sigma-Aldrich, Milan, Italy) to observe fluorescence. All the samples were analyzed under a Leica DMI 6000 fluorescence microscope (Leica Microsystems, Wetzlar, Germany) with a 63× objective, and the images were acquired using the Leica AF 6500 Integrated System for imaging and analysis.

For standardization and comparison of the different groups, only good-quality sections at the same magnification were investigated; at least 10 sections of tissue from each group were evaluated. Two hundred sperm per sample were evaluated. In detail, the images were obtained with HCX PL FLUOTAR 63×/1.25 oil objective; filters for TRIC and FITC were selected. The micrographs were not modified with image elaboration software. The specificity of the antibodies, guaranteed in the datasheets of both antibodies, was also evaluated by omitting the primary antibodies

4.7. *Resolvin (Rv) D1 and Apelin Assays in Testis*

In homogenates of testicular tissue (75% *w*/*v*, in phosphate-buffered solution, pH 7.4), RvD1 and apelin were measured by a quantitative sandwich enzyme-linked immunosorbent assay (ELISA) (MBS2601295 MyBioSource, San Diego, CA, USA, and abx585113, Abbexa, Cambridge, UK, respectively). A biotin-labeled antibody and horseradish peroxidase + avidin were applied for bound biotin-labeled antibody detection. Spectrometric detection of color intensity at 450 nm allowed the determination of testis RvD1 and apelin amounts by comparing the optical density of each sample to the standard curve (respectively ranging from 2000 pg/mL to 31.2 pg/mL RvD1 amounts, ranging from 8000 to

125-pg/mL apelin amounts). In all experiments, measures were performed in duplicate in each sample.

4.8. Statistical Analysis

Lipid oxidation, fatty acids profile, testosterone, and sperm kinetic traits were analyzed with a linear model to evaluate the fixed effect of diet (control, FLAX, FISH; SPSS v28) Least squares (LS) mean, and pooled standard error (SE) are reported. The Bonferroni correction was applied for multiple comparisons. The significance was set at $p < 0.05$. A correlation was built to detect associations among the main variables (apelin, RvD1, MDA, n-3 PUFA, n-6 PUFA, final testosterone, MDA, sperm motility, curvilinear velocity VCL), also within dietary groups (Supplementary Table S1). The correlation was defined as high when the Pearson coefficient was (r) > $|0.5|$, medium when r ranged from 0.3 to 0.5, and low when r < $|0.3|$.

5. Conclusions

The data seems to indicate that dietary supplementation of FLAX increases the amount of apelin in the testis, whereas the RvD1 increased in both the supplemented diets.

Apelin in ejaculated sperm was mainly localized in the tail, and its positive correlations with sperm motility and RvD1 level could suggest an involvement of these molecules in male reproduction and probably a role in the resolution of inflammatory status induced by dietary challenges (acute or chronic).

The proposed diet, by influencing apelin levels, could directly and indirectly, stimulate the resolution of reproductive inflammation.

Further studies on the biological system (in vivo and in vitro) challenged with pro-inflammatory or pro-oxidative molecules could be done to find a relationship with other derived molecules.

Supplementary Materials: The following supporting information can be downloaded at: https://www.mdpi.com/article/10.3390/molecules28176188/s1, Table S1: Correlations (Pearson's coefficient) between all considered variables.

Author Contributions: S.M. wrote the original paper, performed the PUFA profiles, oxidative stress evaluation, testosterone, and animal care; E.M. performed data interpretation, immunochemistry analysis and paper revision; C.C. performed project design, statistical analysis, and paper revision; C.S. performed apelin and resolvin assays; R.C. performed immunocytochemical analysis; E.A. performed oxidative stress and animal care; G.C. was responsible for animals and sacrifice, and performed immunocytochemical analysis, conceptualization, project design, and administration, and wrote the original paper. All authors have read and agreed to the published version of the manuscript.

Funding: The study was supported by the PSR (Plan Support Research, 2267-2022-CGPAR001), Department of Molecular and Developmental Medicine, University of Siena.

Institutional Review Board Statement: Not applicable.

Informed Consent Statement: Not applicable.

Data Availability Statement: Not applicable.

Conflicts of Interest: The authors declare no conflict of interest.

Sample Availability: Not applicable.

References

1. Collodel, G.; Castellini, C.; Lee, J.C.; Signorini, C. Relevance of Fatty Acids to Sperm Maturation and Quality. *Oxid. Med. Cell Longev.* **2020**, *2020*, 7038124. [CrossRef] [PubMed]
2. Castellini, C.; Mattioli, S.; Signorini, C.; Cotozzolo, E.; Noto, D.; Moretti, E.; Brecchia, G.; Dal Bosco, A.; Belmonte, G.; Durand, T.; et al. Effect of Dietary n-3 Source on Rabbit Male Reproduction. *Oxid. Med. Cell Longev.* **2019**, *2019*, 3279670. [CrossRef] [PubMed]

3. Mattioli, S.; Dimauro, C.; Cesarani, A.; Dal Bosco, A.; Bartolini, D.; Galli, F.; Migni, A.; Sebastiani, B.; Signorini, C.; Oger, C.; et al. A Dynamic Model for Estimating the Interaction of ROS–PUFA–Antioxidants in Rabbit. *Antioxidants* **2022**, *11*, 531. [CrossRef] [PubMed]
4. Castellini, C.; Mattioli, S.; Moretti, E.; Cotozzolo, E.; Perini, F.; Dal Bosco, A.; Signorini, C.; Noto, D.; Belmonte, G.; Lasagna, E.; et al. Expression of genes and localization of enzymes involved in polyunsaturated fatty acid synthesis in rabbit testis and epididymis. *Sci. Rep.* **2022**, *12*, 2637. [CrossRef]
5. Elfassy, Y.; McAvoy, C.; Fellahi, S.; Dupont, J.; Fève, B.; Levy, R.; Bastard, J.P. Seminal plasma adipokines: Involvement in human reproductive functions. *Eur Cytokine Netw.* **2017**, *28*, 141–150. [CrossRef]
6. Singh, A.; Choubey, M.; Bora, P.; Krishna, A. Adiponectin and Chemerin: Contrary Adipokines in Regulating Reproduction and Metabolic Disorders. *Reprod. Sci.* **2018**, *25*, 1462–1473. [CrossRef]
7. Shokrollahi, B.; Shang, J.H.; Saadati, N.; Ahmad, H.I.; Yang, C.Y. Reproductive roles of novel adipokines apelin, visfatin, and irisin in farm animals. *Theriogenology* **2021**, *172*, 178–186. [CrossRef]
8. Higuchi, K.; Masaki, T.; Gotoh, K.; Chiba, S.; Katsuragi, I.; Tanaka, K.; Kakuma, T.; Yoshimatsu, H. Apelin, an APJ receptor ligand, regulates body adiposity and favors the messenger ribonucleic acid expression of uncoupling proteins in mice. *Endocrinology* **2007**, *148*, 2690–2697. [CrossRef]
9. Soliman, M.; Arafah, M. Apelin protect against multiple organ injury following hemorrhagic shock and decrease the inflammatory response. *Int. J. Appl. Basic Med. Res.* **2015**, *5*, 195–199. [CrossRef]
10. Than, A.; Zhang, X.; Leow, M.K.; Poh, C.L.; Chong, S.K.; Chen, P. Apelin attenuates oxidative stress in human adipocytes. *J. Biol. Chem.* **2014**, *289*, 3763–3774. [CrossRef]
11. Kurowska, P.; Barbe, A.; Różycka, M.; Chmielińska, J.; Dupont, J.; Rak, A. Apelin in Reproductive Physiology and Pathology of Different Species: A Critical Review. *Int. J. Endocrinol.* **2018**, *2018*, 9170480. [CrossRef] [PubMed]
12. Moustafa, A. Effect of Omega-3 or Omega-6 Dietary Supplementation on Testicular Steroidogenesis, Adipokine Network, Cytokines, and Oxidative Stress in Adult Male Rats. *Oxid. Med. Cell Longev.* **2021**, *2021*, 5570331. [CrossRef] [PubMed]
13. Das, M.; Gurusubramanian, G.; Roy, V.K. Postnatal developmental expression of apelin receptor proteins and its role in juvenile mice testis. *J. Steroid Biochem. Mol. Biol.* **2022**, *224*, 106178. [CrossRef] [PubMed]
14. Salas-Huetos, A.; Bulló, M.; Salas-Salvadó, J. Dietary patterns, foods and nutrients in male fertility parameters and fecundability: A systematic review of observational studies. *Hum. Reprod. Update* **2017**, *23*, 371–389. [CrossRef] [PubMed]
15. Moretti, E.; Signorini, C.; Corsaro, R.; Noto, D.; Tripodi, A.S.; Menchiari, A.; Micheli, L.; Ponchia, R.; Collodel, G. Apelin is found in human sperm and testis and is raised in inflammatory pathological conditions. *Cytokine* **2023**, *169*, 156281. [CrossRef]
16. Serhan, C.N.; Chiang, N.; Dalli, J.; Levy, B.D. Lipid mediators in the resolution of inflammation. *Cold Spring Harb. Perspect. Biol.* **2014**, *7*, a016311. [CrossRef]
17. Duvall, M.G.; Levy, B.D. DHA- and EPA-derived resolvins, protectins, and maresins in airway inflammation. *Eur. J. Pharmacol.* **2016**, *785*, 144–155. [CrossRef]
18. Signorini, C.; Moretti, E.; Noto, D.; Micheli, L.; Ponchia, R.; Collodel, G. Fatty Acid Oxidation and Pro-Resolving Lipid Mediators Are Related to Male Infertility. *Antioxidants* **2022**, *11*, 107. [CrossRef]
19. Marchiani, S.; Vignozzi, L.; Filippi, S.; Gurrieri, B.; Comeglio, P.; Morelli, A.; Danza, G.; Bartolucci, G.; Maggi, M.; Baldi, E. Metabolic syndrome-associated sperm alterations in an experimental rabbit model: Relation with metabolic profile, testis and epididymis gene expression and effect of tamoxifen treatment. *Mol Cell Endocrinol.* **2015**, *401*, 12–24. [CrossRef]
20. Marco-Jiménez, F.; Vicente, J.S. Overweight in young males reduce fertility in rabbitmodel. *PLoS ONE* **2017**, *12*, e0180679. [CrossRef]
21. Fan, J.L.; Castellini, C.; Page, C.P.; Spina, D. The rabbit as a biomedical model. In *The Genetics and Genomics of the Rabbit*; CABI: Wallingford, UK, 2021; pp. 310–325.
22. Estienne, A.; Bongrani, A.; Reverchon, M.; Ramé, C.; Ducluzeau, P.H.; Froment, P.; Dupont, J. Involvement of Novel Adipokines, Chemerin, Visfatin, Resistin and Apelin in Reproductive Functions in Normal and Pathological Conditions in Humans and Animal Models. *Int. J. Mol. Sci.* **2019**, *20*, 4431. [CrossRef] [PubMed]
23. Yuan, C.; Zhang, K.; Wang, Z.; Ma, X.; Liu, H.; Zhao, J.; Lu, W.; Wang, J. Dietary flaxseed oil and vitamin E improve semen quality via propionic acid metabolism. *Front. Endocrinol.* **2023**, *14*, 1139725. [CrossRef] [PubMed]
24. Yuzbashian, E.; Zarkesh, M.; Asghari, G.; Hedayati, M.; Safarian, M.; Mirmiran, P.; Khalaj, A. Is apelin gene expression and concentration affected by dietary intakes? A systematic review. *Crit. Rev. Food Sci. Nutr.* **2018**, *58*, 680–688. [CrossRef] [PubMed]
25. Zhou, Q.; Cao, J.; Chen, L. Apelin/APJ system: A novel therapeutic target for oxidative stress-related inflammatory diseases (Review). *Int. J. Mol. Med.* **2016**, *37*, 1159–1169. [CrossRef] [PubMed]
26. Yan, J.; Wang, A.; Cao, J.; Chen, L. Apelin/APJ system: An emerging therapeutic target for respiratory diseases. *Cell Mol. Life Sci.* **2020**, *77*, 2919–2930. [CrossRef]
27. Agarwal, A.; Rana, M.; Qiu, E.; AlBunni, H.; Bui, A.D.; Henkel, R. Role of oxidative stress, infection and inflammation in male infertility. *Andrologia* **2018**, *50*, e13126. [CrossRef]
28. Tekin, S.; Erden, Y.; Sandal, S.; Etem Onalan, E.; Ozyalin, F.; Ozen, H.; Yilmaz, B. Effects of apelin on reproductive functions: Relationship with feeding behavior and energy metabolism. *Arch. Physiol. Biochem.* **2017**, *123*, 9–15. [CrossRef]

29. Troisi, A.; Dall'Aglio, C.; Maranesi, M.; Orlandi, R.; Suvieri, C.; Pastore, S.; Bazzano, M.; Martínez-Barbitta, M.; Polisca, A. Presence and localization of apelin and its cognate receptor in canine testes using immunohistochemical and RT-PCR techniques. *Vet. Res. Commun.* **2023**, *47*, 937. [CrossRef]
30. Choubey, M.; Ranjan, A.; Bora, P.S.; Baltazar, F.; Martin, L.J.; Krishna, A. Role of adiponectin as a modulator of testicular function during aging in mice. *Biochim. Biophys. Acta Mol. Basis Dis.* **2019**, *1865*, 413–427. [CrossRef]
31. Shultz, T.D.; Bonorden, W.R.; Seaman, W.R. Effect of short-term flaxseed consumption on lignan and sex hormone metabolism in men. *Nutr. Res.* **1991**, *11*, 1089–1100. [CrossRef]
32. Li, W.; Tang, D.; Li, F.; Tian, H.; Yue, X.; Li, F.; Weng, X.; Sun, W.; Wang, W.; Mo, F. Supplementation with dietary linseed oil during peri-puberty stimulates steroidogenesis and testis development in rams. *Theriogenology* **2017**, *102*, 10–15. [CrossRef] [PubMed]
33. Qi, X.; Shang, M.; Chen, C.; Chen, Y.; Hua, J.; Sheng, X.; Wang, X.; Xing, K.; Ni, H.; Guo, Y. Dietary supplementation with linseed oil improves semen quality, reproductive hormone, gene and protein expression related to testosterone synthesis in aging layer breeder roosters. *Theriogenology* **2019**, *131*, 9–15. [CrossRef] [PubMed]
34. Knight, D.C.; Eden, J.A. A review of the clinical effects of phytoestrogens. *Obstet. Gynecol.* **1996**, *87*, 897–904. [PubMed]
35. Martin, P.M.; Horwitz, K.B.; Ryan, D.S.; McGuire, W.L. Phytoestrogen interaction with estrogen receptors in human breast cancer cells. *Endocrinology* **1978**, *103*, 1860–1867. [CrossRef]
36. Apel-Paz, M.; Vanderlick, T.K.; Chandra, N.; Doncel, G.F. A hierarchy of lipid constructs for the sperm plasma membrane. *Biochem. Biophys. Res. Commun.* **2003**, *309*, 724–732. [CrossRef]
37. Nordgren, T.M.; Anderson Berry, A.; Van Ormer, M.; Zoucha, S.; Elliott, E.; Johnson, R.; McGinn, E.; Cave, C.; Rilett, K.; Weishaar, K.; et al. Omega-3 Fatty Acid Supplementation, Pro-Resolving Mediators, and Clinical Outcomes in Maternal-Infant Pairs. *Nutrients* **2019**, *11*, 98. [CrossRef]
38. Zirpoli, H.; Chang, C.L.; Carpentier, Y.A.; Michael-Titus, A.T.; Ten, V.S.; Deckelbaum, R.J. Novel Approaches for Omega-3 Fatty Acid Therapeutics: Chronic Versus Acute Administration to Protect Heart, Brain, and Spinal Cord. *Annu. Rev. Nutr.* **2020**, *40*, 161–168. [CrossRef]
39. Serhan, C.N.; Brain, S.D.; Buckley, C.D.; Gilroy, D.W.; Haslett, C.; O'Neill, L.A.; Perretti, M.; Rossi, A.G.; Wallace, J.L. Resolution of inflammation: State of the art, definitions and terms. *FASEB J.* **2007**, *21*, 325. [CrossRef]
40. Moro, K.; Nagahashi, M.; Ramanathan, R.; Takabe, K.; Wakai, T. Resolvins and omega three polyunsaturated fatty acids: Clinical implications in inflammatory diseases and cancer. *World J. Clin. Cases* **2016**, *4*, 155–164. [CrossRef]
41. Meunier, B.; de Visser, S.P.; Shaik, S. Mechanism of oxidation reactions catalyzed by cytochrome P450 enzymes. *Chem. Rev.* **2004**, *104*, 3947–3980. [CrossRef]
42. Samandari-Bahraseman, M.R.; Elyasi, L. Apelin-13 protects human neuroblastoma SH-SY5Y cells against amyloid-beta induced neurotoxicity: Involvement of anti oxidant and anti apoptotic properties. *J. Basic. Clin. Physiol. Pharmacol.* **2021**, *33*, 599–605. [CrossRef] [PubMed]
43. Chiarella, P.; Puglisi, R.; Sorrentino, V.; Boitani, C.; Stefanini, M. Ryanodine receptors are expressed and functionally active in mouse spermatogenic cells and their inhibition interferes with spermatogonial differentiation. *J. Cell Sci.* **2004**, *117 Pt 18*, 4127–4134. [CrossRef] [PubMed]
44. Zhou, Y.; Ru, Y.; Wang, C.; Wang, S.; Zhou, Z.; Zhang, Y. Tripeptidyl peptidase II regulates sperm function by modulating intracellular Ca^{2+} stores via the ryanodine receptor. *PLoS ONE* **2013**, *8*, e66634. [CrossRef] [PubMed]
45. Moretti, E.; Signorini, C.; Noto, D.; Corsaro, R.; Micheli, L.; Durand, T.; Oger, C.; Galano, J.M.; Collodel, G. F4-Neuroprostane Effects on Human Sperm. *Int. J. Mol. Sci.* **2023**, *24*, 935. [CrossRef]
46. Dobrev, D.; Heijman, J.; Hiram, R.; Li, N.; Nattel, S. Inflammatory signalling in atrial cardiomyocytes: A novel unifying principle in atrial fibrillation pathophysiology. *Nat. Rev. Cardiol.* **2023**, *20*, 145–167. [CrossRef] [PubMed]
47. Zhu, M.; Wang, X.; Hjorth, E.; Colas, R.A.; Schroeder, L.; Granholm, A.C.; Serhan, C.N.; Schultzberg, M. Pro-Resolving Lipid Mediators Improve Neuronal Survival and Increase Aβ42 Phagocytosis. *Mol. Neurobiol.* **2016**, *53*, 2733–2749. [CrossRef] [PubMed]
48. Chen, Z.; Zhou, Q.; Chen, J.; Yang, Y.; Chen, W.; Mao, H.; Ouyang, X.; Zhang, K.; Tang, M.; Yan, J.; et al. MCU-dependent mitochondrial calcium uptake-induced mitophagy contributes to apelin-13-stimulated VSMCs proliferation. *Vascul. Pharmacol.* **2022**, *144*, 106979. [CrossRef]
49. Boiti, C.; Castellini, C.; Besenfelder, U.; Theau-Clément, M.; Liguori, L.; Renieri, T.; Pizzi, F. Guidelines for the handling of rabbit bucks and semen. *World Rabbit. Sci.* **2005**, *13*, 71–91.
50. Shara, M.A.; Dickson, P.H.; Bagchi, D.; Stohs, S.J. Excretion of formaldehyde, malondialdehyde, acetaldehyde and acetone in the urine of rats in response to 2,3,7,8-tetrachlorodibenzo-p-dioxin, paraquat, endrin and carbon tetrachloride. *J. Chromatogr.* **1992**, *576*, 221–233. [CrossRef]
51. Mourvaki, E.; Cardinali, R.; Dal Bosco, A.; Corazzi, L.; Castellini, C. Effects of flaxseed dietary supplementation on sperm quality and on lipid composition of sperm subfractions and prostatic granules in rabbit. *Theriogenology* **2010**, *73*, 629–637. [CrossRef]
52. Folch, J.; Lees, M.; Sloane-Stanley, H. Asimplemethod for the isolation and purification of total lipids from animal tissue. *J. Biol. Chem.* **1957**, *226*, 497–509. [CrossRef] [PubMed]

53. Christie, W.W. A simple procedure for rapid transmethylation of glycerolipids and cholesteryl esters. *J. Lipid Res.* **1982**, *23*, 1072–1075. [CrossRef] [PubMed]
54. Mattioli, S.; Dal Bosco, A.; Maranesi, M.; Petrucci, L.; Rebollar, P.G.; Castellini, C. Dietary fish oil and flaxseed for rabbit does: Fatty acids distribution and Δ6-desaturase enzyme expression of different tissues. *Animal* **2019**, *13*, 1934–1942. [CrossRef] [PubMed]

MDPI
St. Alban-Anlage 66
4052 Basel
Switzerland
www.mdpi.com

Molecules Editorial Office
E-mail: molecules@mdpi.com
www.mdpi.com/journal/molecules

www.ingramcontent.com/pod-product-compliance
Lightning Source LLC
Chambersburg PA
CBHW061443210926
51850CB00008B/1601
9783725811168